Overcoming Fieldwork Challenges in Social Science and Higher Education Research

Abir El Shaban
Innovate Your Classroom (IC) Consulting, Canada

A volume in the Advances in Higher Education and Professional Development (AHEPD) Book Series

IGI Global
PUBLISHER of TIMELY KNOWLEDGE

Published in the United States of America by
IGI Global
Information Science Reference (an imprint of IGI Global)
701 E. Chocolate Avenue
Hershey PA, USA 17033
Tel: 717-533-8845
Fax: 717-533-8661
E-mail: cust@igi-global.com
Web site: http://www.igi-global.com

Library of Congress Cataloging-in-Publication Data

Names: El Shaban, Abir, 1978- editor.
Title: Overcoming fieldwork challenges in social science and higher education research / Abir El Shaban, editor.
Description: Hershey, PA : Information Science Reference, [2021] | Includes bibliographical references and index. | Summary: "This book draws on the experiences of conducting fieldwork in different contexts and world regions that are relevant to social science and education studies by presenting diverse experiences in research processes and contexts with the goal of giving an authentic and realistic description of how research data are collected, the tools needed to envision some of the challenges that they might face and how to effectively solve them"-- Provided by publisher.
Identifiers: LCCN 2020025742 (print) | LCCN 2020025743 (ebook) | ISBN 9781799858263 (hardcover) | ISBN 9781799866046 (paperback) | ISBN 9781799858270 (ebook)
Subjects: LCSH: Social sciences--Fieldwork. | Social sciences--Methodology. | Education, Higher--Fieldwork. | Education, Higher--Research--Methodology.
Classification: LCC H62 .O936 2021 (print) | LCC H62 (ebook) | DDC 300.71/1--dc23
LC record available at https://lccn.loc.gov/2020025742
LC ebook record available at https://lccn.loc.gov/2020025743

This book is published in the IGI Global book series Advances in Higher Education and Professional Development (AHEPD) (ISSN: 2327-6983; eISSN: 2327-6991)

British Cataloguing in Publication Data
A Cataloguing in Publication record for this book is available from the British Library.

Advances in Higher Education and Professional Development (AHEPD) Book Series

ISSN:2327-6983
EISSN:2327-6991

Editor-in-Chief: Jared Keengwe, University of North Dakota, USA

MISSION

As world economies continue to shift and change in response to global financial situations, job markets have begun to demand a more highly-skilled workforce. In many industries a college degree is the minimum requirement and further educational development is expected to advance. With these current trends in mind, the **Advances in Higher Education & Professional Development (AHEPD) Book Series** provides an outlet for researchers and academics to publish their research in these areas and to distribute these works to practitioners and other researchers.

AHEPD encompasses all research dealing with higher education pedagogy, development, and curriculum design, as well as all areas of professional development, regardless of focus.

COVERAGE

- Adult Education
- Assessment in Higher Education
- Career Training
- Coaching and Mentoring
- Continuing Professional Development
- Governance in Higher Education
- Higher Education Policy
- Pedagogy of Teaching Higher Education
- Vocational Education

IGI Global is currently accepting manuscripts for publication within this series. To submit a proposal for a volume in this series, please contact our Acquisition Editors at Acquisitions@igi-global.com or visit: http://www.igi-global.com/publish/.

The Advances in Higher Education and Professional Development (AHEPD) Book Series (ISSN 2327-6983) is published by IGI Global, 701 E. Chocolate Avenue, Hershey, PA 17033-1240, USA, www.igi-global.com. This series is composed of titles available for purchase individually; each title is edited to be contextually exclusive from any other title within the series. For pricing and ordering information please visit http://www.igi-global.com/book-series/advances-higher-education-professional-development/73681. Postmaster: Send all address changes to above address.

Titles in this Series

For a list of additional titles in this series, please visit:
http://www.igi-global.com/book-series/advances-higher-education-professional-development/73681

The Emotional Self at Work in Higher Education
Ingrid Ruffin (University of Tennessee, Knoxville, USA) and Charissa Powell (University of Tennessee, Knoxville, USA)
Information Science Reference • © 2021 • 302pp • H/C (ISBN: 9781799835196) • US $195.00

Higher Education Challenges in South-East Asia
Christian Kahl (Beijing Jiaotong University, China)
Information Science Reference • © 2021 • 320pp • H/C (ISBN: 9781799844891) • US $175.00

Introducing Problem-Based Learning (PBL) for Creativity and Innovation in Chinese Universities Emerging Research and Opportunities
Chunfang Zhou (Aalborg University, Denmark)
Information Science Reference • © 2021 • 150pp • H/C (ISBN: 9781799835271) • US $145.00

Cases on Engineering Management Education in Practice
Despo Ktoridou (University of Nicosia, Cyprus)
Business Science Reference • © 2021 • 197pp • H/C (ISBN: 9781799840633) • US $195.00

Reshaping Graduate Education Through Innovation and Experiential Learning
Toby S. Jenkins (University of South Carolina, USA)
Information Science Reference • © 2021 • 301pp • H/C (ISBN: 9781799848363) • US $195.00

Handbook of Research on Inclusive Development for Remote Adjunct Faculty in Higher Education
Amber Dailey-Hebert (Park University, USA) B. Jean Mandernach (Grand Canyon University, USA) and Emily Donnelli-Sallee (Park University, USA)
Information Science Reference • © 2021 • 333pp • H/C (ISBN: 9781799867586) • US $245.00

IGI Global
PUBLISHER of TIMELY KNOWLEDGE

701 East Chocolate Avenue, Hershey, PA 17033, USA
Tel: 717-533-8845 x100 • Fax: 717-533-8661
E-Mail: cust@igi-global.com • www.igi-global.com

For my sweetheart Jana and Murad- I love you like no one; Adel for his constant support and encouragement; and Mom, Dad and my Guardians whom I miss so much. I know that no one would ever be as proud as you.

For Joy Egbert, my outstanding advisor and friend who always inspires and encourages me to keep learning, publishing and growing in academia regardless of all the barriers.

Thank you all and I love you all.

List of Reviewers

Reima Abobaker, *Savannah Collage of Art and Design, USA*
Denise Brend, *Université de Sherbrooke, Canada*
Rebecca Chism, *Kent State University, USA*
Joseph Ezale Cobbinah, *University of Ghana, Ghana*
Mohamed Elhess, *University of Idaho, USA*
Kula Francis, *University of the Virgin Islands, USA*
Carine Graff, *University of North Texas, USA*
Kenny Hendrickson, *University of the Virgin Islands, USA*
Sonia Lopez-Lopez, *Washington State University, USA*
Peter Sercombe, *Newcastle University, UK*
Seyed Abdollah Shahrokni, *Western Oregon, USA*
Nazlı Ayyildiz Unnu, *EGE University, Turkey*
Yao Wang, *Newcastle University, UK*

Table of Contents

Preface xvi

Acknowledgment xxiv

Section 1
Teacher Education and Education Technology Research Challenges

Chapter 1
Learning Amidst Mayhem: Researching Teacher Development and Technology Use in Second-Grade Classrooms 1
Reima Abobaker, Savannah College of Art and Design, USA
Joy Lynn Egbert, Washington State University, USA

Chapter 2
Handling Research Challenges: Understanding Teachers' Adoption Process of Education Technology 25
Abir El Shaban, Innovate Your Classroom (IC) Consulting, Canada

Section 2
Ethnographic and Qualitative Research Challenges

Chapter 3
A Tale of the Field: Can Ethnographic Challenges Become Opportunities? 46
Yao Wang, Newcastle University, UK
Peter Sercombe, Newcastle University, UK

Chapter 4
It All Works Out in the End: The Experience of Researching Online Language Classes 73
Rebecca Lynn Chism, Kent State University, USA
Carine Graff, University of North Texas, USA

Section 3
Fieldwork Partnerships and Researching Corporate Challenges

Chapter 5
Meaning Making Change: Interpretive Phenomenological Analysis and an Epistemology of Power 96
Denise Michelle Brend, Concordia University, Canada

Chapter 6
Data Collection as a Journey: A Multi-Step Approach to Navigating Access to an Organization 118
Géraldine Bengsch, King's College London, UK

Section 4
Gender and a Cultural Impact on Research

Chapter 7
On the Way to Researching Universities: The Case of Higher Education in Turkey 149
Nazlı Ayşe Ayyıldız Ünnü, Ege University, Turkey

Section 5
Counseling and Psychology Education Research

Chapter 8
When Participants' Trauma Becomes Mine: How I Used "Countertransference" to Enhance the Quality of My Study 175
Sigalit Gal, McGill University, Canada

Chapter 9
Reflections of Challenges and Resolutions in Fieldwork Research on Quality of Life of Cancer Patients: Counseling and Psychology 199
Kavita Gupta, The Maharaja Sayajirao University of Baroda, Vadodara, India

Section 6
Post-Disaster Research

Chapter 10
Reflections on Post-Disaster University Research in the Aftermath of Hurricanes Irma and Maria 219
Kula A. Francis, University of the Virgin Islands, US Virgin Islands
Kenny A. Hendrickson, University of the Virgin Islands, US Virgin Islands

Section 7
Online Gaming Research

Chapter 11
Multiplayer Online Game Research: Navigating an Unreal-yet-Real Terrain ... 243
Seyed Abdollah Shahrokni, Western Oregon University, USA

Compilation of References ... 267

About the Contributors ... 299

Index ... 303

Detailed Table of Contents

Preface xvi

Acknowledgment xxiv

Section 1
Teacher Education and Education Technology Research Challenges

Chapter 1
Learning Amidst Mayhem: Researching Teacher Development and Technology Use in Second-Grade Classrooms 1
Reima Abobaker, Savannah College of Art and Design, USA
Joy Lynn Egbert, Washington State University, USA

The purpose of this chapter is to describe issues with conducting field studies on technology use in elementary schools. The chapter first provides a brief overview of the study that was conducted with two teachers and their 46 second grade students, and then it presents some of the challenges that the researchers faced. These challenges include everything from securing permissions and scheduling to having sufficient personpower to collect all the data and using technology both in the study and as a topic of study. Solutions for future research are included, along with implications for all stakeholders.

Chapter 2
Handling Research Challenges: Understanding Teachers' Adoption Process of Education Technology 25
Abir El Shaban, Innovate Your Classroom (IC) Consulting, Canada

This chapter explores some of the challenging issues that were raised for the author while conducting her first biggest research project as an international doctoral student at one of the United States universities. The main challenge that the author faced was the difficulty of having access to a language center to examine her technology professional development model on language teachers to explore its effectiveness in understanding the teachers' adoption and rejection decision of using education technology and to collect data for her dissertation. After choosing an alternative venue and planning to travel to the United Arab Emirates to conduct the practical part of her research at one of the UAE's universities, an academic networking event changed the course of the latest plans and had a gatekeeper assisting her in experimenting her model and conducting the rest of her study in an American university. The chapter explores some of the challenges and how the author tackled each one of them. The chapter ends with some general recommendations for graduate students and novice researchers.

Section 2
Ethnographic and Qualitative Research Challenges

Chapter 3
A Tale of the Field: Can Ethnographic Challenges Become Opportunities? 46
Yao Wang, Newcastle University, UK
Peter Sercombe, Newcastle University, UK

This chapter discusses a number of fieldwork challenges leading up to and during an ethnographic study undertaken in southwest China. The purpose of this study was to research the circumstances of privately-run migrant schools (PMSs) as forms of educational provision for internal migrant children and the coping strategies that were observed and/or reported to deal with these children. The authors consider the (pre-study) challenge of selecting an appropriate research site, recruitment of willing participants, possible language barriers, and potential issues arising from the role(s) of the researcher and the study population. By explicating the ethnographer's path from fieldwork constraints to potential opportunities, this chapter clarifies the importance of realizing the need for flexibility and adaptability when conducting fieldwork. The chapter is of relevance to current and future ethnographers conducting fieldwork or setting out to conduct fieldwork. Additionally, this chapter provides insight on the unknown eventualities that almost certainly arise during the research process.

Chapter 4
It All Works Out in the End: The Experience of Researching Online Language Classes 73
Rebecca Lynn Chism, Kent State University, USA
Carine Graff, University of North Texas, USA

Qualitative research enables flexible designs unlike quantitative research, but those designs can be modified up to a point. It sometimes happens that the study at hand takes an unexpected turn of events and the researchers have then to find ways to cope with the changes. When the matter being investigated involves online surveys, there is even more leeway for uncertainty, as difficulties such as finding participants and time constraints may arise. This chapter presents a study conducted by two researchers in an American midwestern university and the challenges they encountered after answering a call for papers about technology and language learner psychology. They used an online survey to inquire about students' experience in an online language course. However, the lack of time intertwined with unexpected events involved readjustments and a need for more research and training. In the current article, the researchers share their project timeline, the issues they encountered, and offer some solutions.

Section 3
Fieldwork Partnerships and Researching Corporate Challenges

Chapter 5
Meaning Making Change: Interpretive Phenomenological Analysis and an Epistemology of Power 96
Denise Michelle Brend, Concordia University, Canada

This chapter describes how an interpretive phenomenological analysis (IPA) study generated perceived risk for stakeholders and for participants. Here, perceived risk was interpreted through discourses and practices specific to intimate partner violence contexts that influenced intimate partner violence professionals' subjective experiences. These risk-responses were a fundamental threat to the purpose of the research: to contribute to meaningful change for the participants in their contexts. The clash between the research aim and the risk-responses opened a theoretical space for reflection about power and knowledge relationships in lived experience and meaning-making in IPA research. Specifically, this chapter addresses the question of whether the current epistemological stance grounding IPA research leads to meaning-making that reproduces knowledge in a form that overlooks the omnipresent influence of power and knowledge dynamics. Butler's philosophies of power, knowledge, subjectivity, and performativity are explored as means of expanding the epistemological foundation of IPA.

Chapter 6
Data Collection as a Journey: A Multi-Step Approach to Navigating Access to an Organization 118
Géraldine Bengsch, King's College London, UK

The journey approach for data collection may enable the novice researcher to reflect on their data collection processes, and it aims to point to potential creative solutions that can help create a coherent study deeply rooted in its social environment. Fieldwork becomes a part of the project, rather than an isolated element that needs to be done. Through this approach, even a novice researcher can demonstrate connected insights that are not only relevant to the study itself but also the subject of study. A creative, multi-layered approach to resources and created opportunities may help increase the feasibility of a study by reacting to and interacting with the social context of the field. This chapter reflects on the author's data collection progress during her PhD program and invites readers to discover actionable steps that can be used to overcome the inertia due to inability to secure access to a field site.

Section 4
Gender and a Cultural Impact on Research

Chapter 7
On the Way to Researching Universities: The Case of Higher Education in Turkey 149
Nazlı Ayşe Ayyıldız Ünnü, Ege University, Turkey

This chapter aims to shed light on the significant experiences and challenges that have occurred during a university-funded project, exploring the dynamics of academic employment in the light of gender roles. The data for this mixed methods study comes from a research on Turkish universities, including 505 questionnaires and 46 in-depth interviews with academics from 39 different universities, located in 20 different cities of Turkey. The stratified sample of quantitative research represents the distribution of gender, department, managerial experience, position, and development level of the city, where the universities are located. In-depth interviews are used along with the survey to better understand the nature of academic employment and the implications of quantitative data. This chapter addresses the challenges, such as accessing the universities, ethical and emotional considerations, the effect of Turkish culture, censorship, and the ways that they affect the research process itself and the researchers.

Section 5
Counseling and Psychology Education Research

Chapter 8
When Participants' Trauma Becomes Mine: How I Used "Countertransference" to Enhance the Quality of My Study............................175
Sigalit Gal, McGill University, Canada

This chapter is a reflection on the author's work in the context of trauma-focused qualitative research entitled "Risk and protective factors for the mental health consequences of childhood political trauma (Argentina 1976-1983) among adult Jewish Argentinian immigrants to Israel." By examining the author's emotional reactions during the process of the data collection and analysis of her doctoral study, the author will explore the challenges that she faced, as well as the solutions she employed (both the effective and ineffective). More specifically, using the lens of the psychoanalytical term "countertransference", she will discuss the manifestations of her positionality as a qualitative researcher and its impact on her engagement with her study. The author will elaborate on different strategies that she used for her study, and propose qualitative researchers to use "countertransference" as a way to understand and address the complexity of a researcher's positionality in narrative research.

Chapter 9
Reflections of Challenges and Resolutions in Fieldwork Research on Quality of Life of Cancer Patients: Counseling and Psychology 199
Kavita Gupta, The Maharaja Sayajirao University of Baroda, Vadodara, India

Challenges and resolutions in fieldwork have become a dominant focus of interest for academicians and researchers conducting researches in social sciences. The objective of the chapter is to describe and reflect the researcher's experience in conducting a research study on anxiety, depression, and quality of life of the cancer patients that could be beneficial to novice researchers in the field of cancer studies. The author encountered these experiences in 2016 as part of her doctoral thesis that she had conducted in a tertiary care hospital based in Delhi. The chapter discusses the context of study with respect to the challenges faced in selecting a field site and acquiring access, recruiting, and building trust and rapport with the research participants, concern of confidentiality and emotional and psychological wellbeing of the researcher. Further, the chapter focuses on the resolutions and recommendations to deal with compassion fatigue, burnout, and secondary traumatic stress to avoid and manage great amount of emotional demands on the researchers.

Section 6
Post-Disaster Research

Chapter 10
Reflections on Post-Disaster University Research in the Aftermath of Hurricanes Irma and Maria 219
Kula A. Francis, University of the Virgin Islands, US Virgin Islands
Kenny A. Hendrickson, University of the Virgin Islands, US Virgin Islands

This chapter offered a reflection on a research study conducted at the University of the Virgin Islands, a historically black college and university (HBCU), in the aftermath of Hurricanes Irma and Maria. The research study examined student perceptions of authentic university academic care (AUAC) in a post-disaster setting. AUAC can be viewed as genuine humanistic concern and disciplined nurturing in higher education. AUAC is also an important element of examining the quality of academic services. Thus, this chapter shares the experiences of the researchers, their investigative processes, the challenges faced, solutions used, and recommendations.

Section 7
Online Gaming Research

Chapter 11
Multiplayer Online Game Research: Navigating an Unreal-yet-Real Terrain ...243
Seyed Abdollah Shahrokni, Western Oregon University, USA

This chapter explains the process of collecting data for an ethnographic case study in a Massively Multiplayer Online Role-playing Game (MMOG) called Stronghold Kingdoms, describing the methodological aspects of the study such as the research site, participants, data collection sources, and analysis. Further, the chapter explains the challenges encountered while conducting the study, including entering and becoming a member of the game community, data collection and analysis in an ever-changing world, being a participant-observer and the risk of going native, and participant attrition. Finally, the chapter provides some solutions and recommendations for researchers interested in conducting studies in MMOG settings.

Compilation of References 267

About the Contributors 299

Index 303

Preface

Conducting social science and education research studies that require involvement in fieldwork is not an easy task at all. Many graduate students and novice researchers face difficulties conducting the practical aspects of their research in fieldwork efficiently and effectively. Part of that might due to the lack of finding and/or accessing authentic and realistic descriptions of some of the previously conducted fieldwork experiences and processes in varieties of fields. This could be the case whether the research is going to be on a virtual platform or in a real and actual context.

Therefore, the goal of this book is to shed light on important experiences that portray the successes and the pitfalls of the researchers' personal experiences of fieldwork. The experiences that the academic world chose to be silent about and ignore while focusing solely on literature reviews, research methodologies and research outcomes at the end.

The textbook draws on the experiences of conducting fieldwork in different contexts and world regions that are relevant to social science and education studies. The diverse experiences in research processes and contexts that this book brings, can provide graduate students and novice researchers with an authentic and realistic description of how research data are collected, the tools needed to envision some of the challenges that they might face and how to effectively solve them.

The main body of the book comprises 11 chapters that are reflections of the authors' authentic research experiences and journeys in fieldwork. The authors come from different domains of social sciences and educational fields such as teacher education, education technology, counseling, gender, and ethnographic, and qualitative research. Following a brief description of each chapter and what each tries to cover including the different themes, the nature of the participants, and the contexts, e.g., country, higher education institutes, schools, hospitals, and etc.

The first chapter, for instance, is driven from the current need to explore strategies of implementing coding technology in elementary schools, a case study involving 2-second grade classes with their teachers was conducted to shed the light on and further understand the optimal methods of integrating coding into elementary level classes as well as benefits of context-based learning. This chapter starts with a description of the study goals and producers including the school context, background on the teachers and students and their use of technology. A detailed discussion is presented on the challenges faced by the research team with the different research stages as well as their suggested solutions. The challenges were classified as difficulties with study administration, which included issues related to obtaining permissions (the school district and the university IRB), working with a diverse research team, and collecting and processing both numeric and descriptive data. Another major category of challenges are the ones related to working with teachers and students participants, which entail issues of communication delays, differences in expectations, interviewing the student participants, and dealing with other unpredicted circumstances. Finally, the researchers share technology related challenges that range from introducing and controlling of the coding bots and access to the school technology. The chapter concludes with the researchers' recommendations for themselves, when conducting future research, and others wishing to explore technology use in such authentic classrooms.

The second chapter reflects on the author's challenges in getting access to a language center where she needs to coach teachers to integrate technology into their teaching pedagogy as part of her dissertation. As an international graduate student in the US, the task was not easy at all. So in this chapter the authors reflects on her doctoral research experience focusing on the importance of building trust and rapport with teachers before and while conducting the workshop, expanding the networking umbrella to find doors open when needed, and To be added

The third chapter discusses a number of fieldwork challenges, leading up to and during an ethnographic study undertaken in Southwest China. Consideration here is on the (pre-study) challenges of selecting an appropriate research site, recruitment of willing participants, possible language barriers, and potential issues arising from the role(s) of the researcher and the position of the researched. By explicating the ethnographer's path from fieldwork constraints to potential opportunities, the authors offer a number of coping strategies, such as building a broad set of relationships, developing collaborations, combining different sampling methods, being sensitive to language and researcher positionality, with an emphasis on the idea that fieldwork should be treated as a flexible process in which researchers negotiate their field site, being sensitive to local dynamics.

The authors present a realistic view, the tarnished and messy aspects of a fieldworker's experience amidst the various challenges in the planning and conducting of fieldwork. Unpredicted issues can easily emerge from fieldwork practice that researchers do not necessarily encounter in fieldwork guides. They consider fieldwork to be a unique journey for each ethnographer in which the complexities of the field only gradually emerge by the day. In this vein, the chapter serves as a useful resource for current and future ethnographers conducting fieldwork or setting out to conduct fieldwork. Additionally, it provides some modest insights into the unknown eventualities that almost certainly arise during the ethnographic research process.

The fourth chapter highlights how two world language teacher educators found themselves facing numerous challenges when they began research for an academic article on the psychological and social impact of online classes on lower-level learners. Although both researchers were well-schooled in qualitative approaches, their quest consisted of many twists and turns that, in retrospect, make for an interesting story as they reflected back on their two-year journey from the study's outset to its conclusion. They had to deal with many roadblocks that affected their ability to conduct the type of research they initially wanted to pursue. Some of these included difficulties finding a class to use for the study, attracting participants for the study, and time restraints. As a result of their flexibility and adaptability, they managed to reframe their research, which resulted in a publication, though not on their original topic. Consequently, they are now in a more objective position to describe the unanticipated problems and subsequent solutions based on their experience as well as to offer suggestions and guidelines for future endeavors. It is the objective of this paper to assist graduate students, fellow investigators, and anyone else who seeks to conduct informative research of the potential issues they may face. The scope of the chapter will address the chronological turn of events as a cautionary tale, but one that will offer an awareness, and hopefully, preparation, for such events.

The fifth chapter presents several unexpected issues encountered in the field during the author's doctoral research. Notably, these issues led the author to reflect on how to improve future interpretive phenomenological inquiries. This chapter first provides context for the situation through a literature review that introduces the reader to the author's substantive area of the research and establishes the reasons for this research project. Then she describes how she designed and carried out the research protocol and discusses how she responded and adapted to the unexpected issues that occurred at specific steps in her protocol. Finally, the author draws on her experience of how research methodologies can conceal power and knowledge dynamics, she proposes a mechanism for improving future interpretive phenomenological inquiry for novice researchers and graduate students.

In the sixth chapter, the author describes her approach to overcoming barriers in accessing and researching corporate organizations. The chapter focuses on hotels in the international tourism sector and elaborates on how a researcher may navigate the hierarchy of an organization. The author discusses practices for researching the field prior to data collection. Based on this, she provides ideas for how to approach a prospective organization and make participating in a research project meaningful to them. She suggests a method to overcome inertia in the data collection process. The chapter stresses the importance of communication between the researcher and the researched before, during, and after data collection in the field. The author argues that data collection is a process that, in itself, can add significant value to the study. Especially relevant for new researchers, the chapter encourages readers to reflect on their engagement with the organization they are investigating and learn not only from the data obtained but also their time in the field. The chapter aims to show that data collection is a process unique to each project in which each step contributes to the outcome of the study. All elements, including failed attempts at securing access to a site, shape a piece of research – and the researcher in the process. The author's account shows that data collection in the field is a humbling and grounding experience.

The seventh chapter aims to shed light on the significant experiences and challenges that have occurred during a university-funded project, exploring the dynamics of academic employment in the light of gender roles. Conducting a research on the gendered nature of academia and dynamics of academic employment sheds a new light on some research issues such as consent, censorship, rapport building, ethical considerations, and cultural barriers. This is in addition to a detailed discussion of other challenges and commitments that the research team has to deal with during the data collection process.

The data for this mixed methods study comes from the research on Turkish universities. It included 505 questionnaires and 46 in-depth interviews with academics from 39 different universities. All are located in 20 different cities in Turkey. The stratified sample of quantitative research represents the distribution of gender, department, managerial experience, position, and development level of the city, where the universities are located.

The chapter, specifically, consists of five sections, and each presents a critical aspect of the researcher's journey while conducting this project. The first section covers the general perspective, objectives, and background of the research. It also presents some details regarding the research team and discusses the originality of the study. The second section aims to familiarize the readers with the Turkish higher education system and provides a summary of the system. The third section explores the methodology used. It provides an overview of the research phases, and presents the initial challenges faced by the research team in each phase. The fourth section further presents the challenges faced by the research team throughout the different stages of the fieldwork, such as accessing the participants, dealing with ethical and security considerations. The fifth section reflects on the impact of Turkish culture on conducting research studies and presents the findings of the study. Finally, concluding marks are provided in the last section. The intent is to help graduate students, and novice researchers smoothly conduct their future research studies in the Turkish context.

In the eighth chapter "When participants' trauma becomes mine", the author provides a unique glimpse into the emotional experience of work on trauma-focused qualitative research. In this reflective piece, the author shares her personal journey during her research, using her journal as a reference, describing her emotional and behavioral reactions, and analyzing the impact these reactions have had on her subjectivity.

Given numerous similarities between talk-based therapy settings and the methodology of narrative research, the author uses the psychoanalytical term "countertranference" which relates to therapists' emotional and behavioral responses to their clients, in order to examine the challenges she faced during her research work as well as the solutions she employed.

The chapter begins with an introductory section that introduces the readers to the topic, and to the terms narrative research, psychoanalysis and countertransferance. It then explores some of the dilemmas the author encountered during her research journey when interviewing adult immigrants to Israel who were traumatized as children during the Argentinian military dictatorship (1976-1983), and the manifestation of countertransference in the context of her research work. This exploration illustrates what it is like for a researcher to hear for 30 hours childhood trauma survivors who re-live their trauma during the interviews; what happens when the researcher strongly identifies with the participants' stories; and how can the researcher manage the intersection between her personal history and the traumatic life stories. These dilemmas are followed by a description of the various strategies the author used to manage her reactions and subjectivity during her research work, such as using talk-therapy skills, writing a journal and self-talk techniques. An additional section in this chapter focuses on how the author managed to regulate her emotional and behavioral reactions in the context of her work on her recent research, which was conducted during the stressful time of the COVID-19 pandemic outbreak. The chapter concludes with some "golden rules" and recommendations for self monitoring in relation to handling researchers' subjectivity, such as techniques to develop one's self-awareness to emotional and behavioral reactions during research work, ways of creating a good supportive systems and environments during data collection and analysis and important tips for self-care.

This chapter offers researchers a look at their imbedded subjectivity in a new way, as it urges them to use subjectivity as an important source of knowledge instead of trying to dismiss it. The chapter gives researchers a roadmap to manage their subjectivity by providing concrete and simple tools that are relatively easy to implement, and that ultimately can significantly improve the quality of one

The ninth chapter deals with the challenges posited during the process of acquiring access to the field site, building trust and rapport with the respondents, facing confidentiality issues and, dealing with the emotional and the psychological well being of the researcher. These challenges were definable and observable since research in social sciences is human-centered. The researcher selected cancer patients as the potential research participants. The selection of vulnerable population paved the way with challenges that developed the burnout, compassion fatigue and secondary traumatic stress symptoms in the researcher. Critical thinking played a vital role in combating these challenges and issues. The chapter presents the recommendations that would be beneficial to the novice researchers who are aiming to undertake similar research studies. The resolutions are concluded on the basis of the researcher's experience that would help the researchers to remain motivated and maintain self-confidence. The present chapter is not only a piece of research work but a journey of challenges and resolutions. These challenges give an insight to the researchers to enquire of 'what has happened' with an investigative mind. In the end, this chapter is written with an aim that the reflections of the researcher's experience would be beneficial and valid across the belt of knowledge with an interdisciplinary approach. The implications are further supported by the findings from the literature. Thus, fieldwork considers the researcher's autonomy, courage, perseverance and openness attitude towards research.

The tenth chapter explores how natural disasters can be an unpredictable and challenging area of research. As a result of its complications, there is limited exploration in this area, especially as it pertains to higher education. There is an unquestionable need for research centered on assessing the quality of academic services at universities following disasters. This chapter shares the research experiences of conducting an investigation on Authentic University Academic Care (AUAC) in the aftermath of Hurricanes Irma and Maria. AUAC can be viewed as genuine humanistic concern and disciplined nurturing in higher education. AUAC can also be instrumental in assessing the quality of academic services in a post-disaster setting. This chapter provided the background, purpose, and challenges of the researchers' work, as well as their personal experiences and strategies for data collection. This study was conducted at the University of the Virgin Islands (UVI). UVI is a historically black college and university (HBCU). HBCUs are well-known for their high level of othermothering care, which is a strong exemplification of AUAC. The data collection process for this study utilized an original survey, Caring about Academic Care Inventory (CAACI) that was developed based on the researchers' conceptual framework, a University Academic Caring Carescapes Framework. Ultimately, this chapter provided the researchers' perspectives, approach, and recommendations on conducting post-disaster university research. This chapter is geared to readers and researchers interested in disaster research, student perception research, higher education, and the evaluation of academic services.

The eleventh chapter explains the process of collecting data for an ethnographic case study in a Massively Multiplayer Online Role-playing Game (MMOG) called Stronghold Kingdoms, describing the methodological aspects of the study such as the research site, participants, data collection sources, and analysis. Further, the chapter explains the challenges encountered while conducting the study, including entering and becoming a member of the game community, data collection and analysis in an ever-changing world, being a participant-observer and the risk of going native, and participant attrition. Finally, the chapter provides some solutions and recommendations for researchers interested in conducting studies in MMOG settings.

Each chapter started with a general background description of the nature of that particular research study. It is important to note that authors of this book were given the freedom to share their reflection in fieldwork using their preferable writing style. However, each chapter puts emphasis on two important points: (1) Some of the biggest challenges that each researcher faced during the research journey, emphasizing the data collection process and how they handled them, and (2) recommendations for graduate students, novice researchers, policy makers, administrators, independent researchers or whomever interested in that particular theme and type of research. Readers would find it beneficial learning more about other colleagues and researchers' experiences (accessibility to the fieldwork, traps in the fieldwork, delays because of some controllable and uncontrollable factors in the fieldwork, challenges in the fieldwork ...etc.) Learning about such challenges before entering any fieldwork might reduce the level of anxiety that graduate students and novice researchers might encounter during the process and they might learn from their colleagues' previous experiences how to avoid some of the obstacles and effectively solve them.

It is also important to note that each chapter is a reflection of the authors' journeys, perceptions, and attitudes about the fieldwork experience that might differ from a researcher to another depends on so many different factors such as experience, the type of participants, the nature of the research...etc. However, each shares glimpses that could benefit many researchers and give an overall idea about what some might encounter and a few ways of how to tackle some issues. Each chapter is an attempt to go beyond the methodology section that still pays attention to the number and the types of instruments used to collect the data and the nature of the methods used to analyze them e.g., qualitative, quantitative, or mixed methods. It is very rare to find any textbooks or articles that take the reader through detailed descriptions of the researchers' fieldwork journey. This book does so by providing its readers with a realistic diverse portrait of the untold part of the story of some compassionate researchers who agreed to share their voices and experiences with those who might need it.

TARGET AUDIENCE

The target audiences of this book are:

1. All the social science and education studies' professors who have research as a mandatory part of their curriculum.
2. All graduate students from everywhere in the world and this is why I thought it would be a brilliant idea to include contributors from different regions and cultural backgrounds.
3. All independent researchers who thrive to grow on an industrial level as well as an academic level.
4. Administrators and policymakers who might express their willingness to be of general assistance to help potential future researchers overcome some of the difficulties in fieldwork.

Abir El Shaban
Innovate Your Classroom (IC) Consulting, Canada

Acknowledgment

To the authors of this book who participated in reviewing the book chapters during its first phase, thank you.

To my family, friends, colleagues and fans, I am grateful for your kind wishes and support.

Section 1

Teacher Education and Education Technology Research Challenges

Chapter 1
Learning Amidst Mayhem:
Researching Teacher Development and Technology Use in Second-Grade Classrooms

Reima Abobaker
Savannah College of Art and Design, USA

Joy Lynn Egbert
https://orcid.org/0000-0001-8185-7011
Washington State University, USA

ABSTRACT

The purpose of this chapter is to describe issues with conducting field studies on technology use in elementary schools. The chapter first provides a brief overview of the study that was conducted with two teachers and their 46 second grade students, and then it presents some of the challenges that the researchers faced. These challenges include everything from securing permissions and scheduling to having sufficient personpower to collect all the data and using technology both in the study and as a topic of study. Solutions for future research are included, along with implications for all stakeholders.

INTRODUCTION

After a professional development workshop on technology use in education at a nearby university, a teacher expressed her interest in learning more about using technology in her classroom. The teacher, Ms. N, asked whether the workshop

DOI: 10.4018/978-1-7998-5826-3.ch001

facilitators would work with her around using small robots called "Ozobots" that had been introduced at the workshop. They agreed, and a small university research team was formed to support Ms. N's classroom practice and also explore context-based teacher professional development around technology use, specifically focusing on computer coding using the bots. Ms. N also obtained agreement from a grade-level peer to participate in the project with his class. Her colleague, Mr. R, had no previous experience with the bots, although he was familiar with basic computer coding.

The context for the study, then, was two 2nd grade classrooms in a local elementary school where the teachers wanted to learn about coding by implementing it with the help of the research team. The resultant case study, *Authenticity, autonomy, and just-in-time feedback: Situating teacher and student learning around elementary school coding* (currently under revision for *Computers and Education*), collected data from the two teachers and their 46 second grade students. Data were collected while the teachers taught coding based on lessons from the research team, and the students practiced coding with the bots over a week of instruction in each class. Both numeric and descriptive data were collected from the students and teachers, including videotaped observation of every class and recorded interviews; student data also included two implementations of a task engagement survey, a background survey, and student task documents. Conclusions and implications include what might work when integrating coding into classrooms and the benefits of context-based learning for participants.

The purpose of this chapter is to help researchers reflect on possible challenges when conducting studies that integrate technology and take place in elementary schools. From beginning to end of the project described in this chapter, the research team needed to think and adapt quickly to ensure that all of the data were captured, the teachers met their learning goals for themselves and the students, and all institutional guidelines for working with children were followed. In spite of the challenges, overall the research team was pleased with the data collected, the teachers and students were happy with what they had learned, and the outcomes of the study add to the literature in significant ways.

BACKGROUND

The research team initially consisted of one university professor (the PI), one postdoctoral researcher, and one graduate student assistant (GA), supported by a small amount of university and foundation funding. Although the research team discussed all aspects of the project together, in general the professor was responsible for obtaining support funding and any tools needed, creating instruments and lessons, and obtaining university permission for the study; the post-doc was in charge of liaising with Ms. N (who spoke for both teachers) and the school administration; the GA was tasked with gathering cameras and other research tools, collecting, copying, and filing data, and completing other administrative tasks.

LITERATURE REVIEW

The rising demand for implementing technology in instruction in elementary schools has left teachers in a dilemma (Konokman & Yelken, 2016), as many reflect on the challenges that come with implementing a new technology into their classrooms. These challenges include the availability of the needed technology and school readiness for it; staff, parent, and student understanding, acceptance, and attitude towards it (Vatanartiran & Karadeniz, 2015); the availability of technical support (Al Ghamdi, Samarji, & Watt, 2016; Gupte, 2015; Helm, 2015; Porter & Graham, 2016), and opportunities for effective technology professional development that provides teachers with examples of engaging tasks (Hyndman, 2018; Rebora, 2016).

The project was originally intended as teacher professional development rather than research, so the team's initial goal was only to record as much as possible about what went on during the sessions as an example of teacher situated learning (for more information on this theory in teacher education, see, for example, Arnold, Ducate, & Lomicka, 2007; Egbert, 2006; Korthagen, 2010; McNeil, 2018). Therefore, the general idea of professional development as situated learning grounded the context and process of the original study. As Egbert and Salsbury (2009) note, situating teacher learning is a way to help teachers understand some of the challenges they face, to produce and consider alternatives for integrating new ideas like coding into their curricula, and to collaborate as they reflect on new practices like coding. In other words, teacher development occurs through reflection and participating in activities in real classrooms, both theirs and others (Brown, Collins, & Duguid, 1989; Feiman-Nemser & Remillard, 1996; Lave & Wenger, 1991; Putnam & Borko, 2000). The theoretical framework for the study, therefore, was five of the nine elements of Herrington and Oliver's (2000) instructional design framework that support situated learning; these elements are presented in Table 1. The framework was used to build lessons for the teachers and characterize the findings of the study.

The project goal was to provide teachers with the support they needed to successfully implement coding lessons and tasks into their classroom. The teachers who participated in the project were supported in creating their own context-bound and personalized learning experience in editing and implementing their coding lessons (Knowles, Holton, and Swanson, 2014); in other words, learning to implement a new technology use (coding with the Ozobots and Scratch) into their school curriculum.

Table 1. Herrington and Oliver's (2000) guidelines for instructional design

Element	Guidelines for Implementation
Provide authentic context that reflects the way the knowledge will be used in real-life	a physical environment reflecting real use a non-linear design a large number of resources no attempt to simplify
Provide authentic activities	activities which have real-world relevance well-defined activities a single complex task an opportunity for students to define the task a sustained period of time for investigation the opportunity to detect relevant information the opportunity to collaborate on tasks which can be integrated across subject areas
Provide access to expert performances and the modelling of processes	access to expert thinking and modelling processes access to learners in various levels of expertise sharing of stories access to the social periphery
Provide multiple roles and perspectives	different perspectives on the topics from various points of view the opportunity to express different points of view the opportunity to criss-cross the learning environment
Support collaborative construction of knowledge	tasks which are addressed to a group rather than an individual classroom organization into pairs or small groups appropriate incentive structure for whole group achievement

The small robot (seen in Figure 1) has two modes: 1) it runs on a black track into which specific color codes have been integrated to provide instructions (see an example in Figure 2); 2) it uses the programming language Ozoblokly to follow instructions without a track. Not many studies exist that examine Ozobots in particular (see Fojtik, 2017, for one), but anecdotes across the Internet describe how versatile and engaging bot use can be for student learning (see, for example, a review by Powers at https://www.commonsense.org/education/app/evo-by-ozobot).

The purpose of the project was to describe what the teacher and student learning looked like and to uncover how both the teachers and students worked with the ideas and tools. The initial research questions were:

1. What does teacher situatedness in their learning about coding look like?
2. How do teachers implement tasks and teach elementary school students about coding? What do they consider?
3. How do the teachers and students perceive their participation in the learning experiences?
4. What are the outcomes for teachers?

Figure 1. An Ozobot

Figure 2. A sample Ozobot track with color codes

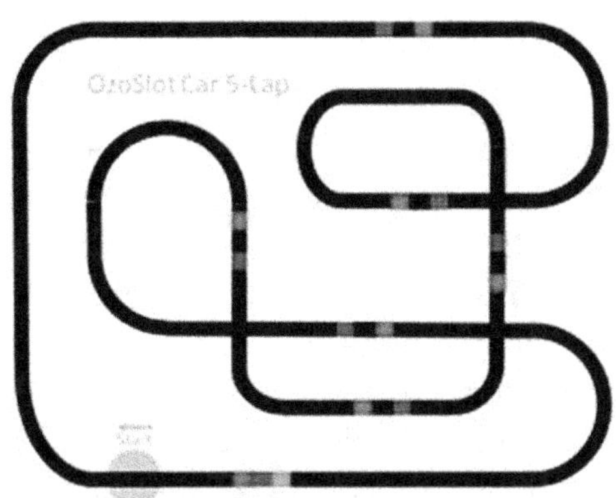

Each class session started with a whole class instruction by the teachers (with Ms. N's students on a rug in the front of the room and Mr. R's at their desk), and then the students were instructed to complete a specific coding task either in groups or individually. However, once students were released from the initial whole group interaction, mayhem (of a creative, loud, and interested sort) set in. Because of the differences in student abilities and their desire to push the bots to see what they could do, no two groups of students were doing the same thing at the same time. This organized chaos was only a small part of the challenges with this classroom-based study.

CHALLENGES AND SOLUTIONS

It is worth noting that both experience and rigorous planning are the key to a successful research study. However, conducting a study in an authentic setting can come with a number of challenges to planning that can create other challenges along the way. The main challenges that the research team faced with this field study are described below, grouped into three categories; these categories are challenges in administration, challenges in working with teachers and students, and challenges with technology.

Challenges in Study Administration

Although small administrative challenges were faced throughout the study period, the 3 main issues with administering this study were getting all the permissions and other paperwork needed in a timely fashion, working with a diverse research team, and dealing with a plethora of data, some of which the research team did not expect to collect.

Obtaining Permissions

Obtaining all the permissions needed was one of the biggest challenges with this study. Once the teachers agreed to participate in the study, the research team had to get approval from both the school district and the university institutional review board (IRB). Working with children requires more than a simple exempt form, so the application was much longer than usual and needed more approvals. At the end of the IRB process, a total of eight IRB related documents had been prepared and submitted; these included the original non-exempt application form, recruitment description, parent consent, student assent, measures, and school/district approval. The research team understood from the outset that the amount of time that the whole process would take was uncertain and depended on quick response from the district and school, the IRB's workload, and any issues the IRB had with the application. Therefore, team members started the original application as soon as possible - in the middle of January -- but all of the paperwork, including permission from the school district, was not submitted until March 31st. IRB permission to conduct the study was granted in May, a week before the study was to begin. This meant that, until right before the teachers wanted to start the study, team members were still unsure if permission would be granted to conduct the study. Once permission was granted, all 46 parental consents and 46 student assents had to be secured within a week because no part of the study could begin until all of the paperwork was signed. Happily for the study, the teachers were able to secure all of the permissions during that week. However, this left little time for the team to get to know the teachers and students and to finish the study preparation before data collection began. This lack of preparation time created other challenges that will be described later in this chapter.

There was no way after the fact to solve this problem, but a couple of partial solutions for future studies may help:

1. Start the IRB process as early as possible, and follow up as often as possible in the way that school partners are most comfortable using and most likely to respond in.
2. Prepare as if the study will be permitted rather than waiting to see if it will be. As much as possible, ask teachers questions about themselves and their students that might impact the study.
3. Be flexible and build in contingencies in case permission comes at the last minute or past that, know what the team will do if dates and opportunities change.

Working With a Diverse Research Team

Not only were the research team members in different positions at the university, they were also from different cultural backgrounds. The PI/professor was an American female, the postdoc a Libyan female, and the GA an Iranian man. Further, after the first day of the study when the initial team understood that it needed more help to capture the data from all of the very active second-grade students, the PI employed two more graduate students (GS), one Libyan male and one Libyan female. Due to the team's variety of cultures, background experiences, English language proficiencies, and genders, members of the research team understood things in different ways. For example, the field notes were really different. Figure 3 shows two different styles and contents for part of one day of Ms. N's instruction. The team member's notes on the left show a few words for each entry without any context connected; that observer conscientiously added a priori codes to her notes. The notes on the right show that that team member took notes verbatim and connected them with arrows to other data but did not add the codes during the initial note-taking.

To ensure that the notes would be usable if the team member who took them was not available to explain them, the team members agreed more specifically about what should be included in the field notes and the format that the notes should take. They also agreed that the codes did not need to be assigned during the observations. A better solution, to avoid this problem in the first place, is to norm the process of taking field notes. This did not happen in the current study because of timing, as discussed previously.

Figure 3. Examples of field notes by different team members

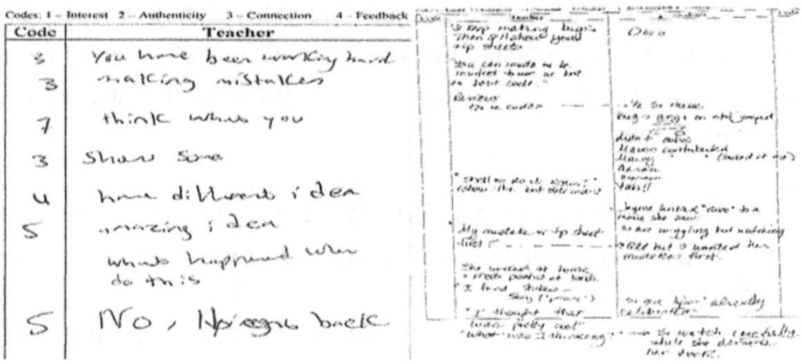

Although a challenge, team diversity was not really seen as a problem. After the second day, when all of the team members were in the classroom, the team agreed to more concrete roles for each member; for example, the GA was responsible for setting up and taking down the cameras and taking roving student shots, and he also posted all the data and conducted data entry; one GS was responsible for monitoring the teacher camera and the other still camera; these two team members also participated in putting all the supplies away at the end of each session. The other GS was asked to take observational notes, help students when the teacher, professor, or postdoc was not available, and she also participated in several of the interviews. As noted previously, the postdoc worked with the teachers, answering questions in each class, making requests for the research team, and helping in any way asked. As the study progressed, team members found additional ways to avoid issues. These included arriving at least 30 minutes earlier than the scheduled class time to set up the bots, disconnect them from their chargers, count the number of bots needed for that day, categorize the materials and worksheets related to the day's activities, check in with the teacher, and set up the cameras and recording devices. After each class session the whole team met informally to discuss what members had seen that day and how the process had worked out; team members also talked about any changes needed to make for the next day and who would bring which supplies.

Solutions to the initial challenges of working together can support any research team, whether it is as diverse as this set of researchers or not. These included, then, having frequent brief meetings with the whole team where everyone noted their perceptions of how the day went and expressed ideas about what could be done differently or better. In addition, dividing the responsibilities helped each team member to know that nothing was being missed and it would all get done. Further, the PI frequently checked the data, including the interview recordings, field notes, and student documents, as soon as possible after collection to address any issues with how they were collected. By providing oversight and having one person who understood the big picture, this helped the process to run as smoothly as possible.

Collecting and Processing All the Data

The team experienced several challenges with data collection and analysis. First, the research team did not know the number of students who would actually be in the classes for each session, because different students were pulled out of class to work with different specialists on different days during the time that the study would be conducted in the school. This was a challenge because it was difficult to figure out how many bots, handouts, or surveys would be needed for any day and how much data the team would be collecting from whom. One solution was to make enough handouts and prep enough bots for the whole class each time, using the extras either for future data collection or for "scratch" paper in the classroom. Another solution was to check that all study documents had names on them and then have the GA let the other team members know each day if something was missing so that, if there were an opportunity, the data could be collected the next day.

A second challenge was not having enough time to interview all the students. Because of how the school day was constructed, the students had a short recess after the daily research period and then they had to move on to other things. In the four days that we had planned for interviews for each class, it was impossible to interview all of the students individually. The team devised two solutions to this issue. The first was to add one of the additional GAs to the interviews and train her how to ask the questions. This took extra time but turned out to be an effective solution for reaching more of the students. The second solution, which actually appeared to help the students be less reticent to talk, was to interview the students in small groups. In this way most of the students were able to talk to the interviewers.

Third, in spite of some student absences, the sheer amount and variety of data collected was daunting. For example, the team was not aware ahead of time that Ms. N was going to allow her students to draw their own original bot tracks with codes on paper the second day of the study, and the research plan did not initially include collecting this kind of original data. However, the team noted that the students were very creative in creating original tracks (see an example in Figure 4) even after the brief period of their teacher's instruction on how to use the bots, and collecting their tracks as data might be one useful way to describe student outcomes. With the teacher and students' permission, the team decided to collect, scan, and organize this "new" data and then return the papers to the students/teacher the following day. This happened four days out of the five in Ms. N's class and several times in Mr. R's. This added many new pieces of data.

Figure 4. A student's original track with color coding for the Ozobot

The original study as proposed was based mainly on the data from the pre and posttest and an engagement survey (explained more in another section), in addition to teacher interviews. However, once in the class the researchers saw the need to collect photos, documents, incidental interactions, and student group videos instead of whole class videos. For example, the team noticed that when the students worked together to solve coding problems, they had different solutions and attacked the problem in different ways that were independent of how the teacher had taught them to address them. In order to capture as much of that diversity and problem-solving data as possible, the "roving" GA was tasked with taking photographs and videos of students' solutions as they worked. However, some of the data were missed because, during the class, one team member might see something interesting and signal wildly for photographs to be taken while the GA was already in another spot filming something else.

That means, by the end of the study, we had oral, written, numeric, anecdotal, pictorial, video, and other data to store, analyze, and synthesize in some way. The storage problem was solved with the decision to have one GA collect, scan or input, and organize it all; this helped the team not only to know where the data were but who to go to if a team member could not find something. In addition, the group decided in discussions to use individual student data like the coding tracks to illustrate findings in the numeric data, thus avoiding having to code or otherwise categorize the drawings in some way.

Overall, the challenges of finishing paperwork on time, working with a variety of team members, and figuring out how to manage both the planned and unexpected data were overcome quickly and mostly efficiently by the flexibility of the research team and the willingness of the team to adapt to changing circumstances.

Challenges in Working With Teachers and Students

In addition to challenges with setting up the study, conducting the study in an elementary school at the end of the semester meant entering an environment in which the participants had already established roles and routines; without enough time for the team to get to know the context and participants well before the study started, this was bound to create some challenges. The research team was faced with multiple unexpected situations that varied from teacher-oriented challenges to those involving students, as noted below.

Communication Delays

Ms. N initiated a chain of emails with the research team, inquiring about the possibility of collaboration in implementing the bots in her classroom. Ms. N was a busy teacher; in addition to her teaching responsibilities, she was also working on other projects and participating in professional development around reading. The research team had hoped to spend several weeks in the school collecting data. An agreement was eventually reached four months before the research commenced, and the specific dates and times of the study sessions were set by Ms. N (in collaboration with Mr. R) based in part on the school testing schedule. This points out that, when working with schools, the administration and the teachers determine the study, not the researchers. As with many of the challenges the team faced, it was important to be open-minded and work with whatever was given.

Once the dates were set, working with the busy teachers to plan the actual tasks for the study also met with time-consuming delays. Finding a mutual method of communication was an issue. For example, the teacher participants did not answer their emails often, but it was the researchers' favored way of communication. Two solutions were implemented. First, the post-doctoral researcher had to request in-person meetings and visited the teachers on the school campus several times to obtain the needed materials and answers. Second, not knowing what the teachers' specific goals were prompted the researchers to request a letter of professional expectations and goals for the proposed study. This letter was written by Ms. N, but it was not completed until immediately before the study was set to start; this minimized the time the researchers had to specifically prepare the study materials and procedures.

However, the written document did allow the research team to do what they could ahead of time and was more effective than not having any indication of expectations.

Communication delays also forced the research team to reconsider or change some of the tentative plans they had made. For example, the corresponding teacher, Ms. N, confirmed the participation of two of her grade-level colleagues, therefore including all three of the school's 2nd grade classes in the study. However, before the study was set to start, one of the teachers withdrew her participation from the study, decreasing the number of participants by 1/3, changing the duration of the study, and eliminating the possibility of collecting additional enriching insights. Such changes might be expected in studies involving human participants, but the notice of this change came at such a late date that there was nothing the research team could do. Therefore, the team focused on collecting as much data as possible from the other two classes and adapting their initial plans to obtain the needed insights and findings. In other words, the solution was to revise the plans quickly and to discover what we could from the remaining two classes.

Differences in Expectations

While the research team members were well-versed in the use of the bots and coding languages, Ms. N expressed her concern of her limited knowledge of using the bots as well as creating lessons incorporating the bots to fulfill her students' learning outcomes. The data showed that, with the basic lessons that the team prepared to outline possible steps for the class sessions, Ms. N had very little trouble adapting and implementing focused lessons based on her students' needs and abilities; in other words, Ms. N, a very experienced teacher, was by far the expert in teaching second graders, which most of the research team had little experience with. On the other hand, Mr. R was a fairly new teacher with no previous experience with the bots, and his expectations of the research team were quite different. He appeared willing to follow the lesson plans verbatim and stopped to ask questions of the team members during class any time he did not quite understand what to do. Because of this inconsistency, a couple of challenges emerged.

One difference in expectations involved the lesson plans. Initially the lessons prepared by the research team were based on the instructions and worksheets that came with the bots. Each of the five lessons created for the teachers before working with the students included goals, objectives, and task steps in a preview/ do/ review format. However, Ms. N slightly adapted the lesson plans on multiple occasions, using her 17 years of teaching experience to tailor its content to fit her class conext and student progress. Such changes were not communicated to the research team before implementing them so team members could not follow along on their copies and make comments. The team therefore requested copies of the adapted plans and filed them with the other data to help examine the adaptation process, adding an interesting and unanticipated data set for the research team to analyze.

Another challenge around expectations was the nature of the lessons themselves. While the lessons that the research team created focused on using the technology to learn about coding, Ms. N expressed right before the study started that she hoped to mainly focus on creating an environment for her students to reflect on problem-solving (emphasized in the process of debugging their codes) and the idea that making mistakes was part of life. In her modelling and questioning of students she focused on making solving a problem into a thinking journey full of trial and error to find the optimal solution to the problem at hand. Taking the cue from Ms. N, this modification led the research team to participate in the class sessions with questioning, guiding, and modeling the tasks for the students as opposed to only teaching the bot operation and collecting data. On the other hand, Mr. R stuck fairly closely to the original lessons, teaching the bots and coding as the topic of the lessons. This led the team to reconsider how to frame their approach to the study and their findings; while originally the study was focused on the teachers and their professional development, the team realized that the students' feedback and outcomes were a part of the larger classroom picture that needed to be understood and reported. Fitting these pieces together with the large amount of data took several brainstorming sessions.

Based on the aforementioned challenges, the research team had to adapt their approach to allow for multitasking, providing the necessary support for the teacher and the students as well as making sure that no important, unplanned data were missed. Interview questions for both teachers and students, as well as the student posttest, were reformatted to address the updated focus. The research team could have continued with the initial study focus, but the data would not have been as rich or as explanatory. This is yet another instance where the solution to the challenge is to adapt as quickly as possible, while retaining the rigor and theoretical soundness of the overall process.

Changes in Researchers' Roles

The research team did not expect to have to help the students' needs. The 2nd graders, introduced to a new, exciting technology, wanted a lot more support more quickly than the researchers had anticipated. The teachers could not address all of the students' constant questions or their desire to show them what they had accomplished, so the teachers enlisted the research team to listen to and check on students. The team had to decide whether and how such participation would influence the study outcomes and decided that it was more important to understand and do what the teachers needed than to watch them and their students flounder and spend time being discouraged. The team had to figure out ways to balance helping students and collecting data; one way was to take notes while helping students and ask the students and/or teacher questions about their process/progress after the session was over. This actually helped the team better focus the interview questions and learn more specifically about the teachers' thoughts.

Another challenge originating from the change in the researchers' role from passive data collectors to active classroom visitors was figuring out what appropriate interactions were. For example, students sometimes asked the research team members about topics unrelated to the study, such as permission to do something that the teacher had not mentioned at the beginning of class to solving peer conflicts. Again, the team did not want to assume or detract from the teachers' established classroom roles, so they consulted with the teachers and established boundaries for the team members' interactions with students. Part of this meant letting the students know that they did not have the authority to attend to certain types of inquiries and that they would have to wait for the teacher. For some students, this meant that they ended up waiting impatiently or became discouraged briefly, but it also helped the research team interfere less in the authentic data collection.

There may be no way to avoid some of the aforementioned challenges, but previously-mentioned solutions might help. For example, a research team may consider postponing their study if planning time is too short and it is not possible to work with the teachers to agree on important aspects, from goals and roles to procedures. Or, understanding that these aspects might pose challenges, the team can provide teachers with a questionnaire or a proposal so that everyone is on the same page at the beginning of the study. Further, agreeing on a suitable method of communication and a response deadline for all collaborators may ease some of the pressure and prevent misunderstandings. At the same time, being flexible enough to work with changing expectations and considering them part of the study data can mitigate what might otherwise be a frustrating experience for all involved.

As noted above, working with the very excited, energetic, and chatty students also posed challenges. Excluding the senior researcher, the research team members had limited experience interacting with and teaching a 2nd grade class of 7-9-year-old children. Coming from cultures where elementary education was more sedate, teacher-centered, and quiet, the international researchers were surprised by what they perceived as a chaotic, loud, and constantly moving classroom dynamic. This lack of initial understanding of the environment led to several issues.

Introducing the Bots

On the study's first day, after some whole class explanation by the teachers, the teachers and research team distributed the Ozobot boxes to teams of students. Each box contained a bot, four different colored markers, 2 decoration skins, 25 task

Figure 5. A student interacts with the contents of the Ozobot carton

worksheets, 2 code stickers, and 3 cardboard detachable decoration sets (see an example of a student interacting with the contents of an Ozobot box in Figure 5).

Then the teachers started introducing the bots, describing steps for opening the boxes and setting up for the day's task. However, unwrapping and unboxing took the students quite a bit longer than anticipated because they wanted to look at every item and understand what it was. The students were very excited to open the boxes, take everything out, and attempt to operate the bots despite the teachers' repeated instructions to only take out the materials needed for the day. The researchers and Ms. N realized the mistake of allowing the students to use the decorations in the box to dress their bots - student teams not only disagreed, but they wanted to keep changing the decorations. This used up a lot of the first session, and Ms. N asked the research team to remove all the decorations from the boxes after the first day. Before starting to work with Mr. R, the team explained to him about the situation with the decorations that had occurred in Ms. N's class, and he decided to keep the decorations separate and use them as needed.

This same type of situation occurred in some way during each day of the study. For example, the students were distracted to the point that some of them did not pay attention to the teacher's instructions on how to calibrate the bots (described in the next section) and apply the codes for the first lesson. This led to students' frustration, as many of them could not successfully get the bot to work or apply the codes. The researchers and the teacher had to reteach some individual groups of students and walk them through the steps again. By the time that most of the students had accomplished the task (and some had moved far ahead), the class was almost over. It was time for clean up and organization, which the students not only did not want to do but had a hard time doing because there was so much to put back in the box. The research team discussed this issue with the teachers and decided to reorganize the materials to only include the worksheets needed for each lesson and to store most of the decorations in another place in the classrooms. This led to smoother lessons afterwards, although some students still exhibited excited and anxious behaviors while they were supposed to be listening and interacting with the rest of the class.

Unpredicted Circumstances

As in any elementary classroom, data collection meant that the research team had to consider that there were lots of student pullouts for special services during the coding lessons. Some students were sick some of the days, minimizing the pool of candidates for student interviews. In addition, one student in Ms. N's class had special needs and became upset enough to throw everything off his desk when he could not get the bot to work as he wanted; the student shut down and refused to talk to the researchers during or after class. These circumstances are common in any classroom, but the disadvantage for the research team was being unaware of them and their potential impact on study procedures. The researchers asked the teachers to identify students who might have different needs, such as English language learners, those with special education needs, and any others that the teachers wanted to provide information on, and the team also decided to identify the students who were present each day as potential candidates for interviews.

Class Layout

The layout of the class promoted collaborative interactions between students, as their desks and chairs were grouped to form tables, each seating 2 pairs of students. The initial challenge was with setting up the recording equipment. With only 2 cameras available initially to capture footage of the teacher and all 23 students' interactions, it was difficult to place the camera's tripods in optimal spots to clearly record student's voices, conversations, facial expressions, and movements without them being occasionally and accidently knocked over by the students. With the addition of 2 more GAs and extra cameras (explained in the administration section above), this problem became less of an issue. However, using a recording program such as Panopto or Camtasia for each set of students might have allowed the research team to collect student data more easily.

Interviewing the Student Participants

Interviewing participants means that a researcher needs to both stimulate genuine responses without leading the participants and consciously control a biased interpretation of the observed activities. Interviewing the 2nd graders, who were quite voluble in class but not nearly as outspoken during interviews, made this difficult.

First, the research team had struggled to set a time to interview the students. The one-hour class was not enough time to conduct a lesson, tasks, and interviews. Running over the set time during almost each class led to interviewing student participants during their recess, moving to group interviews after conducting a couple of individual ones, and minimizing the overall number of interviews by only meeting with students who showed an interesting behavior or idea during the lesson. These solutions allowed the researchers to interview about half of the students in each class.

Further, the teacher's unexpected modifications of the lesson plans led the researchers to add interview questions about the changes. The researchers needed to take extra care in preparing the additional questions to ensure that they were phrased in age-appropriate language and carefully crafted to generate a clear response. The original set of questions asked about student understanding of the purpose of using the bots, their level of difficulty, and tasks liked and disliked, as well as ways that the students aspired to learn about the bots and coding. Adding questions to an interview that was probably already too long for wiggly 2nd graders meant that the interviews would take an extra couple of minutes. This added to the time constraints, meaning that students who the team were interested in interviewing because of their behaviors or contributions on one day were sometimes not interviewed until the following day. To remedy such delay, the researchers used stimulated recall for many of the questions by showing the students their work or providing reminders of the observed behavior, but this did not always result in useful responses.

Further, researchers sometimes received a brief response to an interview question for which details were needed but not provided. In this case, the question had to be restated in age-appropriate language or explained and paraphrased in order to encourage a meaningful response. Students were regularly asked to give examples and illustrate their intent, but their responses, for the most part, were very brief. However, adding data of student conversations and behaviors during their tasks helped the researchers make sense of the student interview data. While interviewing the students the researchers also needed to be aware of their tone, posture, facial expressions, and body language to project friendliness and create a comfortable environment. The students had never been interviewed by outsiders before, and this allowed them to more freely open up and share their thoughts. Further, PI and the female GA who was assigned to doing interviews (but had not previously done so) teamed up at first so that the GA could learn how to conduct interviews with children and control bias and leading questions. This led to consistency across interviews.

While waiting for teachers or administrators to respond to requests, or planning other aspects of a study, researchers can avoid some interview challenges by norming the questions across interviewers and practicing the interviews so they elicit the most data in the shortest amount of time. In addition, listing and providing solutions for a variety of "what-ifs" can also mean saving time and trouble when the study actually begins.

Technology Challenges

It is probably the rare classroom study employing technology where the technology works and is used perfectly, and the current study was no exception. From storage and set up to access and student excitement, the challenges with the technology were the most crucial because if the technology could not be used, the study could not be conducted. Four challenges in particular are worth noting.

Control of the Bots

One challenge with using the technology was that, because we initially left the materials, including the bots and cameras, at the school site, the team did not have control over what happened to them when it was not there. For example, even after purchasing additional power plugs for the Ozobots to make sure that all of them would be charged when needed, the team would find the bots unplugged upon arrival at school. In addition, several times one or more of the cameras ran out of storage space or power, and that meant that that camera could not be used for the rest of that day. To solve this problem, the team asked the teachers to make sure that the bots were charged, and the study PI purchased additional memory cards for the cameras that could be swapped for the full ones immediately. Further, one GA took the cameras to his office every day to plug them in there and make sure that they would have sufficient power for the next day's data collection. These seem like simple solutions to problems that should have been avoided at the study's outset, but the team's initial lack of familiarity with the two classrooms and lack of sufficient planning time created challenges that otherwise may not have existed.

Access to the School Technology

Often the issue in technology-based studies is that students do not have access, or equitable access, to the technology. In the current case, it was a problem that we did not have access to the school's technology. The 2nd grades had a rolling cart of

Chromebooks that they shared, and Ms. N scheduled them to be in her classroom during our sessions. However, we did not find out until the lesson plans had been written and the study was about to commence that the school technology rules and lack of sufficient school technology support meant that the team could not install, or have installed, the software that was planned for use after the students had mastered the Ozobots. Instead of moving on to the Scratch programming language, the students continued using the Ozobots and Chromebooks with the web-based programming language OzoBlokly, which did not require installation. This was a satisfactory solution because both Scratch and OzoBlokly are block coding languages that have similar functions.

However, the use of Ozoblokly did pose a challenge in itself. To use the software, the students first put their codes together in the programming software. When their short programs were finished, they had to hold the Ozobots on the screen so that the bots could intake the instructions. However, because the Chromebook screens were not bright enough, the bots had a hard time reading and accessing the programs that the students had developed. One GA figured out the problem, and then the team had to find out how and then instruct the students in turning up the brightness on the Chromebooks. This solved the problem for the most part, and it turned the students' discouragement at the work they thought they had wasted into excitement. That was part of another challenge that the team faced, student excitement about the bots.

Student Perceptions and Use of the Bots

The first day in each of the classrooms when the students saw the boxes of bots, their excitement was easily observed. Student enthusiasm grew as they were allowed to open the boxes and they saw the small, round bots; a crescendo of "awww" was heard across the classroom. In both classes, the students' eagerness to make the bots do what they wanted them to do grew over the week, and it led to a variety of challenging situations. For example, several groups of boys were engaged in seeing what would happen if they let their bots fall off their desk. Although the research team appreciated the students' curiosity, it was also concerned that using the bots in this way would render some of them inoperable. However, it was not the role of the research team to discipline the students, so a team member discussed the issue with the teachers. The teachers then set the consequences for the students of doing activities that could harm the bots. Although there was some delay time between the students' act and being given this warning by the teacher, supporting the teacher's role as classroom leader was important not only to the purpose of the study but also to working with her as the expert on her classroom.

Another result of student excitement was that they became agitated quickly when they could not get the bot to work. The Ozobots required a short process to calibrate them before they would follow instructions, but the students often forgot to do this (or how to do this). In their eagerness to program and see what the bots would do, the students did not want to wait the 3 seconds that calibration took or to retry when they calibrated incorrectly. One solution in Ms. N's classroom was to discuss this issue with her students at the beginning of each session, where she reminded the whole class of what they had already learned before teaching them a new concept. Then, research team members made a practice of asking students whether they had calibrated before answering other questions that the students had. This practice carried over into Mr. R's class during the second week of the study.

RECOMMENDATIONS

Some literature suggests that students may act differently or be distracted by the researchers/data collection methods, but our team did not see this at all, and the teachers agreed that the students were too engaged with the technology for this to be a problem; this made one challenge that we did not have to deal with! Although some of the challenges faced were easier to solve than others, by following our own recommendations below we will be able to conduct our next field work more easily. The overall recommendations include getting to know the participants and context, planning for unplanned situations, having clear, shared expectations, and, above all, being flexible.

Get to Know Participants Ahead of Time

Unless it is problematic for the study focus, As time allows, observe in the class and talk more extensively with the teachers before the study period begins. We were not allowed to collect data early, but we could have asked more questions that were not considered "data." We would have but couldn't, but this would have allowed us to do many things differently, from providing student support to asking more specific questions to better understanding how these students could best understand the surveys.

Plan

Two weeks is a relatively short time to explore classroom processes and outcomes, but it is important to get the best data out of the time that the teachers can spare, even if only two weeks. Again, any time spent planning is time well spent that can help make the most of the time on site. During the planning time, it can be useful for researchers to brainstorm possible challenges, layout alternative scenarios, and make sure that all team members understand their roles and tasks well. Norming on instructions, interview techniques, and other study activities can save time and help the data collection period flow smoothly.

Make Expectations Clear for Team Members and Participants

As part of their planning (and when appropriate for the study), researchers conducting studies in natural classroom settings can work with the teachers to set out expectations in writing. This will help to make sure that the goals and process are agreed on and everyone understands how changes or differences will be worked out. Because teacher professional development means building a relationship with the teacher rather than dictating what the teacher should do, respecting teachers' roles in their classrooms is paramount. It is worth mentioning that providing opportunities for teachers to cooperate in the design of the professional development can not only support teacher outcomes but also help the research team gain valuable insights into both the teacher's thinking and her knowledge of her students.

Be Flexible

If the study would be ruined if something went wrong, it is probably not the study to do in a fieldwork situation, particularly with young children. On the other hand, if the researchers are open to the possibilities that can arise and have planned in extra time and resources just in case, interesting ideas and new insights might emerge.

CONCLUSION

Any authentic education setting can pose challenges for researchers, due to the "messiness of the real world" (McKinley, 2019, p. 2) and the related "methodological messiness" (p. 6) that might ensure. However, as also he notes, without seeing them in their actual context it would be far more difficult to understand the "complex issues that teachers deal with in their daily practices" (p. 2). In the project described in this chapter, the researchers experienced both the mayhem and the complexity, and by being flexible and willing to bring in additional help, the team addressed many of the challenges that the elementary school context presented. If we did this project again, we would certainly heed our own advice, as offered in this chapter, to plan better, understand the context more thoroughly, and provide clear roles for all the participants. In this way, we might be able to work better within the mayhem that can be part of researching in authentic contexts.

REFERENCES

Al Ghamdi, A., Samarji, A., & Watt, A. (2016). Essential considerations in distance education in KSA: Teacher immediacy in a virtual teaching and learning environment. *International Journal of Information and Education Technology, 6*(1), 17–22. doi:10.7763/ijiet.2016.v6.651

Arnold, N., Ducate, L., & Lomicka, L. (2007). Virtual communities of practice in teacher education. In M. A. Kassen, R. Z. Lavine, K. Murphy-Judy, & M. Peters (Eds.), *Preparing and developing technology-proficient L2 teachers* (pp. 103–132). CALICO.

Brown, J. S., Collins, A., & Duguid, P. (1989). Situated cognition and the culture of learning. *Educational Researcher, 18*(1), 32–42. doi:10.3102/0013189X018001032

Egbert, J. (2006). Learning in context: Situating language teacher learning in CALL. In P. Hubbard & M. Levy (Eds.), *Teacher Education in CALL* (pp. 167–181). John Benjamins. doi:10.1075/lllt.14.15egb

Egbert, J., & Salsbury, T. (2009). Out of complacency and into action: An exploration of professional development experiences in school/home literacy engagement. *Teaching Education, 20*(4), 375–393. doi:10.1080/10476210902998474

Feiman-Nemser, S., & Remillard, J. (1996). Perspectives on learning to teach. In F. Murray (Ed.), *The teacher educator's handbook: Building a knowledge base for the preparation of teachers* (pp. 63–91). Jossey-Bass.John.

Fojtik, R. (2017). The Ozobot and education of programming. *New Trends and Issues Proceedings on Humanities and Social Sciences, 5*, 8–16.

Gupte, M. (2015). Does information and communication technology have an important role to play in Indian higher education? *Abhinav International Monthly Refereed Journal of Research in Management & Technology, 4*(1), 41–46.

Helm, F. (2015). The practices and challenges of telecollaboration in higher education in Europe. *Language Learning & Technology, 19*(2), 197–217. http://llt.msu.edu/issues/june2015/helm.pdf

Herrington, J., & Oliver, R. (2000). An instructional design framework for authentic learning environments. *Educational Technology Research and Development, 48*(3), 23–48. doi:10.1007/BF02319856

Hyndman, B. (2018, August 13). Ten reasons teachers can struggle to use technology in the classroom. *The Conversation*. Retrieved from http://theconversation.com/ten-reasons-teachers-can-struggle-to-use-technology-in-the-classroom-101114/

Konokman, G. Y., & Yelken, T. Y. (2016). The development of the resistance scale towards technology supported instruction: Exploratory and confirmatory factor analysis. *International Online Journal of Educational Sciences*, *8*(4), 199–214. doi:10.15345/iojes.2016.04.016

Korthagen, F. (2010). Situated learning theory and the pedagogy of teacher education: Towards an integrative view of teacher behavior and teacher learning. *Teaching and Teacher Education*, *26*(1), 98–106. doi:10.1016/j.tate.2009.05.001

Lave, J., & Wenger, E. (1991). *Situated learning: Legitimate peripheral participation*. Cambridge University Press. doi:10.1017/CBO9780511815355

McKinley, J. (2019). Evolving the TESOL teaching-research nexus. *TESOL Quarterly*, *53*(3), 875–884. doi:10.1002/tesq.509

McNeil, L. (2018). Situated learning in CALL teacher education. In J. Liontas (Ed.), *The TESOL encyclopedia of English language teaching*. John Wiley & Sons. doi:10.1002/9781118784235.eelt0438

Ozobot & Evollve, Inc. (2020). *Ozobot: Robots to code and create with.* Ozobot. https://ozobot.com/

Ozobot & Evollve, Inc. (2020). *Ozoblockly*. Ozobot. https://ozobot.com/create/ozoblockly

Porter, W. W., & Graham, C. R. (2016). Institutional drivers and barriers to faculty adoption of blended learning in higher education. *British Journal of Educational Technology*, *47*(4), 748–762. doi:10.1111/bjet.12269

Putnam, R., & Borko, H. (2000). What do new views of knowledge and thinking have to say about research on teacher learning? *Educational Researcher*, *29*(1), 4–15. doi:10.3102/0013189X029001004

Rebora, A. (2016). Teachers still struggling to use tech to transform instruction, survey finds. *Education Week*, *35*(35), 4–5. https://www.edweek.org/ew/articles/2016/06/09/teachers-still-struggling-to-use-tech-to.html/

Vatanartiran, S., & Karadeniz, S. (2015). A needs analysis for technology integration plan: Challenges and needs of teachers. *Contemporary Educational Technology*, *6*(3), 206–220. Retrieved from https://files.eric.ed.gov/fulltext/EJ1105647.pdf

Woodcock, J. (2015). *Coding with Scratch: Workbook*. New York, NY: DK. https://scratch.mit.edu/

KEY TERMS AND DEFINITIONS

Calibrate: Calibration is a short process that helps to initiate the Ozobot's code and line-reading accuracy. For example, a button on the bot is pushed and held, when the bot's light flashes white, the bot is set on a black dot. When the light on the bot flashes green and it moves a short distance, it has been calibrated and is ready for use.

Debugging: Debugging means finding and solving problems with code.

Elementary: In the United States, elementary schools generally start at kindergarten (students aged 5-6) and then include 1st through 5th grades. Some elementary schools also include 6th grade. After elementary school, students move on to junior high/middle school.

Institutional Review Board: A group, commonly known as IRB, that approves the study, making sure that all legal and ethical responsibilities are considered and carried out. This group reviews the study methods, all instruments and measures, and the wording of consents and assents, in addition to other study aspects.

Instructional Design: This term refers to using the elements of instruction that ensure its quality; this process starts with understanding student needs, abilities, and goals and then working to develop instruction with them as the basis.

Ozoblokly: This is a coding language that uses blocks of text that can be loaded on the Ozobots to provide them with instructions (available via http://https://ozoblockly.com/).

Ozobot: An Ozobot is a small round robot made by Evollve (ozobot.com) that can follow instructions that have either been color-coded or programmed with the block script Ozoblokly.

Professional Development: Also known as "PD," it refers to teacher learning after their initial certification. Public school teachers in U.S. schools are required to participate in a certain amount of PD in order to keep their teaching certificates valid.

Scratch: Another type of block coding that is generally considered easy for young students to use (available free from https://scratch.mit.edu/).

Technology: In this study, "technology" refers to digital tools that can be used for educational purposes.

Chapter 2
Handling Research Challenges:
Understanding Teachers' Adoption Process of Education Technology

Abir El Shaban
Innovate Your Classroom (IC) Consulting, Canada

ABSTRACT

This chapter explores some of the challenging issues that were raised for the author while conducting her first biggest research project as an international doctoral student at one of the United States universities. The main challenge that the author faced was the difficulty of having access to a language center to examine her technology professional development model on language teachers to explore its effectiveness in understanding the teachers' adoption and rejection decision of using education technology and to collect data for her dissertation. After choosing an alternative venue and planning to travel to the United Arab Emirates to conduct the practical part of her research at one of the UAE's universities, an academic networking event changed the course of the latest plans and had a gatekeeper assisting her in experimenting her model and conducting the rest of her study in an American university. The chapter explores some of the challenges and how the author tackled each one of them. The chapter ends with some general recommendations for graduate students and novice researchers.

DOI: 10.4018/978-1-7998-5826-3.ch002

INTRODUCTION

"Today, I will defend my dissertation entitled 'In-service ELL Teachers and Technology Adoption: Exploring Diffusion of Innovations in English Education'. In this presentation, I will give an over all idea of the work I have done during the last three years". This is how I, the author of this chapter, started the moment that each graduate student is aiming to; however, reaching to that moment was not an easy task at all. For a researcher to be called a practionars scholar, he or she need to master some important skills and know very well the components of running a research project. During the research process, we all have certain goals to achieve and some challenges to deal with; some of the challenges are difficult to manage, while others are manageable. It is our research, yet it necessitates the involvement and management of many different factors to succeed in reaching the intended goals. Some of the very common, yet manageable challenges that each graduate student face starts in choosing a research topic, finding a true gap in the literature, writing a clear proposal, and getting approval of an advisor and committee members. From my experience, those mentioned challenges are the easiest to manage, however, challenges that are associated with getting access to a particular institute, recruiting participants, and having to deal with each participant in person are the two main difficult to manage challenges, especially for an international student, like me, whose research study involved seeking access to language institutes and coaching its teachers on how to integrate technology into their teaching pedagogy.

As someone who is coming from a very low tech area and who is first heard about the term education technology and its value in an American university, I became curious to know how the successful diffusion of educational technology can be achieved. I wanted to know as an individual how to help teachers and students in low tech context to be prepared for the 21st century needs and challenges including the use of technology. Unfortunately, this dilemma is not only in developing countries, but it does exist in developed countries as well. For instance, research showed that regardless of the many benefits of using educational technology in classroom settings and the amount of money institutions spend on hard ware and software, teachers still resist using technology and that adequate examination of the process that leads to the adoption of educational technology by teachers in general and language teachers in particular is still lacking (El Shaban, 2017; Kessler, 2006).

Regadless of the technology gap that exists where I used to study and live in the US and regerdless of me offering free services to coach teachers on how to integrate technology effeiciantly and effecitively into their teaching pedagogy, receiving an adminstrative approval to conduct my research in a language institute was the toughest challenge that put a very heavey burden on my shoulders. Without an access to an institution, I will either end up stoping working on my dissertation or changing the technology intevention (the workshops) part that I was and I am still so passionate about. The last was not an option at all. Thus, in this chapter, I will reflect on my research journey including the challenges that I faced during the research process and how I solved them, but first, in the background section, I will give a brief introductory idea about my research topic.

BACKGROUND

The ubiquitous use of computer technology in our daily life has influenced the educational field (McGrail, 2005). General educational reforms to institute innovations have been a goal of the U.S. federal government for more than a century; the overall goal is to enhance students' academic achievement and educational technology has been considered as part of the reforms (Fullan, 2007). The Association for Educational Communications and Technology (AECT) defines the term "educational technology" as the ''ethical practice of facilitating learning and improving performance by creating, using and managing appropriate technological processes and resources'' (Januszewski, & Molenda, 2008, p. 1). Many educators believe that the use of educational technology in learning may bring about significant positive outcomes to students' learning. For example, Drexler (2010) found that technology use could transform the traditional classroom environment. Specifically, the teacher shifts from knowledge provider where the students are the recipients of knowledge to implementing a differentiated classroom where students collaboratively work with the teacher to achieve individual, pair, and group learning objectives. Also, technology use may lead to learners' autonomous learning (Terrell, 2011). Furthermore, the use of technology can help teachers create an engaging and interactive learning environment. It can provide second language learners with the opportunity to interact via speaking and writing in the target language and may motivate "learners to produce more language than they otherwise might have done" both outside and inside classrooms (Stanley, 2013, p.2). Despite these potential benefits, adequate examination of the process that leads to the adoption of educational technology by language teachers is still lacking (Allen, & Seaman, 2013; Brahier, 2006).

Thus, this study had two goals. First, it aimed to understand the process of decision making in educational technology by in-service language teachers, using Rogers' Diffusion of Innovations Theory (DIT) as a theoretical framework. Second, it strived to support professional development facilitators in providing context-appropriate professional development in education technology (El Shaban & Egbert, 2018). Based on the idea that the integration of educational technology facilitates changes in teaching and learning, this research study sought to draw an overall picture of the decision to adopt educational technology among in service English language teachers in an intensive English language program. The main goal was to understand how Diffusion of Innovations Theory reflects the decision-making process of the in-service language teachers. To accomplish this goal, the study featured a qualitative design with semi-structured interviews, three on line surveys, and anecdotal data to explore language teachers' decision making around educational technology adoption within a context that applied scholarly recommendations for teacher professional development training. More importantly, it involved four technology workshops that were delivered by the author to introduce language teachers to education technology and to understand the adoption process of technology occurs. This study accomplished this goal by identifying the participants' self-reported innovativeness and their perceptions of the characteristics of educational technology as an innovation. Their responses on where they perceive themselves in the innovativeness categories that Rogers proposes in his theory (2003) and what kind of attributes (also based in DIT) they desire in an innovation helped to understand how Diffusion of Innovations Theory informs the process of teacher development in education technology after participating in workshops and other environmental enablers that include administration support, face-to-face and digital communications.

Based on the problem statement and purpose noted above, this study was designed to respond to the following broad question: To what extent does the Diffusion of Innovations Theory explain the adoption decisions of in-service ELL teachers?

Findings of the research study revealed that language teachers' innovativeness and its link to their perception of the attributes of technology, mainly, relative advantage, ease of use, and compatibility were what shaped their decision to adopt technology in the context of this study. This is in addition to some other factors such as time, lack of resources, nostalgia, and over trusting technology. Further, findings showed that female teachers were more frequently significant adopters of educational technology more than their male counterparts. In short, the study found that DIT explained the decision-making process of in service language teachers fairly well. Further, as the model for professional development indicates, educational technology adoption is not as likely to occur without expert and administrative support. The research concludes with implications of the findings for administrators, facilitators, and researchers.

CHALLENGES AND SOLUTIONS

The research premise and the passion that you have as a researcher towards exploring it are the triggers that would push you to tackle all sorts of challenges you might face in the way. So, whether you are a graduate student or a novice researcher, conducting research studies is an essential part of embracing the world of academia and becoming a reputable scholar who thrives to add value to his or her domain of study and bring about a positive social change. However, entering the world of research in academia has its ups and downs like all other professions and more. I personally faced a few of them along my research journey. In the following sections, I will reflect on some of the main challenges that I faced as a novice researcher. They are challenges in finding a focused research topic, in limiting the literature review information, in accessing institutes and recruiting research partcipants, in building a rapport with the teachers, with collecting the research data, and in dealing with diverse population of participants.

A Challenge in Finding Focused Research Topic

I learned about the terms of education technology and Computer Assisted Language Learning after I had joined my university in 2011 and I became so passionate about them the moment I learned about their values. I saw first hand how can they transform the teaching and learning experiences of teachers and students. I thought this is what schools are missing in my home country, Libya and other similar contexts. This is part of what these countries need to help reinvent their education systems. Therefore, in every class, academic gathering, and conference, I add, 'implementing technology in low tech areas' as one of my research interests when introducing myself. I did so as part of introducing myself to my new professors, colleagues and audiences. I kept doing so for two consecutive years while working on my course work to fulfill my PhD study requirements. During which I published two education technology papers and participated in a few conferences and among all the different topics presented and papers published, technology was present. However, when the time came to seriously work on my PhD proposal and put into action the words and beliefs that I had for two years and kept repeating in every new class and academic event, I found myself lost. I was lost between the ocean of information and resources that were available for me online and in the library and my passion in finding a solution that would help me implement technology in low-tech areas. I realized that for two years, I used to claim conducting a very broad and unmanageable project. Saying something is completely different from truly doing it, so I devoted most of my time after finishing my coursework, doing nothing but digging deeper in the literature review trying to find the best framework and theory that might help me make my dream in finding a solution to implementing technology come true and I did. Finding the appropriate framework that would help me lead the research question felt like a breeze. "Now, I am ready for what comes next". That is what I told to myself that night around 4:00am at the library and I was sure that my advisor would see its value the way I saw it.

Therefore, my advice to graduate students and novice researchers is to be patient and detail driven when you start your search to find enough resources that support your dissertation and help you focus your topic. It will be challenging to research a topic that is very broad or very narrow. My advice is try to limit your framework to two or three theories or models to choose from as soon as you realize that that is the topic that you are passionate about and willing to spend the rest of the years researching. Also, be flexible, as you might need to adjust it to what is available. Make sure to entertain the following points:

- Develop an overall idea about your field of study, find out what interests you the most and try to read everything relevant to that topic to help put your knowledge into action.
- Have in mind that finding the appropriate theoretical framework to support your endeavor and help lead your research question is a crucial component that you need to consider early on before starting the practical part of putting your proposal together.
- Avoid narrowing or broadening your topic. It should be reasonably manageable and doable, e.g., materials, resources, establishments, time, funds, participants, …etc.
- They say a good dissertation is a done one, please don't pay attention to such a claim and thrive to bring about a positive social change. You are spending at least 5 years of your life as a graduate student and at least one year as a novice researcher; add a value to your field of study.
- Putting your topic in the form of a question will definitely help you focus on the type of data you need to gather.

A Challenge in Limiting the Literature Review Information

After spending ample time searching for resources such as books and articles that are related to my research study and support the framework selected, I felt that each piece of information is as important as the other and I wished if I can find way to include all of those information in my introduction and literature review chapters. Now, I lough at myself every time I remember how I struggled to treasure and include all of the notes I highlighted while working on my dissertation. In such scenarios, following your advisor's guidance can save you time and reduce your stress level. Your advisor is there to help you see the complete picture objectively and advice you to narrow your themes to serve what is important and what supports your claims. Therefore, learn that it can be quite challenging to deal with the unlimited amount of literature that you as a novice researcher might have retrieved. Soon, you will find yourself acquiring new scholarly skills that would help you access the necessary information that supports and informs your theoretical framework efficiently, and manage the literature through interpreting the information well, synthesizing it and positioning it professionally to serve the context of your study.

A Challenge Accessing Institutes and Recruiting Research Participants

Very often recruiting research participants entails going through institutions. This could mean more barriers and challenges any researcher can have while trying to recruit the research participants. In my case, accessing a language center that would allow me to coach its teachers and collect data from them was my biggest challenge during conducting my research study. I was an international student in the US and my home country was going through a brutal civil war that eliminated it as an option to experiment my model with some of its teachers. Therefore, as soon as I defended my proposal and became a PhD Candidate, I started contacting the directors of some of the language centers, proposing my willingness to train their language teachers on how to integrate technology into their teaching pedagogy in exchange of allowing me to collect data from the teachers before and after the technology workshops. I emailed my proposal to a few, but all that I heard was rejections and excuses as my research requires time and commitment from the teachers to attend the planned technology workshops which was a crucial part to test the effectiveness of my model. I remember having a meeting with one of the program directors to convince her to let me in. During the meeting she showed willingness to have me coach her teachers and asked me to provide her with some time until the next meeting with her faculty members. That raised my hopes, but unfortunately the response that I received was not in my favour. The message implied that teachers have no time to learn how to use new technology tools. However, the rejection did not stop me and I decided to pursue an alternative venue.

As an alternative, I started searching for universities outside the United States and I chose the United Arab Emirates for a few personal reasons. I started communicating with some universities in the United Arab Emirates. From five emails I sent to various universities, I heard from two only. The first conditionally accepted my proposal to train their teachers, but as part of an academic exchange program. In other words, the university would host me as a visiting researcher, but as an exchange, my American university has to host a scholar from that Emirate university, so I ignored that option. The second university was Zayed University. I received a positive email from the dean of the College of Education at that time in less than 24 hours. The dean who was also American, saw my proposal as promising and told me that Zayed University invested a lot in technology including iPads and that she would support having me to train its pre-service and in-service teachers and collect my data. After that the dean kindly connected me with the Human Resources representative and the research board committee to review my proposal and evaluate its suitability to the needs of the university. I was honored and lucky to have my proposal for training teachers approved, which expediated the process of my visiting scholar visa to the UAE. I was beyond grateful that I finally found a suitable environment to coach teachers and see the effectiveness of my model (see, El Shaban & Egbert, 2018). However, deep inside, I was worried and in a state of uncertainty. I was asking myself what if I leave the United States to train teachers in the United Arab Emirate, but then I fail to obtain a visa back to the USA to join my family and continue my study? My home country was in a civil war and I had a colleague who waited 12 months to get her visa approved again after she decided to visit Libya to see her parents. I was afraid to go through the same experience even though I was not planning to visit Libya. However, I had no choice, but to take the risk and keep the plans in place to travel to UAE early Fall 2014 to coach teachers and collect my data.

In March 2014, a few months before my travel time, I went to TESOL Convention in Portland, Oregon. My colleague and I presented an online interactive English lesson on a website and emphasized how such creative technology tools can help teachers create more engaging courses and exercises for their students. Most of the attendees in our presentation found the idea doable and interesting. Luckily, one of the attendees was the director of one of the Intensive English Programs at one of the midwest universities. The director showed interest in our work and I found that to be an excellent opportunity to share more about my future research and my willingness to coach the teachers in his language center if the administration allows me to. During our networking and conversation time, I received a positive response from the director and I was asked to email the proposal to him and to the provost to see the kind of commitment teachers need to show and if that would really benefit them and would not conflict with their regular daily teaching tasks. And yes, the power of networking can open many closed doors and I joined that Intensive Program, coached its teachers, taught its students, collected my data in 5 months and went back to my university in Washington State. I was ready for the next following phase of my research journey which is data analysis.

If there are factors that I would love to highlight to assist novice researchers in their journey, they are perseverance, networking and selling oneself. A combination of the three helped me accomplish my goal and conduct my research in the USA. To explain each to those who might need to understand them, perseverance helped me continue in the course of action to pursue my dream and experiment the effectiveness of my model through training teachers on how to integrate technology into their teaching pedagogy regardless of the many challenges and barriers I faced trying to get access to language centers and their teachers. On the other hand, networking was a key factor for unlocking many closed doors, for instance, it provided me with the opportunity to get access to a gatekeeper who allowed me to coach the language teachers at one of the US Midwest universities. There is no doubt that acquiring the needed skills for building long lasting relationships through checking on people vie sending emails and making phone calls should be a common sense to any professional who is hoping in building and sustaining his/her relationships with others. Furthermore, developing the skills of selling myself to gatekeepers, e.g., the intensive English program director and the dean of the College of Education at Zayed University, had them perceive me as an individual with a potential positive impact that was ready to bring about to their organizations.

Challenges in Building a Rapport With the Teachers

Access to the research participants, as mentioned earlier, was provided by the director of the intensive English program. The director stated that he was very willing to introduce his teachers to educational technology. In early Summer 2015, an email that included a description of the study, proposed workshops, and consent information was sent to the program director. Eight participants (2 males and 6 females), the total number of faculty members in the intensive English program, agreed to participate in this study except one female participant who decided to withdraw after the first interview. Therefore, she was excluded from the research participants. Six of the participants were over 40 years old and the other two were between 20 and 30 years old. All of the participants except one had MA degrees. The participants had varied experience in teaching English as a second language that ranged from 2 to more than ten years. Table 1 shows the background information of each of the participants based on pseudonyms provided for the study.

Table 1. Participants' Background Information

Participants	Gender	Age	Educational Experience	Teaching TESOL Experience
Ken	Male	Over 40	MA degree	>10
Dan	Male	Over 40	MA degree	>10
Kayla	Female	Between 20-25	BA degree	0-2
Jace	Female	Over 40	MA degree	>10
Sarah	Female	Over 40	MA degree	7-10
Kate	Female	Over 25	MA degree	3-6
Helen	Female	Over 40	MA degree	>10
Denisha	Female	Over 40	MA degree	0-2

(El Shaban, 2017)

The night that I had put my foot on the campus where I had to meet my participants, brought two conflicting emotional states in me. I felt so excited that I was so close to fulfill one of my dear dreams and I was so scared that I would not be able to deliver my workshops effectively. I had experience training K-12 teachers and introduced a few to some interesting technology tools, but I was part of a team that was lead by my advisor, Dr. Joy Egbert. I was not the sole facilitator who is responsible for all that teachers might need. To make my task easy, I chose to be a research participant and a coworker with my participants. Considering my experience in teaching English as a second language, the director offered me a few courses to teach. So I was teaching English as a second language and at the same time, I trained the in-service language teachers, my participants and colleagues, on how to use education technology. This dual role helped me break some of the barriers with my participants who I purposely leaned to a few of them to learn more about my students' performances in their classes and to ask them to share the syllabus and their expertise with me. Still, I faced few challenges during the training and the data collection process. For instance, I could not gain the trust of one of my participants. She responded to two surveys and had me conduct the first interview with her to understand her needs, perception and familiarity with the use of education technology in classrooms, so I can better assist her and tailor my choices of technology tools to meet each one of my participants' needs. However, unlike the rest of her colleagues who showed enthusiasm in learning more about technology, she chose to withdraw and not to participate in the technology workshops. Honestly, hearing that news did not make me feel good and had me reflect on my interaction with her and wonder if I unintentionally said or did something that had her change her mind and withdraw from the training and the research. Unfortunately, I failed to reach to any answer considering the very limited times I interacted with her and I had to respect her decision and chose not to ask her why she decided that. I chose not to ask her to avoid offending her and myself. Later, I heard from the director that she was planning to move to a middle school and teach there and that could be the reason behind her decision. So, whether you are a skilled or a novice researcher, go to your field of study while expecting losing some of your participants for reasons that could be unexplained and totally out of your hand. Move on and learn to get the best results with what you have in hand.

A Challenge With Data Collection, the First Interview

Data used in this study were collected from three online surveys, two sets of interviews, and informal anecdotal sources such as emails, observations, and brief hallway talks as the researcher made herself approachable and available to respond to the participants questions and reelections on their use of educational technology.

In order to understand more about the participants' decisions to adopt or reject technology and to reach an in-depth understanding of the adoption of education technology, all of the participants took part in semi-structured interviews. Each participant was interviewed individually. Two interviews, an initial and a follow-up, were conducted. The initial interviews were conducted before the first workshop and the follow up ones were conducted after the last workshop. The questions addressed in the first interview were determined by considering the participants' responses on the innovativeness scale (IS). For example, one of the questions asked about how they saw themselves in the eyes of their colleagues. Other questions addressed in the follow-up interview were determined by considering the participants' responses on the Perceived Characteristics of Innovating Scale (PCIS) and their decisions to adopt. For example, one of the questions asked the participants about their decision about whether to adopt or not to adopt educational technology in your teaching practices. The interview questions also addressed the participants' ways of teaching and types of technology they used, if any; this interview gathered data to understand the level of the teachers' familiarity and use of educational technology while teaching English as a second language and to create an overall picture of their teaching style. Only one participant missed the first interview as she traveled to Russia for ten days' vacation and joined the study for the first workshop. Both of the semi-structured interviews were audiotaped and transcribed.

One of the biggest pitfalls I faced as a novice researcher had to do with the initial interview. I mismanaged it when I allowed teachers to speak and cover so many topics and deviate from the main idea that made me feel overwhelmed when I had to set and transcribe each word verbatim. Yes, that open conversation provided me with more detailed and rich information about my participants, still such unrestricted quantity of information can also mislead novice researchers and leave them stuck and overwhelmed. I remember when I shared the manuscript of the verbatim transcribed interviews with my advisor, her first response was, 'you can make a whole another different research paper from this data'. So my recommendation to novice researchers is to go prepared to the fieldwork knowing very well how to manage the interview with your participants and direct and redirect them to that specific point to save time and energy. If the questions were set for a specific time slot of fifteen minutes, then be it fifteen minutes not forty. In my case, each interview supposed to last for a maximum of fifteen minutes, however, each of the semi-structured interviews lasted between fifteen to thirty minutes. Creswell's (2009) steps for qualitative data analysis were adopted to analyze the interview data: Transcribed data were organized and coded based both on ideas from the literature such as challenges, decision categories, and so on and codes that emerged from the data in order "to develop a sense of categorical, thematic, conceptual, and or theoretical organization" of the collected data (Saldana, 2009, p. 149). Then, codes were grouped into themes that were integrated and interpreted with regard to the research questions. The interview provided detailed information about which of Roger's decision-making stages applied to the participants.

A Challenge Introducing Technology to a Diverse Population

The English program's classrooms were equipped with a TV, computer, document camera, and projector. iPads were available upon request. Also, a computer lab at the university was assigned to the program for use during this study. The program coordinator and director with support from the university's Provost covered all funding related to the study, including incentives for the teachers for their participation (pay for the hours they attended workshops) and for the cost of registration for any needed technology tool. This implies that this context has no issue with the accessibility to the required technology for training teachers. However, as a novice researcher you should not expect the availability of most needed resources elsewhere. In many contexts, the lack of access to the required resources could be one of the biggest research barriers.

Nevertheless, I sensed a sort of resistance from some of the participants during the initial workshop. Such issues could rise especially when you have a diverse population of participants, e.g., diverse in age, authority level, experience and perception. Before I elaborate on that let me give you a brief idea about the technology workshop. The intensive English program teachers participated in four technology workshops. The purpose of the workshops was to introduce tools, provide examples of integration, and provide support for practice as a basis for their decision-making about educational technology. All participants except one participated in all of the workshops; that participant experienced time constraint that caused her to miss parts of the second and the third professional development sessions.

The implementation of the model for the professional development began during the fall semester with three formal workshops to meet stage one requirements and a fourth workshop to give an example of stage two requirements. Each workshop lasted for about two hours. In each workshop, the researcher introduced three software packages. Rogers' decision-making process was considered while introducing the idea of educational technology. In order to engage teachers in tool exploration, brainstorming, and reflection, participants had hands-on training for each tool. The next section provides more details about the workshop map.

The First Workshop

The first workshop took place in week 1 of the fall semester and it lasted for 2 hours. This workshop was conducted as follows:

- The workshop commenced with a *Jeopardylab* (2013) game as an icebreaker to build on teachers' prior experiences of technology and give them an overall idea of what educational technology in a classroom could look like.
- Second, because the researcher viewed herself more as a change agent rather than simply a facilitator, in the introduction to this workshop she highlighted three reasons for why it is necessary for teachers to consider integrating technology into their teaching practice. She presented these ideas using Popplet software (2018). The reasons, as stated previously, included self-efficacy, marketing and students. The intention was to convince the participants to integrate technology into teaching - "the why" (persuasion). After addressing why teachers should learn to use technology (persuasion), an explanation of educational technology and task engagement that served as "the what" (knowledge) was presented. In addition, an educational technology tool, *JeopardyLabs* (n.d.), was presented as an example. A set of diverse questions was first set. It included language, history, and art questions. During the workshop, the participants were divided into two teams with instructions on how to participate. A small incentive was given to the winning team. Because chocolate was the incentive, the other team was invited to enjoy theirs as well. Further, the presentation was delivered using the software *Popple*t (2018) as well. This tool was used to show the participants how it can be used as an instructional tool. Also, a preliminary map of the upcoming technology workshops was presented.
- For this workshop, the "how" (practice) section was the longest as it involved hands on approach used for the rest of the first workshop and the other workshops.
- Third, hard copies and online supporting materials with the necessary technology tools were shared to the participants.
- Fourth, the researcher demonstrated how to use *Jeopardylab* (2013), *Popplet* (2018), and *Glogster* (2020).
- After each tool, the participants were given ample time to try the tools themselves. Then, following a lesson plan template, they were asked to create a lesson using any tool that interested them the most, focusing on the objectives of any of the language contents that they were teaching.
- Finally, *Socrative* software (2020) was used to wrap up the workshop. In order to give the participants, the opportunity to experience the tools themselves, the researcher provided a class code. This will also enable them to experience the tool themselves, examine their understanding of some of the CALL principles and collect their feedback about the workshop.

- Additionally, the participants agreed to use Facebook (Facebook, 2020) as a digital form of communication in the same workshop. The digital form of communication is part of stage two (the enabling environment) of the teachers' professional development model.

The Second Workshop

The second workshop took place in week 3 and it was lasted for two hours. This workshop was conducted as follows:

- Teachers were engaged in an informative argument about their experiences when they integrated the tools they have learned in the previous workshop.
- The researcher continued with "the how" (practice) and presented *English Central* (2020), as an interactive tool for English language learning and the specific language skills that could be enhanced using this tool. Then, the participants were given ample time to try on its different levels on their own and reflect on how they felt about using it.
- A co-worker, volunteered to present *Breaking News English* software (2020), and the participants were given the time to try the tool and to reflect on it as well.
- There was a quick navigation of *Google* (Google, 2020) and exchange of ideas about *Google image*, search, and *YouTube*. Then, the researcher paid special attention to how to use Google document to engage students in collaborative projects and writing. All the participants were invited to the Google document and given the opportunity to experience how it feels to be on the same page at the same time.

The Third Workshop

The third workshop took place in week 2 of the second term of the Fall semester and it was lasted for two hours. This workshop was conducted as follows:

- It started with an introduction of the three tools that will be presented and how each tool can be integrated into teaching English as a second language. These tools included *VoiceThread* (2020), *TEDxESL* (Jones, 2014), and Story jumper (2020).
- *VoiceThread* (2020) was the first tool presented and the participants were advised to create their accounts in advance. The researcher shared an email with all the participants containing instructions on how to create their accounts two days prior to the workshop.

- The researcher demonstrated how to use the tool and navigated it with the participants. Then, she asked them to access a link that she shared with them five minutes before the workshop started. She posted an illusion photo as a media and asked the participants to comment using either microphone, telephone, or writing to express what they thought bout the photo. The participants were given time to read and listen to the comments of each participant and also reflect on them, either on the software or in the lab.
- Then, the researcher introduced *TEDxESL* (Jones, 2014) and shared one of the tool's already existing lesson plans. After watching one of the clips, the researcher facilitated a meaningful discussion, considering the content of that clip and their overall evaluation of it.
- Last but not least, *StoryJumper* (2020) was introduced. The researcher asked the participants to create their accounts and explained how teachers can create their classes and invite their students to work under their guidance. The participants observed that this software is one of the very engaging ones. After the workshop, some of them decided to stay in the computer lab to finish creating their stories.

The Fourth Workshop

The fourth workshop took place in week 5 of the second Fall semester. This informal workshop was conducted as follows:

- A fourth informal workshop led by the participants was conducted as an example to sustain teachers' professional development in using technology. The goal was to provide them with the opportunity to collaborate and share their experiences.
- During the same workshop, some participants shared other educational technology tools that are similar to the ones they had learned about in the previous workshops such as *Kahoot* (2020) and *Factile* (n.d.).
- The goal of this workshop was for teachers to brainstorm and share their experiences on recent research in the field of ESL and technology. This informal meeting is scheduled to take place twice or once a month. Also, as mentioned earlier, the participants agreed to use Facebook as an online platform for communication starting after the first workshop.

For more details about that technology teacher professional development model, see (El Shaban, Egbert, 2018). The resistance that I noticed during the first workshop was basically from one teacher. She was the program coordinator. She was the oldest and the most experienced in the teaching field. I can still recall her during that first technology workshop commenting more than once, ‘I’m not a technology person’, and she repeats, ‘I am not going to learn technology’. Honestly, as a technology facilitator and someone who is using the Diffusion of Innovation Theory as part of my theoretical framework and knows very well the impact that peers and colleagues can have over one another, see (El Shaban, 2017) for more details about the theory, I felt so frustrated during that first workshop, but of course I could not show that as a courtesy and just kept smiling whenever I hear a similar comment during those two hours. After the workshop, I decided that this should be stopped. This teacher is the program coordinator this means that she is an authority figure and a trusted person among her colleagues and whatever she says or believes in might directly or indirectly impact them. I spent that whole day after the workshop figuring out how to motivate this teacher and had her change her perception towards the technology. I know enough about her to use in my favour. For example, she taught writing for the majority of her life. This means that if I introduce her to a tool that she can use and help her see how easy it is to use, this might help me reach the outcome I desire. So in the second workshop, I introduced Google Document and invited all of the teachers to write a summary and help one another on the same document and I had them see their interaction on the data show. Interestingly, she showed interest in that technology tool and wanted to learn more about it and about similar writing technology tools, so she can use them with her advanced international students. Not only that, this same teacher collaborated with me and a third teacher to present Google Document as an effective tool for enhancing the students’ writing, engagement and collaboration skills in TESOL Convention 2016. Our presentation paper was accepted. The following other workshops, this same teacher showed interest in learning more technology tools and brought about a positive impact to the rest of her colleagues. Understanding your participants and working smart to meet their needs could work wonder and would remove any possible barriers.

RECOMMENDATIONS

Whether you are a graduate student, or a novice researcher and regardless of the type of methodology you plan to use to collect and analyze your data, conducting a research study is not as well ordered and organized as it seems to be in published scholarly research articles and reports. Furthermore, the investigation, the exploration and the examination of different topics in your field of study is an integral part of being a scholar-practitioner who is capable of making an effective social change and you can be a one. In this section, I will summarize some valuable practical recommendations with those who might need them.

- **Make sure that your research idea is doable.** To do so, you need to carefully calculate the availability of the required resources, e.g., humans, materials; you need to guarantee the success of your journey.
- **Do your work.** You are a researcher after all. Make sure to read and read until you feel that you are a master in that specific topic. This will help you craft an excellent proposal and close a true gap in the literature and add a true value to your field of study. Trust me, a good dissertation is not a done dissertation. It is a useful one. Do not forget the time and energy you invested towards it. Yours should bring about a positive social and academic change in your field.
- **Make sure that you are passionate about your research.** You have to be burning with an idea, or a problem, to give it the attention it needs. Remember that you are going to spend at least one year thinking about and working day and night on your project.
- **Do not be in a rush.** Spend enough time planning for your project and do not feel afraid to say, 'I don't know' or seek experts' knowledge and guidance. You can start small by doing a pilot study. I saw first hand how that saved time and money for a few of my colleagues. Also, you might end up publishing two papers or more instead.
- **Embrace the power of networking and building relationships with people you do not know.** This would open up some closed doors as what happened with me. Go to conferences, attend, participate and network whenever possible. Search for other social and professional events near you, register, attend and network. Leverage the power of social media, open an account on LinkedIn and make sure to send genuine messages to people who you might one day need to work for or with. Be consistent and send emails and do phone calls occasionally and when necessary to keep the connection circle a live. You should not connect only to request something. If you do so, the answer would most likely be no response or no and you do not want to hear that. So, learn to reach out to people when you need them the least to find them when you need them.
- **Get to know well your participants.** When I used to train teachers in the US, I always reminded them to see their students as humans first. As a researcher, I encourage you to see your participants, as humans not merely subjects or participants to collect your data from and go. By doing so, you will be able to solve many of the unknown obstacles that you might face in the way. Also, this will help you avoid jumping to conclusions and work smart to address their needs.

- **Be flexible in all of your approaches.** As you progress with your research, have in mind that your ideas might progress too and your questions might need slightly to be changed. There might not be a real gap in literature, or there might be a lack of knowledge and resources that are related to that specific topic, be flexible to deal with such a shift professionally. Be open to explore all possible academic opportunities and directions.

CONCLUSION

In this chapter, I tried to highlight some of the biggest challenges I faced and figured out as a graduate student while working on my PhD dissertation, hoping that by doing so to guide and empower other graduate and novice researchers. I want you to know that you are not alone. Even the best experienced researchers have their share of fieldwork challenges, but unfortunately, they rarely talk and publish articles about it. This might due to the different uncontrollable variables that each researcher might have while conducting his or her research study. We need to think of fieldwork and research studies as no size fits all. My experience in that particular context with the same exact participants could be completely different from yours if it happened to conduct your research in the same context and interact with the same participants. Regardless, I believe that showing our vulnerability through publishing about the untold part of our research journeys reflects strength and could be empowering to many. With that being said, sometimes, in the course of conducting research studies, the challenges can be internal and/or external as the ones listed above. I believe it might be easy to control the internal challenges if we are passionate enough about the research topic. Internal challenges might have to do with the lack of passion, perseverance, curiosity and motivation. You need to have each of the four works in your favor to keep progressing despite the other external barriers in your research. By doing so, you will be able to handle all of the other challenges.

REFERENCES

Allen, I. E., & Seaman, J. (2013). Changing course: Ten years of tracking online education in the United States. Newburyport, MA: Babson Survey Research Group and Quahog Research Group, LLC: Pearson.

Brahier, B. R. (2006). *Examining a model of teachers' technology adoption decision making: An application of diffusion of innovations theory*. University of Minnesota.

Breaking News English. (2020). Retrieved from https://breakingnewsenglish.com/

Creswell, J. W. (2013). *Research design: Qualitative, quantitative, and mixed methods approaches*. Sage publications.

Drexler, W. (2010). The networked student model for construction of personal learning. *Language Learning & Technology*, *26*(3), 369–385.

El Shaban, A. (2017). I*n-Service ELL Teachers and Technology Adoption: Exploring Diffusion of Innovations in Language Education* [Doctoral dissertation, Washington State University]. Retrieved from https://research.libraries.wsu.edu/xmlui/bitstream/handle/2376/13005/ElShaban_wsu_0251E_12122.pdf?sequence=1&isAllowed=y

El Shaban, A., & Egbert, J. (2018). Diffusing education technology: A model for language teacher professional development in CALL. *System*, *78*, 234–244. doi:10.1016/j.system.2018.09.002

EnglishCentral. (2020). Retrieved from https://www.englishcentral.com/videos

Facebook. (2020). Retrieved from https://m.facebook.com/?refsrc=http%3A%2F%2Fwww.google.com%2Fsearch&_rdr

Factile. (n.d.). Retrieved from https://www.playfactile.com/

Fullan, M. (2007). *The new meaning of educational change*. Routledge.

Glogster. (2020). Retrieved from http://edu.glogster.com/?ref=com

Google. (2020). Retrieved from https://www.google.com/

Google Docs. (2020). Retrieved from https://www.google.com/docs/about/

Januszewski, A., & Molenda, M. (Eds.). (2013). *Educational technology: A definition with commentary*. Routledge. doi:10.4324/9780203054000

Jones, A. (2014). *TEDxESL*. Retrieved from https://tedxesl.com/

Kahoot. (2020). Retrieved from https://getkahoot.com/

Kessler, G. (2006). Assessing CALL teacher training: What are we doing and what could we do better. *Teacher Education in CALL*, 23-42.

McGrail, E. (2005). Teachers, technology and change: English teachers' perspectives. *Journal of Technology and Teacher Education*, *13*(1), 5–25.

Popplet. (2018). Retrieved from http://popplet.com/

Rogers, E. M. (2003). *Diffusion of innovations* (5th ed.). Free Press.

Saldana, J. (2009). *The coding manual for qualitative researchers*. Sage Publishing.

Stanley, G. (2013). *Language learning with technology: Ideas for integrating technology in the classroom*. Cambridge University Press.

StoryJumper. (2017). Retrieved from https://www.storyjumper.com/

Terrell, S. S. (2013). Integrating online tools to motivate young English language learners to practice English outside the classroom. In *Explorations of Language Teaching and Learning with Computational Assistance* (pp. 184–192). IGI Global. doi:10.4018/978-1-4666-1855-8.ch012

Section 2

Ethnographic and Qualitative Research Challenges

Chapter 3
A Tale of the Field:
Can Ethnographic Challenges Become Opportunities?

Yao Wang
https://orcid.org/0000-0002-2559-738X
Newcastle University, UK

Peter Sercombe
Newcastle University, UK

ABSTRACT

This chapter discusses a number of fieldwork challenges leading up to and during an ethnographic study undertaken in southwest China. The purpose of this study was to research the circumstances of privately-run migrant schools (PMSs) as forms of educational provision for internal migrant children and the coping strategies that were observed and/or reported to deal with these children. The authors consider the (pre-study) challenge of selecting an appropriate research site, recruitment of willing participants, possible language barriers, and potential issues arising from the role(s) of the researcher and the study population. By explicating the ethnographer's path from fieldwork constraints to potential opportunities, this chapter clarifies the importance of realizing the need for flexibility and adaptability when conducting fieldwork. The chapter is of relevance to current and future ethnographers conducting fieldwork or setting out to conduct fieldwork. Additionally, this chapter provides insight on the unknown eventualities that almost certainly arise during the research process.

DOI: 10.4018/978-1-7998-5826-3.ch003

INTRODUCTION

When entering a research field, ethnographers who are eager to explore a particular area of social life may come across obstacles for which their academic training has not adequately prepared them (Hammersley & Atkinson, 2007). This chapter cannot address all types of such challenges. Rather, it offers an aspiring ethnographer's insights based on experience accruing from the fieldwork of one of the authors and reflections about how she dealt with stumbling blocks she encountered in the course of her fieldwork.

The chapter starts by outlining the research project as a means to contextualize the four main fieldwork challenges and constraints that the researcher encountered. These include the (pre-study) challenge of selecting an appropriate research site, the challenge of recruiting willing participants, possible language barriers, potential challenges arising from the role(s) of the researcher and the study population, and the coping strategies that emerged to deal with these challenges. By documenting and highlighting the various ways in which ethnographers can accommodate challenges, the authors provide a perspective that attempts to clarify the importance of flexibility while conducting ethnographic fieldwork. In this vein, we hope this chapter can serve as a useful resource for current and future ethnographers conducting fieldwork or setting out to conduct fieldwork. Additionally, we hope this chapter will provide insight into the unknown eventualities that almost certainly arise during the research process.

BACKGROUND

One of the main features of ethnographic study is that it takes place "in the field" (Hammersley & Atkinson, 2007, p. 3), meaning that research occurs in an environment occupied by those being researched; hence, ethnography is immersive. Due to the inevitable differences between sites, any fieldwork is unique in certain respects. To help a novice prepare for ethnographic study, many researchers have shared hands-on experience in various research settings. Emphasis tends to be placed on methodological and technical aspects to provide procedural guidelines (see Hammersley & Atkinson, 2007; Murchison, 2010). The four main fieldwork challenges in this chapter are concerned with the research site, participant sampling, possible language issues, and the researcher's role. Researchers can also find advice about these issues in various textbooks (e.g., Bryman, 2016; Fetterman, 2010; Madden, 2017). However, principles from such texts or a research methods course do not necessarily translate seamlessly to fieldwork contexts because ideas are not the same as practices. Besides these, unpredicted issues can easily emerge from fieldwork practice that researchers do not encounter in fieldwork guides.

Regarding a research site, writers of ethnographies suggest minimizing the geographic distance from the researcher's working base for practical reasons (Madden, 2017; Murchison, 2010). However, few writers discuss the process of "selecting a research site." There is no ideal research setting; rather, there are always likely to be considerations and constraints to take into account. Political sensitivities, language barriers, and physical access might well be issues that researchers need to overcome or that make an investigation of a research topic difficult or even an unrealistic prospect. Therefore, it is important to acknowledge these and have potential strategies for coping with them.

The matter of a research site is best considered at the planning stage, prior to fieldwork. Concerning sampling and recruiting participants in qualitative research, common strategies include letters, email, flyers, telephone calls, or word of mouth to solicit interest in participation. These somewhat "standardized" means seem more applicable to situations in which physical sites are not necessarily an issue. For example, this might be the case in non-ethnographic work, such as surveys, when researchers may not know where participants are situated, or this being unimportant. The third challenge, identified here, is that of language(s). In this case, the challenge of language relates to varieties of Mandarin, and this was unexpected for the researcher investigating in her home nation. Language(s) can also be a key dimension of reflexivity in many ethnographic studies. For example, studies about reflexive language in an international collaboration team (Creese & Blackledge, 2012) or in multilingual community settings (Creese et al., 2008; Ganassin & Holmes, 2013) have contributed to understanding the complexities of researching when more than one language (variety) is involved. It is easy to overlook language barriers in studies that take place in a local or native context. It may be assumed, especially in a country such as China, that the officially recognized common language, Mandarin, is shared by citizens across the nation.

Finally, other researchers have discussed strategies that can influence the role of the researcher and the researched, which can influence field dynamics (Reyes, 2018) and human interrelations during the research process. Rather than focusing on demographic characteristics, as in previous literature (Flores, 2016; Venkatesh, 2006), we focus on how to make use of available tools and skills in order to build relationships with informants and aid data collection. The four issues are not independent but rather interconnected. To address how these challenges influence research and possible ways to adapt to them, this chapter elaborates on fieldwork conducted as part of the first author's doctoral research, in which the researcher is referred to in the first person.

The study concerns privately-run migrant schools (PMSs) which constitute the main form of educational provision for the children of internal migrants in China (Chen & Feng, 2013; Lin, Ren, & Yang, 2019; Shi & Sercombe, 2020). Most Chinese migrant workers are employed in manufacturing and construction industries (NBS, UNFPA, & UNICEF, 2019) with low pay and low social status. Together with these workers, there are around 14.07 million migrant children at school age. Among these children, 21.3% are studying at PMSs (MOE, 2017). Therefore, PMSs, which are low-cost private schools, cater to millions of children. PMSs are, generally, the sole providers of educational opportunities for this (21.3%) minority who would, otherwise, very likely be "left-behind" children (Hu, Lu, & Huang, 2014, p. 1).

This study explores how these schools operate and it explores their educational and social circumstances. The research included a three-month fieldwork period in Guiyang City in southwestern China (Figure 1). Previous studies have mainly considered PMSs in highly developed cities, such as Beijing and Shanghai (Chen & Feng, 2017; Lai et al., 2014; Ling, 2015). In contrast, this project focuses on schools in a relatively impoverished yet burgeoning city, Guiyang, which is the capital of one of China's least developed areas – Guizhou Province. Grassroots privately-run migrant schools are widespread in cities across China, although the precise number of these is unknown. Indeed, researchers conducted this study in a context that is in the throes of considerable infrastructure development. This has resulted in the presence of many marginalized internal migrants in search of a better life. These migrants and their children raise challenges for local governments in terms of social and education policy. As a result, PMSs have sprung up to fill educational gaps.

Figure 1. Maps of the research site
Source: Map 1&2: Wang, 2020; Map3: All-free-download.com, 2020

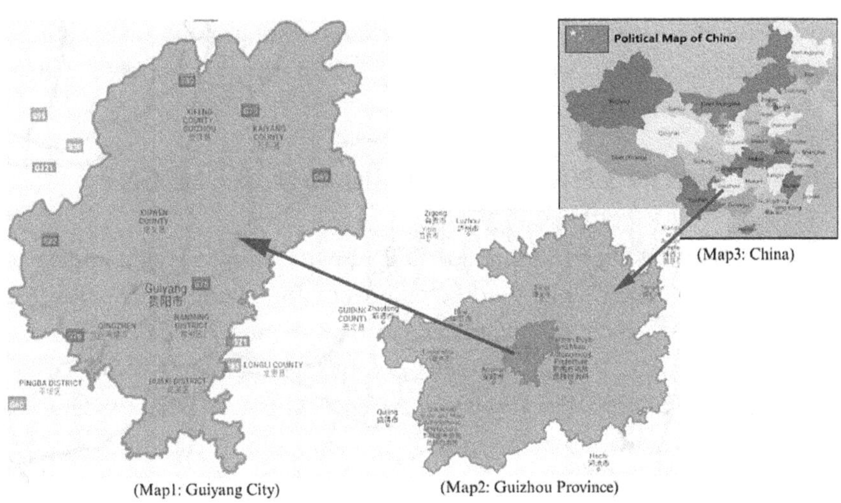

I adopted an ethnographic approach on the basis that this would be the most productive way to address my research questions about the social and educational characteristics of PMSs in Guiyang, the role(s) being played by PMSs in the education of migrant children in the city, the kinds of challenges that PMSs face, and the kinds of improvements they aspire to, according to the main stakeholders. A broad but fundamental aim of the research was to try to make sense of the experiences of students and staff at PMSs. Therefore, I applied multiple methods to investigate PMSs and their stakeholders' views.

First, I collected the perspectives of a selection of local NGO staff (among whom I undertook semi-structured interviews with 3 NGO directors who were working with PMSs). Second, the perspectives of head teachers (15 of whom completed questionnaires, the purpose of which was to gather baseline data about the social and educational circumstances of PMSs). Third, I undertook participant and non-participant observation, which involved observing school meetings, coordinating extracurricular activities, spending time in staffrooms interacting with staff, spending some stints working as a substitute teacher, as well as visiting the homes of students (recorded in field notes and photographs). Finally, I conducted 22 semi-structured interviews with head teachers, and teachers, to explore schools' challenges and aspirations, from their perspectives. I have examined data to establish the types of role that PMSs are playing in the education of migrant children and, more broadly, in their lives.

Overall, this research aims to shed light on the challenges and needs of PMSs in Guiyang that might otherwise remain invisible from the public and officials. A larger purpose of this research is also to highlight the circumstances of these migrant minorities, bearing in mind that a nation's citizens are its most valuable resource. If this resource is unable to participate in society to its fullest potential, both citizens and the nation are losing out. The data used in this chapter are based on field notes that documented the researcher's ethnographic observations and reflections. Additionally, interview data as further information sources are included.

ACCOMMODATING ETHNOGRAPHIC CHALLENGES

The four issues emerging from this doctoral research project might appear in any study of a similar nature. While there is a tendency for a budding researcher to pay attention to questions such as "what is ethnography?" and "how does one conduct ethnographic research?", much formal or published guidance offers principles and strategies to avoid pitfalls or bypass barriers without sufficiently detailing the process of "what if a real problem arises?" For example, a research fieldwork site almost certainly contains some degree of unpredictability. This is inevitably part of an ethnographer's journey of discovery. By knowing very little on arrival but being filled with curiosity, one is in a good position to learn about the research site and those who populate it, if circumstances allow.

Instead of presenting "polished, theorized, sanitized, and carefully selected parts" (Rothenberg, 2016, vii) of the fieldwork, this chapter attempts to humanize the fieldwork experience, raise some of the main challenges encountered by a person new to ethnography based on personal experience, and reveal some of the strategies found useful to accommodate the challenges. Not selecting an appropriate research site is a potential danger that can probably be avoided if a researcher has considered perceived barriers (e.g., political concerns, language issues, and access to a site) beforehand. In my case, these included resistance to my entrance into some schools, skeptical attitudes towards my research intentions, and negative or aloof responses to inquiries. On the other hand, some obstacles revealed issues that otherwise would not have been noticed, as referred to in subsequent sections below. However, facing up to and accommodating these field challenges sometimes generated new opportunities regarding the research site, relevant social networks (among those being researched), a more comprehensive understanding of these networks, and a greater sense of empathy towards those who acted as my informants.

SOLUTIONS AND RECOMMENDATIONS

This section exemplifies some tentative recommendations and solutions in dealing with the four challenges identified and foregrounded through the experiences of the first author in the field.

The Choice of Research Site

The first fieldwork challenge discussed is the choice of research site. As mentioned, fieldwork took place in Guiyang; however, Guiyang was not my original choice of the research site. As Lichterman (2017) suggests, in order to make ethnography more transparent, ethnographers need to present how they came up with their research choices and interpretations, how they made mistakes, and what arose as a result of good fortune. Therefore, this chapter continues with its first constraint. As a result of making an invaluable contact, this constraint led to a change of research site – from Beijing to Guiyang, as the authors explain below.

I selected Beijing as the initial research site for very pragmatic reasons. First, I volunteered to teach in a Beijing PMS for one year, and this both motivated me and shaped my research interest towards the education of China's internal migrant children. Based on previous experience, I naively believed there were locally available schools to be investigated. Second, I had studied and worked in Beijing for over six years by the time of planning my fieldwork, so I was confident about researching a place with which I was familiar and that was close to family and friends.

Before finalizing my fieldwork site, I undertook two pre-fieldwork visits. The first visit, which was to Beijing, convinced me to abandon this setting and turn my attention to Guiyang. On the other hand, the second visit, which was to Guiyang, confirmed the accessibility, feasibility, and value of conducting the study in this city in the southwest of China. Finally, the shift of research site then allowed for a productive match between my aims and the planned research context. This match significantly benefited from having a confirmed research setting exploring the research topic beforehand, and I partly achieved this by building relevant social networks in advance of fieldwork.

Building relationships was a gradual process that included helpful initial contacts, unplanned meetings, and further useful contacts as a result of these (meetings). My initial visit to Beijing to seek out a research site took place at the end of 2018. However, my connection with the original (Beijing-based) research site happened earlier. I read a paper (Sui & Song, 2017) that was co-authored by a leading scholar, Professor Yingquan Song, who had conducted a number of studies about migrant children's education in Beijing and elsewhere. After reading this paper, I contacted him and received a positive reply. Visiting Professor Song was the first practical stage of my research. Without a plan and a clear design for my research, I took what I prefer to call a "networking trip." During this trip, I became acquainted with people who could help, or perhaps not help me or even hinder my access to PMSs (as I later discovered).

For a researcher new to a particular area of investigation, it is very beneficial to find a suitable person(s) who can help smooth the path to tasks such as contact potential informants from the outset. In conversation with Professor Song, I learned from his research experience about PMSs, and this helped as a kind of "informal socialization" (Cerwonka & Malkki, 2008, p.3). In addition, being familiar with my area of research, he introduced me to two head teachers and one NGO director through whom I was able to approach potential informants. This meeting provided me with more than expected and I was offered a number of useful sources by domestic researchers about migrant children's education. Additionally, I was able to get in touch with other stakeholders of PMSs in Beijing to whom I had been referred. It would otherwise have been difficult to access these without some form of introduction.

Unfortunately, later interactions with the two head teachers introduced by Professor Song turned out to be unsuccessful. Both were reluctant to participate in my research. The first head teacher rejected my request to talk in person only allowing a phone call. He rejected my research ideas for two main reasons. First, he had participated in studies previously but none of those had really helped his school. He said, "research like yours is of no value, at least for me and my school (Translated by the first author, same below)." This comment made clear that my inquiry would fail to bring any tangible or substantive change or benefit to the context being studied. Second, he told me his school had been closed and he was jobless in his home city. He also added, "You are not able to stop the closure so your research is meaningless for me!"

In contrast, during his phone call with Professor Song (the call being on speaker phone, so I was able to hear what the head teacher said), he mentioned that he was working as the head teacher at a PMS in another district of Beijing. Likewise, another head teacher refused my research request directly. She told me on social media, "If you insist, I have to ask for permission from the local education officials to allow your research, but it will be very hard and time-consuming." I saw this response as a polite decline. Both head teachers showed their frustration with research even though someone they trusted introduced them to me. This caused me to worry about future meetings with the NGO director and my subsequent intended fieldwork.

Nonetheless, the meeting with the NGO director in Beijing went smoothly. He was open-minded, helpful, and generous, for which I was very grateful. Working on the front line to serve migrant children's education for years, he had witnessed and recorded the development of PMSs in Beijing. Two themes emerged from our more than two-hour conversation. First, the pattern for PMSs was generally one of closure, and this was rarely documented in academic literature. According to his investigations, the number of PMSs in Beijing had declined from over 300 before 2006 to 127 in 2014 (only 65 of them being officially recognized by the local education authority). With the closure of migrant schools in recent years, he estimated that there were, at the time of our meeting, 111 migrant schools in Beijing accommodating about 50,000 students.

In contrast, four years ago, the number of students was more than 100,000 (NGOCN, 2018). The reduced number of students and PMSs represented a growing trend in Beijing for reasons that the authors outline below. By the time we met, the NGO director told me that several schools had already received closure notices from the local education authority. I also learned from him that the school at which I had previously volunteered was now closed.

Second, a rigorous policy on the migrant population corresponded with the closure of PMSs. Since 2014, the Beijing government had taken control of migrant population movement by relocating industries that relied on migrant labor to nearby surrounding areas such as Tianjin and Hebei (People's Daily Online, 2014). Some students had to move again (having already moved from the countryside to a city) with their parents due to a change of job location. This meant that some PMSs had to close without enough students even if they had not been told to close by the local education authority. This experience suggested to me that Beijing was not an ideal research setting given the closure of many PMSs, the difficulty of access that I encountered, and the orientation of municipal policy towards PMSs.

The result of this exploratory trip to Beijing was, in some respects, frustrating. However, further interactions with the NGO director to whom I had been introduced in Beijing allowed me to see the potential for transferring my focus to Guiyang. The NGO director kindly introduced me to Mr. Zhang, another director of an NGO serving PMSs and migrant children in Guiyang City. Contact with Mr. Zhang smoothed my access to those PMSs in Guiyang with which he was familiar and that would otherwise have remained inaccessible to me, as with PMSs in Beijing.

The point here is that the practical requirements of research may well trump other priorities. My first priority was finding regular access to a research site (Murchison, 2010). After obtaining access, I immersed myself in reading and becoming familiar with Guiyang, a fast-developing city with a concomitantly increasing volume of migrants. Guiyang was somewhat different than the highly developed cities to which migrant workers had previously tended to gravitate, such as Shanghai and Beijing.

A city such as Guiyang is representative of newly burgeoning urban centers facing mass rural-to-urban migration and high demand for education of migrant children across China. In fact, Guiyang had 1,339,288 migrant workers at the most recent estimate, which accounts for 26.18% of the total population of the city, according to the Guiyang Local Chronicles Compilation Committee (2018). Besides, among the 789 schools in Guiyang at the primary or junior high school level, the compulsory education stages in China (Guiyang Education Bureau, 2019), more than 150 (19%) of these are PMSs that provide rich possibilities for conducting research. PMSs in Guiyang are mainly located in the two central districts – Nanming (南明区) and Yunyan (云岩区), each of which has over 60 PMSs. It was also affordable and convenient to travel into the two districts, given their city center positions. I thus proceeded to conduct a pilot study in Guiyang four months after my initial explorations in Beijing.

As it turns out, constraints can also be changed to opportunities. My investigation started with reading a paper written by a scholar in Beijing and it, then, led me to Guiyang, over 1200 miles from the nations' capital, Beijing. Additionally, constraints led me to various head teachers, teachers, and students in PMSs, as well as migrant parents, who were the main stakeholders and who would be my main research focus. Finally, the choice of Guiyang as a research site turned out to be in line with my research questions in terms of their relevance to the circumstances of PMSs in the city. Furthermore, it seemed that the study would be able to contribute to research gaps regarding PMSs in less developed parts of China. In so doing, the research would ultimately provide some modest contributions to debates about the roles and value of PMSs in Guiyang as well as other, less-developed parts of the country.

Sampling and Recruitment Challenge

After deciding where to conduct the research, my next challenge was in considering who might be included in the project as part of the sample and how to recruit suitable participants. Purposive sampling is often a significant strategy in qualitative studies (Bryman, 2016). Compared with probability sampling, purposive sampling can help a researcher gain access to a wide range of individuals relevant to their research in a deliberate manner, so that the researcher is able to achieve the desired focus and cover a range of different perspectives and activities. Within this strategy, I first adopted the "snowball sampling method" (Bryman, 2016, p. 415) to recruit 8 PMSs – schools A-H (as shown below).

Figure 2. PMSs in Guiyang recruited by snowball sampling method
Source: Wang, 2019

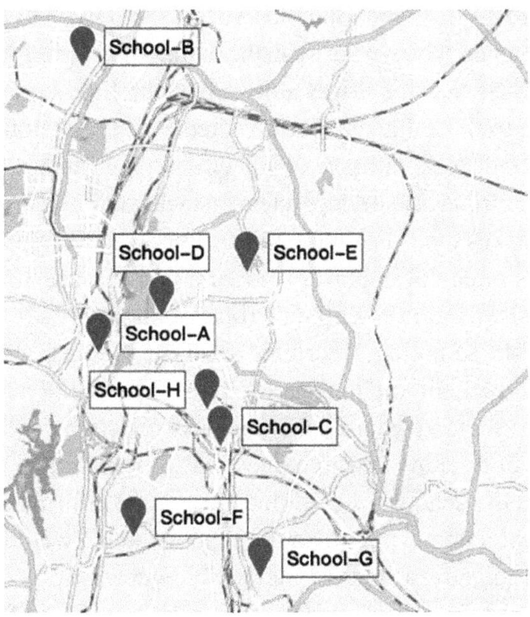

Mr. Zhang suggested 6 out of the 8 schools, while 2 participant head teachers recommended the other two schools (B & H). Gaining and maintaining access to the 8 schools was feasible to realize, unlike in Beijing. Successful access made me ambitious to recruit 30 PMSs in order to give me as wide of a sample as possible to investigate the circumstances and challenges of PMSs in Guiyang. However, recruiting participant schools also had its constraints and uncertainties, including a number of "faceless" and hasty rejections before I had the chance to introduce my proposed research ideas to those who could allow me access to local PMSs.

To begin with, I needed to determine which schools were PMSs from the list provided by the local education bureau, as these were not specified in relation to government schools and the high-cost private schools. Second, each school maintained a rigorous safety and security policy that restricts individuals from entering unless they were already involved with the school in some way (Guiyang Education Bureau, 2010). The head teacher is the person who bears responsibility for safety management (ibid., 2018). In order to access a PMS, I first required permission from a head teacher, which was what I tried to obtain.

To identify the location of other PMSs, I used opportunistic sampling, which means to "capitalize on opportunities to collect data from certain individuals, contact with whom is largely unforeseen but who may provide data relevant to the research question study" (Bryman, 2016, p. 409). I visited 23 schools in person to enquire whether or not they would participate in my study. Without an accurate official list of addresses of PMSs, I was still able to devise two ways to locate 23 PMSs. A participant head teacher, who wrote down the names of several PMSs, identified and suggested some PMSs. By this, I was able to find the PMSs on a local map. Other schools were close to those I had already visited, and I took note of their locations when I passed them to seek entrance at a later date. As Sarsby (1984) pointed out, ethnographic research can also be dependent on luck, although one can equally argue that a researcher makes their own luck. Subsequently, I accessed another 8 schools (schools I-P) by a strategy of opportunistic sampling (Figure 3). Likewise, one can see schools where I was refused entrance (black buttons).

With opportunistic sampling, I recruited another 8 schools (schools I-P), while 15 out of the 23 targeted schools did not agree to participate in the research. Sometimes, the locked gate and security guard led to faceless rejections without providing me the opportunity to outline my research questions. For example, with a locked door and no guard found in a school, I would find the contact information on their student admission posters or a notice put up on the school gates. I would then call the head teacher on the phone and was almost inevitably rejected, sometimes being cut off by whoever answered the phone immediately following my self-introduction.

Demanding negotiations were not only required to gain initial access, but also to maintain such access throughout the data collection process (Høyland, Hollund, & Olsen, 2015). For some of those schools that granted me the chance to talk to the person in charge, I was still sometimes declined. Reasons generally included one or more of four points outlined by those who spoke to me. First, my study was linked to an overseas university and caused head teachers to worry about how data might be used. Second, some did not want to participate in research except that the local education authority had explicitly approved or proposed. Third, some of those

Figure 3. Location relation of PMSs
Source: Wang, 2019
Key:
- *Blue buttons: schools accessible by snowball sampling*
- *Red buttons: schools accessible by opportunistic sampling*
- *Black buttons: schools I could not access*

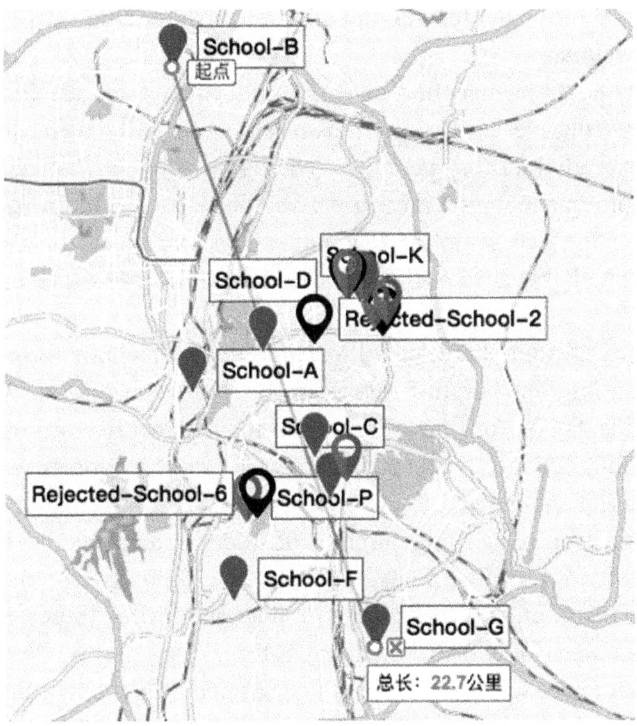

working in PMSs thought my study was without merit and would not bring any substantial benefit to their school. Fourth, school representatives said they would have to seek permission from the local education authority in order to permit any outside study of their institution. Finally, in some instances, no reason or explanation was offered as to why I was denied access.

Negotiating access to a research site is an important practical consideration (Murchison, 2010) but can often come “as a rude surprise” (Feldman, Bell, & Berger, 2003, p. vii) to researchers who may not have anticipated difficulties associated with access. Nonetheless, how one attempts to access a fieldwork site can itself be a way of generating knowledge about the field, its constraints, the concerns of others, and the kinds of information privy to these concerned others (Murchison, 2010). In my case, I regarded the negotiation of access as a way of sifting through potential participants to identify those who were more open to the idea of engaging in research, which is an important consideration in the recruitment of participants (Edwards & Holland, 2013). Also, I discussed the resistance I came across in some schools with participant head teachers. Their responses and the reasons provided by schools from which I had been rejected helped to further explain the delicate paths that PMSs follow as private institutions that help compensate for the deficiencies in state educational provision for migrant children, besides being a means of income for their proprietors.

At least, by having a selection of willing informants, my research was able to achieve a representative sample, a criticism sometimes raised regarding qualitative studies, which academics may perceive as being insufficiently adequate to derive a valid pattern of representation among a given population (Dudovskiy, 2019). The combination of the two sampling techniques was evidenced to work effectively, and this was largely because of the specific context of PMSs. The authors discuss these below.

Geographic proximity between a PMS and a student's home is one of the main concerns when migrant families select schools for their children. With a falling number of students due to some returning to their hometown or, gradually, being admitted to public schools, head teachers from adjacent schools may compete with each other. This means there is a tendency to avoid sharing information. For example, this is especially true concerning numbers of students, teachers' qualifications and salaries, and a school's future plans. Likewise, the NGO director, Mr. Zhang, also acknowledged that there is a real competition among PMSs. Hence, the PMSs that they referred to are widely distributed across three districts.

On the other hand, schools for which I negotiated access opportunistically were those surrounded by schools recruited through the snowball method. Even though attempts to enter some PMSs failed, the selection of a school as a potential research site still took into consideration either the geographical setting of a PMS (e.g., the furthest distance between two visited PMSs was 14 miles from school B to school G, as shown in Figure 3) or the distance to PMSs access to which was gained by snowball sampling (Figure 3). As a result, the researcher did not need to spend excessive amounts of time traveling between schools.

With these coping strategies, I finally recruited 16 schools to participate in the study. Snowball sampling, complemented by opportunistic sampling, emerged as the most pragmatic choice in my situation. I adopted them to ensure a wide spectrum of different PMSs in Guiyang City and thus, a heterogeneous sample (as far as possible), in order to maximize the variety of data that I could obtain (Patton, 2002).

The Use of Languages

Having some understanding of the cultural parameters in which human actions take place is an important consideration for an ethnographer (Madden, 2017). This may be obvious for a study taking place in a setting unfamiliar to a researcher. However, it can be even more important when studying a topic in an apparently more familiar context. Indeed, when a setting is familiar, presumptions about a local environment may blind one to issues that would be more obvious in an unfamiliar location. This was especially true regarding the use of language in my fieldwork, giving rise to my third challenge.

Growing up in Northeast China and studying and working in Beijing, I have spent almost all of my life in parts of China where Mandarin is the default code for communication and people speak Mandarin without facing problems of mutual unintelligibility. I assumed Mandarin to be the "universal" Chinese language. Thus, I had several underlying beliefs about the language of the field in the southwest: First, I believed that all local people could speak Mandarin in a mutually intelligibly manner, even though they may have unique accents. Second, participants from schools would speak Mandarin because it is an essential aspect of a teacher's qualifications as well as a language requirement in education.

In my pre-fieldwork visit in Guiyang, interaction with several head teachers and local NGO staff without intelligibility issues reconfirmed my belief that Mandarin had been promoted and implemented as a "universal" language of China. When conducting research in my home country, the "culture and macro-language bloc" (Madden, 2017, p. 45) gave me a feeling of being home in the sense that I would be easily able to interact with whom I came into contact, wherever they were located in the country. Consequently, my confidence in studying a context within China led me to downplay my supervisor's inquiry before entering the field (Supervision meeting, 2019):

My supervisor: "Will you have any language barriers? Do the people in Guiyang speak the same language as you do?"

I: "Yes, of course! Although they have their own variety, they can all speak Chinese and we are Chinese! Also, this study will mainly take place at schools where everyone is able to speak Mandarin – it is regulated by the Ministry of Education."

While carrying out fieldwork, it turned out that the use of languages was more complex and subtle than I had previously realized. Indeed, there were three layers of language used by the informants in the fieldwork – Mandarin, Mandarin with a local accent, and Southwestern Mandarin, a variety considered to be only about

50% mutually intelligible with Standard Mandarin (Cheng, 1998). For example, Southwestern Mandarin does not possess the retroflex consonants (zh, ch, sh, r) of the standard variety (Zhang, 1989). As a (standard) Mandarin speaker, I came across certain intelligibility issues that were not reported in the literature I had read about Guizhou Province (much to my surprise).

Mandarin, also known as Putonghua (普通话), is the mandated standard variety of Chinese, that is, the official language of China. In 2000, the law of the People's Republic of China regarding the standard spoken and written Chinese language came into force. This stipulates a particular variety of Mandarin as China's national language. As Article 19 regulates (National People's Congress, 2000):

All staff members who need to use Putonghua as their working language shall have the ability to speak Putonghua. The Putonghua level of those who use Putonghua as their working language, such as broadcasters, program hosts and hostesses, actors and actresses of films, TV series and plays, teachers and State functionaries shall reach the respective standards set by the State; those who have not yet reached such standards shall receive different training, as the case may be. (Translated by Asian Legal Information Institute)

Education bureaus in each province of China demand a certain standard of Mandarin-speaking in the Putonghua Proficiency Test (普通话水平测试) in order to acquire a teaching qualification. I was awarded the Certificate of Putonghua Proficiency at Level 1-B (meaning "92% accurate"), which is the second-highest level, while the highest level is required for presenters on national and provincial radio and television. Therefore, one can see my Mandarin as very close to the officially specified, most accurate variety. Although everyone has an accent, I emphasize Mandarin with an accent here because, as a tonal language, it can be easy to misunderstand utterances due to tone differences that arise as a result of dissimilarities in accent among varieties of Mandarin. This proved to be the case in my fieldwork. I specify Mandarin with a regional accent as the second layer of language here. This was particularly true when participants communicated in Mandarin, whether talking to me or teaching classes. Regarding the Southwest variety of Mandarin, this is the conventional language variety that the people of Guiyang use in their day-to-day lives.

The three layers of language I found in the fieldwork were certainly not what I had anticipated. This made data collection more complex and difficult than initially anticipated. Among all 25 interviewees, 12 could speak Mandarin in a way that was comprehensible to me. Twelve of the interviewees spoke Mandarin with a local accent. This brought me the first language challenge. There were 12 informants with whom I was able to interact with mutual intelligibility, thus avoiding missing or misinterpreting information. However, I still needed to ask interviewees to slow their pace of speaking, and I occasionally asked them to repeat what they said when necessary. However, one interviewee only spoke Southwestern Mandarin. In this case, I invited Mr. Zhang, who the interviewee trusted and who agreed to explain what the interviewee said if necessary.

Non-participatory observation in the form of interaction outside the school, such as conducting home visits with teachers and chatting with street peddlers, was nearly all in Southwestern Mandarin. In other instances, such as when I audited meetings of school or NGO staff, there was a mix of Southwest Mandarin and standard Mandarin. People in these meetings ostensibly tried to adopt standard Mandarin to accommodate me, but they sometimes fell back to the local variety seemingly because of people's unconscious lapse into Southwest Mandarin. Below is a scene that depicted this and illustrated the fluidity of the three layers of language and how there were shifts among them in different scenarios (Field notes 09/10/2019, translated by the first author):

Today, I audited the first school meeting for the new semester at school E. In the beginning, the head teacher said, "Let's use Mandarin to make sure Yao can understand because she is nonlocal."

The stumbling block arising as a result of language was obviously related to mutual intelligibility on some occasions. To cope with this, I first learned something about the differences of pronunciation between standard and Southwestern Mandarin, especially the pattern of tone changes. Second, I spent time with local friends and consulted them about the Southwestern Mandarin variety. These friends provided me with direct exposure and opportunities to practice and adapt to local forms of speaking in a relatively short time. Third, in situations in which I needed to be sure of what I had heard, such as during an interview with one head teacher who spoke Southwestern Mandarin (but not the standard variety), a third person was on hand to help.

In ethnographic research, researchers conceptualize language as a social practice that is shaped by social and historical forces. Language also constitutes individuals' lives by specifying, creating, maintaining, and thus shaping the frames of their actions (Eriksson & Kovalainen, 2008). This also applied to the researcher. The language(s) used by ethnographers can influence how they are received and treated in different contexts. Although my study was mainly based on a group of people connected to PMSs, fieldwork was not only limited to primary participants. I interacted with NGO staff, parents waiting outside schools, shop owners and their customers, as well as migrant workers in a wide range of jobs, all of whom would be more likely to be speakers of Southwestern, rather than standard Mandarin.

However, language differences did not hinder my interactions with participants; rather, the differences facilitated conversation. For example, participants applied certain common Chinese hospitality customs, by which they treat guests with kindness and respect by hosts, to me as a guest visitor in Guiyang. People could easily notice my outsider identity because I spoke standard Mandarin rather than the local variety. In random interactions with street vendors, shop owners, and parents waiting to pick up students, standard Mandarin seemed to elicit more tolerance and patience from local people instead of being a social barrier. As a result, locals were more likely to answer seemingly "silly" questions that had obvious answers to them.

Positionality

The last challenge is positionality. When talking about how an ethnographer's positionality influences or shapes their access, interactions, and conversations within the field, many scholars have shared experiences more related to demographic traits, such as gender or nationality (Flores, 2016; Venkatesh, 2006; Zuberi & Bonilla-Silva, 2008). Positionality is connected to how a researcher "relates to research participants and what can/cannot be done vis-à-vis the research within the context of institutional, social, and political realities" (Sultana, 2007, p. 376). Beyond characteristics pertaining specifically to researchers, the following section describes three tools generated by the research, selective use of which helped to sustain access to fieldwork sites and a reasonably smooth process of data collection. The three tools (described below) include: using a file folder; sharing stories; and trying to provide an emic (or insider's) perspective, despite being mainly an outsider in Guiyang. These helped to position the researcher in ways such that she appeared professional, trustworthy, and empathetic during the research.

Using a File Folder

The first "tool" I used was a file folder, which is a common means of carrying documents. However, this apparently trivial item aided me when knocking on the doors of some schools as part of opportunistic sampling. These doors may not have been opened otherwise. When attempting to gain entrance, appearing like a professional person was important, and I believe that this was achieved by carrying a file folder. This idea had been inspired by my experience in one school, as described below.

At every school, the main door or gate would be locked. The first thing to do was to seek permission to enter from the gatekeeper, normally a security guard. If things went smoothly, the guard would agree to allow me entry, but I would then have to remain in the guard room. The guard would then provide me with the phone numbers of head teachers or call them to the gate so that I had the chance to inquire about their willingness to accommodate me. This brought me to the second level of gatekeepers – head teachers.

On one occasion, a young teacher inside the school walked towards me and looked through a crack on the door. She asked me "Who are you looking for?"

I replied, "I am coming to see the head teacher; I am a…".

Instead of waiting for my self-introduction, she opened the door and let me in, saying, "Well, you come to see the head teacher. You can wait here for a moment and I'll tell her to come and see you."

I was surprised because it took me plenty of time to outline my identity and research purpose in other schools before I was either allowed in or before my request was rejected. Several minutes later, the head teacher came out from a classroom and walked over to me. Obviously, the head teacher did not know me and was surprised that I was inside the door as well. Before talking to me, she turned to the young teacher and asked with an intention to blame her, "Why did you let her in?"

The young teacher replied, "Because she's carrying a file folder in her hands. I thought she was from the education authority or an official who had come to inspect the school."

I realized that, with a file folder, people were more likely to regard me as someone arriving with serious intentions. In other cases, they might think I was from the local education authority to undertake some official business. Since this occurrence, every time I visited a new school, I would always bring a plastic file folder with me that could easily be seen to have many documents inside. It turned out that I was granted access to more schools than before as a result of this strategy. Holding a file folder functioned as a tool to indicate my professional status. A result of this was that gatekeepers were more likely to allow me to state my purpose. Similarly, other researchers can, with some foresight and good fortune, find relevant tools s/he needs by paying attention to certain details in the field that might encourage others to be predisposed towards them.

Sharing the Researcher's Story

If relying on tools to give PMSs an impression of my professional status and intentions helped me to open the official doors of some schools, it was the sharing of my own story that helped open the doors of the hearts of participants. Entering a school was just the first step in this process. The next step(s) entailed: accessing the second level of gatekeeper, the head teacher; reassuring informants of my intentions; and encouraging stakeholders to divulge their stories. These steps remained challenging matters. To mitigate people's hesitance and to build trust, I shared my volunteer story at a PMS in Beijing after a head teacher asked why I had traveled thousands of miles to undertake research in Guiyang.

Researchers need to show that they come to do more than build their own careers. For instance, when interviewing the head teacher in school B, he asked me about the reason for self-funding this research, which took place far from my hometown and the university where I was enrolled. I told him that I was determined to improve education for migrant children, a sentiment that arose in me when I volunteered as a teacher for one year. My familiarity with the educational experiences of migrant children, their educational challenges and the lack of social mobility opportunities available to them were, and continue to be, strong motivations for me. Soon afterward, the head teacher started to reveal how he established his school 22 years ago. This followed requests from migrant parents from the same area as Bijie (毕节) in Guizhou Province, whose children were unable to enroll in public schools because of the high fees at that time, stated (Interview data 09/28/2019, translated by the first author):

I was a substitute teacher at another PMS… One day, in my spare time, I spent time in this area. Here, a lot of migrant workers from my hometown (Bijie City, 毕节) were working as peddlers in a very big vegetable market. I went to the market and many people came to me. They said, "Well, compared to work as a teacher in another school, why don't you open a PMS in this district? We will all send our children to your school! It is so hard to enroll our kids in school here!" ... They all stopped what they were doing and stood surrounding me, saying "Teacher Xu, Teacher Xu, it would be fabulous if you could open a school here! We won't plead anyone anymore [to admit children to school]" … Indeed, I did some investigation later, very hard for their children to go to school… Back then, the public school charged very high fees from the migrant children, which euphemistically known as "commission education charge."

我那时在另一个民办打工子弟学校做代课老师……然后闲暇间,我就到这边来玩嘛。在这边嘛,我的很多老乡(毕节人)都在这里,我就去菜市场上,很多人在这里说"哎呀,你在别的学校上课,你不如就在咱们这个区办一个学校,我们把孩子送到你那里,这里读书太困难了!一大堆人围着我啊,他们生意都不做了。"徐老师,徐老师,你要是在这个地方就非常好。我们就不用去求人!" "确实,我一调查,很困难…… 因为他们当时要交高费,那时候高收费美其名曰叫代培费。

I believed that he felt comfortable telling me about his experiences because of his empathetic feelings regarding my own involvement with PMSs and a shared sense of good intentions to improve migrant children's educational experience, as he told me. Sharing my stories prompted similar accounts from informants, following accounts of my time at a PMS in Beijing.

Presenting Emic Perspectives

Besides my own PMS story, accounts and information passed on by other participants significantly facilitated my data collection and allowed me to relate these to my own PMS experiences. The juxtaposition of these facilitated the building of rapport fairly quickly. Additionally, they helped to demonstrate my understanding of PMS circumstances and thus, those of my teacher participants. Sometimes, besides sharing my own PMS experiences, I mentioned what I had observed in other PMSs in Guiyang, discreetly and ethically of course. I did so while adopting the jargon that head teachers would use. The kinds of stories that I shared depended on the ways in which I was received. For example, I spoke with the head teacher in school J and a teacher in school L about their recent work but referenced information gathered from other schools. In my field notes, I recorded these successful attempts at developing rapport (Field notes 11/13/2019, translated by the first author):

This week, school E has been very busy with their annual inspection. Every time I went there, the head teacher was extremely busy with "doing materials (做资料)". "Doing materials (做资料)" means preparing all kinds of documents for the inspection, such as the record of school events. It is a jargon that only a person who deal with school business will use.

When I was granted access to school J, the first thing I asked the head teacher was "How are you? Are you busy with the annual inspection? It's a bit time-consuming, isn't it?" She replied, "Yes! I wish I could have 25 hours a day!" I followed, "It must be due to 'doing materials (做资料)'!" she smiled, "Aha, it's driving me crazy!"... Adopting some specific expressions or jargons used by other head teachers helped me get into the participants' world and maintain the relationship with informants.

Likewise, as some participants told me about some situations they came across, I referred to some relevant comparable cases, but I did so selectively. For instance, I asked a teacher in school L, "Have any students or parents asked you to deal with some issues unrelated to school life? Because one teacher told me before that she had to pick up a child from a police station after he was lost!"

He answered, "Yes, very often!" He then introduced a lady waiting in the next-door room to me. She was a student's aunt who was taking care of the child after his parents' divorce and, later, his father's passing away. The aunt came to discuss her family hardships but stopped when I arrived. Instead of keeping her waiting, they reopened the topic and I naturally joined their conversation as someone who was considered eligible to take part.

Information acquired in the fieldwork process sometimes worked as a form of elicitation before a teacher raised a specific topic. Sometimes, this information worked as a complement for the mentioned topic, such as the aforementioned examples in school J and L, which I used strategically to show that I was also entitled to hold an emic position if needed. Referencing the participants indicated two things, covertly. First, their peers in other PMSs had verified my trustworthiness such they could entrust me with inside information. Some new participants would, therefore, ask me things such as, "Have you come across this in other schools?" Other participants would state, "You have definitely known this! It's very common in schools like ours!" Second, my familiarity and professionalism concerning PMSs further evidenced my prior experience of PMSs.

FUTURE RESEARCH DIRECTIONS

In this chapter, a number of strategies have emerged that can allow one to accommodate the challenges that may otherwise threaten (the robustness of) an ethnographic study and that may even turn these challenges into opportunities. From selecting a research site and sampling to recording fieldwork observation and reflecting on these, the complexity of language(s) and the role(s) of the researcher and the researched can offer solace to those fledging ethnographers. These ethnographers may be planning their fieldwork with numerous unknowns or who may be doing fieldwork with hesitancy, self-doubt, or a sneaking sense of possible failure. By providing this account of fieldwork experience, the authors hope to contribute to the construction of fieldwork strategies and foster the "development and transmission of ethnographic research skills" (Hirvi & Snellman, 2012, p. 8).

Besides guide books on how to conduct ethnographic studies and the tidying-up of research findings and their presentation, it is crucial to pay attention to ways in which one can accommodate the challenges, obstacles, and constraints that are prevalent in any research field. It is also important to develop strategies to overcome these. While some scholars have suggested generic and standardized ways to cope with challenges in specific social or cultural contexts (Reyes, 2018; Rothenberg, 2016), personal, close-up accounts of both successful and unsuccessful means developed by fieldworkers can offer insight to help deal with basic yet common practical issues. In addition, this chapter has focused more on the actual process of undertaking fieldwork; and there remains a paucity of accounts of ways in which fieldworkers have developed as researchers through their fieldwork. Regarding future research, taking account of budding researchers, their personal experiences, and what can be learned from these is a worthwhile consideration.

FURTHER IDEAS AND RECOMMENDATIONS FOR NOVICE RESEARCHERS

By using a doctoral-level study, we have presented how the (pre-study) idea of fieldwork for a novice may change amidst the real field that we are able to access and investigate. These sometimes unexpected changes can bring new challenges to the researcher. However, by arguing that challenges can be accommodated or even adapted to opportunities, we recognize that fieldwork should be treated as a flexible process in which researchers negotiate and perceive the dynamic of their field. We thus suggest the following ideas for novice researchers:

- Choice of **research site** is significant and denotes the extent of a space in which sociocultural events occur. This means what is included or not, the boundaries of research, by which the field site is established and how it relates to a larger social space, such as nation. It may be the case that one cannot obtain access to a chosen research site. However, one can explore other options by building relationships broadly, seeking relevant people's suggestions and finding key contacts, by which one looks for possible gateways to a potential site.
- **Sampling and recruitment**: In order to acquire a diverse sample, one can combine different sampling methods, e.g. adopting both snowball and opportunistic procedures, as we did. Researchers need to make additional effort to develop collaborations with trusted local stakeholders, including leaders (enhancing credibility), as was done in this case with those in local non-government organizations, thus enhancing the possibilities of recruitment and participation.
- The **use of language**: A native researcher conducting research in their home country can easily be assumed to be an insider, one who is familiar with the local vernacular. Given the diversity between as well as within languages, language varieties used in the field may diverge from each other in form, function and other ways. This can be significant as ethnographic descriptions are inevitably mediated through discourse. A challenge for the ethnographer is to bring together different stories, from the same context to produce coherent representations of people and phenomena under study, who are implicitly or explicitly ascribed expertise. Theory comes from description, whereby indigenous meanings emerge from findings.
- **Positionality**: As researchers, we need to be sensitive to characteristics of ourselves and our subjectivities (reflexivity) as well as the environment that may influence, facilitate or affect the research process. One emphasizes the importance of positioning towards participants' own rationalities, while acknowledging our interpretive capacities are fundamental in making sense

of observations (etic). Avoiding ethnocentricity, being dialogic (such that participants' ideas are considered on the basis of their legitimacy rather than by social position), being inductive (such that theory is derived from local meanings an emic position (not pre-imposed ideas), and being holistic (by taking account of aspects of the broader macro context).

CONCLUSION

In this chapter, we discussed the challenges an ethnographer may encounter when conducting research. This chapter places an emphasis on a number of practical challenges, from the stage of selecting an accessible research site and building relationships to recruiting willing participants, possible language barriers, and potential challenges arising from the role(s) of the researcher and the researched. By exemplifying these through reference to a three-month period of fieldwork, the chapter shows how these challenges emerged and influenced the research process. More importantly, this chapter demonstrates how coping strategies emerged from fieldwork. Consequently, it is suggested that it is important to provide tarnished and messy aspects of a fieldworker's experience amidst the various challenges in the planning and conducting of fieldwork.

Researchers can consider fieldwork as a journey in which the complexities of the field emerge and become clearer by the day. This is done by engaging with gatekeepers, participants, and key informants and by involving "a myriad of encounters and events" (De Neve & Unnithan-Kumar, 2016, p. 87). Each of these aspects have helped inform this study. As a researcher who was new to ethnographic fieldwork, I realized the importance of flexibility in my negotiations and interactions. Gradually, I increased my awareness of a number of uncertain challenges and how these might be mitigated. In other words, the more I found out, the more I realized I did not know.

REFERENCES

All-free-download.com. (2020). *China political map.* Retrieved May 1, 2020 from https://all-free-download.com/free-vector/download/china-political-map-vector_587825.html

Bryman, A. (2016). *Social research methods* (5th ed.). Oxford University Press.

Cerwonka, A., & Malkki, L. H. (2008). *Improvising theory: Process and temporality in ethnographic fieldwork.* University of Chicago Press.

Chen, Y., & Feng, S. (2013). Access to public schools and the education of migrant children in China. *China Economic Review*, *26*, 75–88. doi:10.1016/j.chieco.2013.04.007

Cheng, C. C. (1998, May). *Extra-linguistic data for understanding dialect mutual intelligibility*. Paper presented at the meeting of Annual Conference of the Pacific Neighborhood Consortium (PNC). Taipei, Taiwan.

Creese, A., Baraç, T., Bhatt, A., Blackledge, A., Hamid, S., Wei, L., Yağcıoğlu-Ali, D. (2008). *Investigating multilingualism in complementary schools in four communities*. Final Report for the ESCR for Project No: RES-000-23, 1180.

Creese, A., & Blackledge, A. (2012). Voice and meaning-making in team ethnography. *Anthropology & Education Quarterly*, *43*(3), 306–324. doi:10.1111/j.1548-1492.2012.01182.x

De Neve, G., & Unnithan-Kumar, M. (2016). *Critical journeys: The making of anthropologists*. Routledge. doi:10.4324/9781315575032

Dudovskiy, J. (2019). *Snowball sampling*. Retrieved May 1, 2020 from https://research-methodology.net/sampling-in-primary-data-collection/snowball-sampling/

Edwards, R., & Holland, J. (2013). *What is qualitative interviewing?* Bloomsbury.

Eriksson, P., & Kovalainen, A. (2008). Ethnographic research. In P. Eriksson & A. Kovalainen (Eds.), *Introducing qualitative methods: Qualitative methods in business research* (pp. 138–153). Sage. doi:10.4135/9780857028044.d77

Feldman, M. S., Bell, J., & Berger, M. T. (2003). *Gaining access: A practical and theoretical guide for qualitative researchers*. AltaMira.

Fetterman, D. M. (2010). *Ethnography: Step-by-step*. Sage.

Flores, G. M. (2016). Discovering a hidden privilege: Ethnography in multiracial organizations as an outsider within. *Ethnography*, *17*(2), 190–212. doi:10.1177/1466138115575660

Ganassin, S., & Holmes, P. (2013). Multilingual research practices in community research: The case of migrant/refugee women in North East England. *International Journal of Applied Linguistics*, *23*(3), 342–356. doi:10.1111/ijal.12043

Guiyang Education Bureau. (2010). *Guiyang Education Bureau on further strengthening the private primary and secondary schools (kindergartens) safety prevention work of the relevant provisions*. Retrieved May 1, 2020 from http://www.fsou.com/html/text/lar/172412/17241244.html

Guiyang Education Bureau. (2018). *Measures for the management of private schools in Guiyang city (Trail)*. Retrieved May 1, 2020 from http://www.xifeng.gov.cn/zxfw/bmlqfw/jypx/xxgl/201811/t20181130_1953320.html

Guiyang Education Bureau. (2019). *Table of basic information of education in Guiyang in 2019*. Guiyang Open Government Data Platform. Retrieved May 1, 2020 from https://data.guiyang.gov.cn/city/toListResContent.htm

Guiyang Local Chronicles Compilation Committee. (2018). *Guiyang Yearbook 2018*. Retrieved May 1, 2020 from http://www.guiyang.gov.cn/ztzl/ztzlsjtj/ztzlsjtjgynj/

Hammersley, M., & Atkinson, P. (2007). *Ethnography: Principles in practice* (3rd ed.). Routledge. doi:10.4324/9780203944769

Hirvi, L., & Snellman, H. (Eds.). (2012). *Where is the field? The experience of migration viewed through the prism of ethnographic fieldwork*. Studia Fennica Ethnologica. doi:10.21435fe.14

Høyland, S., Hollund, J. G., & Olsen, O. E. (2015). Gaining access to a research site and participants in medical and nursing research: A synthesis of accounts. *Medical Education*, *49*(2), 224–232. doi:10.1111/medu.12622 PMID:25626753

Hu, H., Lu, S., & Huang, C. C. (2014). The psychological and behavioral outcomes of migrant and left-behind children in China. *Children and Youth Services Review*, *46*, 1–10. doi:10.1016/j.childyouth.2014.07.021

Lichterman, P. (2017). Interpretive reflexivity in ethnography. *Ethnography*, *18*(1), 35–45. doi:10.1177/1466138115592418

Lin, Q., Ren, M., & Yang, M. (2019). Identity crisis among rural-to-urban migrant children in China: A proposal for school and government interventions. *Journal of Asian Public Policy*, *12*(2), 144–159. doi:10.1080/17516234.2019.1588841

Madden, R. (2017). *Being ethnographic: A guide to the theory and practice of ethnography*. Sage. doi:10.4135/9781529716689

MOE. (2017). *Education in China*. Retrieved May 1, 2020 from http://www.moe.gov.cn/jyb_sjzl/s5990/201810/t20181018_352057.html

Murchison, J. (2010). *Ethnography essentials: Designing, conducting, and presenting your research* (Vol. 25). John Wiley & Sons.

National People's Congress. (2000). *Law of the People's Republic of China on the standard spoken and written Chinese language* (final rule Jan. 1, Asian Legal Information Institute, Trans). Retrieved May 1, 2020 from http://www.asianlii.org/cn/legis/cen/laws/ssawcll414/

NBS, UNFPA, & UNICEF. (2019). *Domestic migrants in China (2018): Trends, challenges and recommendations.* Retrieved May 1, 2020 from https://china.unfpa.org/zh-Hans/node/42245

NGOCN. (2018). *Beijing's largest Privately-run migrant school may be demolished after its 20th anniversary.* Retrieved May 1, 2020 from https://chinadigitaltimes.net/chinese/2018/08/ngocn-%E4%BA%8C%E5%8D%81%E5%91%A8%E5%B9%B4%E6%A0%A1%E5%BA%86%E5%88%9A%E8%BF%87%EF%BC%8C%E5%8C%97%E4%BA%AC%E6%9C%80%E5%A4%A7%E6%89%93%E5%B7%A5%E5%AD%90%E5%BC%9F%E5%AD%A6%E6%A0%A1%E6%88%96%E5%B0%86/

Patton, M. Q. (2002). *Qualitative research & evaluation methods* (3rd ed.). Sage.

People's Daily Online. (2014). *Beijing proposes to control the size of the population to and strengthen the residing management.* Retrieved May 1, 2020 from http://news.eastday.com/eastday/13news/auto/news/china/u7ai865159_K4.html

Reyes, V. (2018). Ethnographic toolkit: Strategic positionality and researchers' visible and invisible tools in field research. *Ethnography*. Advance online publication. doi:10.1177/1466138118805121

Rothenberg, C. E. (2016). *On doing fieldwork in Palestine: Advice, fieldnotes, and other thoughts.* Springer. doi:10.1007/978-3-319-34201-6

Sarsby, J. (1984). The Fieldwork Experience. In R. Ellen (Ed.), *Ethnographic research: A guide to general conduct.* Academic Press.

Shi, J., & Sercombe, P. G. (2020). Poverty and inequality in rural education: Evidence from China. *Education as Change, 24*, 1–26. doi:10.25159/1947-9417/4965

Sui, H., & Song, Y. (2017). Does the left-behind experience affect the exam behavior, drop-out behavior, and intention to colleges of secondary vocational students? Based on longitudinal data from Zhejiang and Shaanxi Provinces. *Chinese Education & Society: The education of China's rural children: Chinese migrant children and left-behind children, 50*(4), 350-367.

Sultana, F. (2007). Reflexivity, positionality and participatory ethics: negotiating fieldwork dilemmas in international research. *ACME, 6*(3).

Venkatesh, S. A. (2006). *Off the books: The underground economy of the urban poor*. Harvard University Press.

Zhang, F. (1989). The initials of Chengdu speech as compared with Standard Chinese. *Journal of the International Phonetic Association*, *15*(2), 59–68. doi:10.1017/S002510030000298X

Zuberi, T., & Bonilla-Silva, E. (2008). *White logic, white methods: Racism and methodology*. Rowman & Littlefield Publishers.

Chapter 4
It All Works Out in the End:
The Experience of Researching Online Language Classes

Rebecca Lynn Chism
Kent State University, USA

Carine Graff
University of North Texas, USA

ABSTRACT

Qualitative research enables flexible designs unlike quantitative research, but those designs can be modified up to a point. It sometimes happens that the study at hand takes an unexpected turn of events and the researchers have then to find ways to cope with the changes. When the matter being investigated involves online surveys, there is even more leeway for uncertainty, as difficulties such as finding participants and time constraints may arise. This chapter presents a study conducted by two researchers in an American midwestern university and the challenges they encountered after answering a call for papers about technology and language learner psychology. They used an online survey to inquire about students' experience in an online language course. However, the lack of time intertwined with unexpected events involved readjustments and a need for more research and training. In the current article, the researchers share their project timeline, the issues they encountered, and offer some solutions.

DOI: 10.4018/978-1-7998-5826-3.ch004

INTRODUCTION

Qualitative research strives to find answers by examining every aspect of a focus setting or dilemma in order to answer questions, dig deeper, and further understanding. Fieldwork is considered standard practice in such investigations. Pole and Hillyard (2015) define fieldwork as an "intellectually and technically challenging total experience aimed at capturing meaning" (p. 1). Fieldwork in social sciences (Douglas, 1976) and higher education (Mligo, 2013) has been well documented and generalized in qualitative research circles (Brockington & Sullivan, 2003). Certainly, language teaching and learning falls under this purview, as noted by Benson et al. (2009). Standard investigative protocols in the field may include interviews and surveys, action research, observations, etc. in order to truly understand the dynamics of the intended subjects and as a means to answer research questions. Teacher education is dependent on such research for the articulation of best practices and support for professional development as indicated by its long history (Buchmann, 1984; Richards, 2009). The teaching and learning of world languages has also benefited from such investigative research (Bernhardt & Hammadou, 1987), even more so with the advent of technology in the latter 20th century to the present day.

Nevertheless, despite the traditions and procedures associated with such investigations, it is often imperfect. Studies point to the challenges associated with fieldwork within social sciences and higher education, with particular attention given to interview techniques (Partington, 2001; Opdenakker, 2006; Alshenqeeti, 2014), biases (Norris, 1997; Mehra, 2002; Roulston & Shelton, 2015), and ethical considerations (Rodham & Gavin, 2006; Rahman, 2017; Sim & Waterfield, 2019), amongst others. In addition, because of their very nature, such investigations can be unpredictable with obstacles not planned for or anticipated (Hobbs & Kubanyiova, 2008; Baker & Lee, 2011; Ballamingie & Johnson, 2011).

Such was the case when two world language teacher educators found themselves facing numerous challenges when they began research for an academic article on the psychological and social impact of online classes on lower-level learners. Although both researchers were well-schooled in qualitative approaches, their quest consisted of many twists and turns that, in retrospect, make for an interesting story as they reflected back on their two-year journey from the study's outset to its conclusion. They had to deal with many roadblocks that affected their ability to conduct the type of research they initially wanted to pursue. As a result of their flexibility and adaptability, they managed to reframe their research, which resulted in a publication. Consequently, they are now in a more objective position to describe the unanticipated problems and subsequent solutions based on their experience as well as to offer suggestions and guidelines for future endeavors. It is the objective of this paper to assist graduate students, fellow investigators, and anyone else who seeks to conduct informative research of the potential issues they may face. Baker and Lee (2011) found that "too few research reports reveal or discuss the difficulties investigators may have encountered during processes of data collection, thus leaving novice researchers with an impoverished perspective on complications that may arise when conducting classroom research" (p.1445). The scope of the paper will address the chronological turn of events as a cautionary tale, but one that will offer an awareness, and hopefully, preparation, for such events.

THE NARRATIVE

The narrative began in Fall 2017 when a Ph.D. candidate in the modern language department at a large, midwestern university visited an associate professor during her office hours to talk about research, deciding they should do something together. Different ideas were considered until the associate professor noticed a call for papers on the topic of technology and language learner psychology:

...although technology undoubtedly has a direct effect on the cognitive and psychological processes of language learners, research about such processes sorely lags their implementation. Hence, there is clearly a need to uncover how these learners are being affected by the technologies that are being practiced. This edited volume's primary objective, then, is to provide a clearer picture of how technology influences the behavior and affects the mindset of second language learners (Zarrinabadi, 2017).

The researchers were initially interested in the outcomes of online world language classes on proficiency in the interpretive, presentational, and interpersonal modes of communication (ACTFL, 2012). Interpretive communication refers to comprehension from listening, reading, or watching. Presentational communication involves speaking to an audience, writing, or producing a tangible product. Lastly, interpersonal communication is what we think of as conversation; that is, two-way spontaneous dialogue between two interlocutors. As interpersonal communication is considered to be one of the less practiced of the three modes of communication in world language classroom settings, the researchers were particularly interested in seeing how it might be promoted and facilitated in an online environment.

Because the associate professor's areas of interest include computer-mediated communication and world language teaching, she thought it would be a good topic to pursue, knowing the world language department at the university offered hybrid (a combination of online and face-to-face) classes for beginning world language levels. She mentioned this call for papers to the graduate student and they decided to submit a proposal in October 2017 asking the following research questions:

1. How can instructors increase opportunities for online active learning in a lower-level world language class?
2. Does an online active learning approach improve students' skills across the three modes of communication (presentational, interpretive, and interpersonal)? Do students display more willingness to communicate (WTC)?
3. How do students perceive the interactions with their peers after participating in online active learning?
4. What are the thought processes of students? Does online active learning encourage a growth mindset?

The researchers were informed of the editors' acceptance of the proposal in December 2017. In preparation for the fieldwork needed to conduct the study, the associate professor discovered that the hybrid program at the university was no longer in existence and had been shelved due to staffing and spatial issues. Thus, there were no longer online lower level world language courses available for the study. Since the home university no longer offered online lower-level world language classes, they had to find a willing host instructor to help carry out the study. In an attempt to find an instructor, the associate professor reached out to a friend of hers from graduate school who teaches at a large, southern university that offered online lower-level world language classes. However, since he was not teaching online courses at that time, he suggested his colleague, who would be teaching an online class in the Spring 2018 semester to begin in January. During the winter break, the associate professor contacted the recommended colleague, who gave verbal agreement for her course to be used for the study. She readily shared her syllabus and course materials, although she was concerned about the overall ability and motivation of the students. That is, based on her previous experience teaching the course, she found that the majority of the students were taking the course to fulfill a university requirement and therefore only did the minimum amount of work necessary, if that. Also, previous attempts on her part to engage students with technology online in the form of Voicethread produced little to no response (personal communication, February 2, 2018). Nevertheless, the researchers were glad to have found a class and hurriedly set about securing approval from the researchers' home university's Institutional Review Board (IRB) in order to proceed. In the beginning, the researchers envisioned developing activities and assessments for each of the academic units that corresponded with the course textbook in order to promote and measure student proficiency in the three modes of communication, as well as to see if the activities themselves encouraged interpersonal communication in an online format. Since the course was already built within Blackboard Learn, a university-sponsored learning management system, the researchers thought it best to stick with the same platform to maintain consistency and to simplify usage. Additionally, it was a platform with which the researchers were also already familiar. Blackboard has a feature for synchronous chat, Blackboard Collaborate, which the researchers felt would meet the requirements for facilitating interpersonal communication. Users are able to communicate in real time as well as see and hear each other via video and audio. The researchers set out to develop information gap activities that were consistent with the themes presented in the course with the intention to maximize interpersonal communication scenarios. The themes were spread out over six chapters and included the following: shopping, daily routine and personal hygiene, foods and meals, celebrations and personal relationships, parts of the body and medical conditions, and the car and computer products. The researchers considered ways to ensure that the students could meet up at certain predetermined times via the learning management platform in order for them to engage in the interpersonal tasks. However, because the initial discussion with the host instructor took place in early January, the researchers had very little time to develop and prepare the activities before the onset of the class later that month. The researchers wanted to keep within the predetermined themes of the course and to design activities that would promote the three modes of communication, particularly the interpersonal mode, by an integrated performance assessment, or IPA. The IPA is designed to measure student progress in attaining the competencies described in the World Readiness Standards for Learning Languages (W-RSLL) and the ACTFL Performance Descriptors for Learners. The IPA was not to be included in the course grading, but served as the format for the activities. The researchers found that they were rushing to develop materials and were not able to keep up with the pace of the course. They decided that they would instead focus on one chapter, parts of the body and medical conditions, instead of all of them so that they could adequately prepare the materials. This particular chapter was chosen as it was to be covered later in the semester and the researchers felt they could create meaningful activities with the topic by then. As the researchers were preparing the activities and assessments for the chapter, they had already sent the host instructor the letter of recruitment for the study and the background survey. Mid-way through the semester (March 2018), the host instructor informed them that no one in the class had consented to participate, despite encouragement from the instructor. At this juncture, the researchers offered to hold a drawing for a gift card (from their own pocket) for those students who agreed to participate in the study; nonetheless, still no one agreed to take part. The host instructor speculated that they were unwilling to participate due to the ubiquity of online surveys and their tendency to get overwhelmed with requests, resulting in their dismissal of them. Even the promise of the opportunity to win a gift card was not enough to persuade them. As such, in April 2018, the researchers decided to post a call for participants on a variety of listservs hosted by professional organizations and other teacher educator forums for potential leads. Responses to the inquiry did produce some willing host instructors; however, when the researchers followed up with them, they were either unresponsive, or the populations and other parameters did not fit the profile needed for the study. For instance, several high school teachers offered use of their classes. Both researchers were working at the same university at the time of the study and had filed an IRB (Institutional Review Board) application within their institution that focused on higher education participants. If they had wanted high school students to participate, they would have had to file a different IRB mentioning that a different population was going to be used and waited for approval to be able to have high schoolers participate. Given the IRB timetable and the lack of time the researchers were experiencing, that was not an option. Thus, the researchers found themselves at the end of the Spring semester without any data. Hope emerged in May 2018 when the friend of the associate professor mentioned he was going to be teaching a beginning world language class online beginning in June and that we could use his course if needed. At this point, as the researchers were no longer able to investigate their original proposal, they decided to pursue another angle by extending the questions on the original survey in order to glean additional information on the students' perceptions of the efficacy of online beginning world language classes. They redesigned their research questions to fit the new scope of the study. These questions were as follows:

1. How do students describe their experiences in an online lower-level world language class?
2. How do these experiences influence their willingness to communicate (WTC) in the target language?
3. How do these experiences shape students' growth mindset or self-perception as world language learners?
4. How do these students perceive their overall experience in this format?

The researchers put the survey online via Qualtrics and sent the link to the instructor for distribution. Because he highly promoted participation in the study, the researchers were finally able to obtain some data to use. The associate professor sent him a gift card to thank him for his assistance with the study; only when he went to use the card, there was no money on it. Upon further investigation, they discovered that it is common for thieves to enter into establishments that sell gift cards (such as a grocery store) and to somehow scan the bar number and sell it. This particular card number had been used in another state. Again, the researchers could not have imagined these events at the onset.

The resulting manuscript summarized how students described their online class experiences, including what they enjoyed and did not enjoy as well as what they perceived as the best and worst aspects of online learning. The study also considered the influences of online lower-level world language classes on their growth mindset as language learners and their overall willingness to communicate (WTC) in the target language. Results were evenly divided between positive and negative descriptions of their class and its influence on their growth mindset and willingness to communicate. As the majority of respondents were older students who acknowledged that interaction with their peers would likely improve their proficiency, they instead preferred to work alone. The researchers concluded that any online interaction must be productive and user friendly, with definitive teacher and peer presence to encourage engagement. They noted that the increase of online course offerings provides many advantages to today's students in that instruction can take place outside of the university setting, providing more flexible and convenient access to information and communication. However, because of this, it is more important than ever to consider the pedagogical and psychological impact of remote learning environments.

The manuscript was submitted for peer review to the editors in July 2018. The manuscript came back with reviewers notes in December 2018 and the researchers submitted a revised version in January 2019, which came back for minor tweaking in March 2019. In June 2019, the editors sent an email stating that a publishing company had agreed to distribute the edited volume; the researchers gave the final approval of their chapter contribution in July 2019. The editors sent periodic emails concerning the status of the edited volume and its trajectory toward publication. Finally, in April 2020, the edited volume appeared in print and a complimentary copy was sent to each of the authors. The resulting volume featured a chapter entitled *The impact of online lower-level courses on world language learners'self-perceptions, mindset and willingness to communicate* (Chism & Graff, 2020).

Challenges: Problems and Solutions

Before even starting the study, the graduate student was trying to publish and that was not such an easy endeavor. There is a controversy about PhD students publishing articles. Even if there is no specific written rule about it, some journals are against it while some others do not see any issues with it. In the article by Colleen Flaherty "Should grad students publish?" (2017), the problem of too many articles being published is raised in the humanities and in the natural sciences: there are too many publications with a problem of research quality and graduate student publications could be the culprit. Would reducing publication access to graduate students help alleviate the problem? Actually, that might make things more difficult for students who are ready to go on the job market. Indeed, even if PhD students write a long and challenging dissertation and finally obtain their diploma, that is not enough to find a job in academia (Van Cott, 2005). During their studies, graduate students should behave as if they were professors already: similarly to assistant or associate professors, they need proof that they are active as researchers. In other words: publish! However, those students already have to deal with many different situations such as working on the side and/or teaching, and taking care of a family while mainly having to concentrate on their dissertation—even the process of finding an acceptable topic can be very arduous, as the student is not the only person involved in that choice. Staying motivated to write is difficult for those students (as professors do know because they also have issues staying motivated to write regularly themselves) especially in an environment with little or no support. Graduate life is very lonely, notably after the course work is over, one has to organize their own schedule to stay on track with their writing. Nobody teaches that skill in graduate school and students are left to their own devices trying the best they can. To avoid the problem of being rejected by journals for being a graduate student, PhD candidates could ask professors to publish with them. Collaborating with an established professor helps "getting one's foot" in the publishing world. It also helps with staying motivated and eases the navigation of the publication process. Sharing ideas also broadens perspective and diversifies the publication agenda. As Hafernik et al. (1997) would say, "Researchers working together enrich the field with their combined insights and wisdom as well as with their ability to grapple with more complicated problems" (p. 31). However, there are two main issues with that collaboration request when it comes from a student. The first pertains to the curriculum: as publications are often not required in PhD curricula (as part of the course work for certain courses, for example), it can be difficult for graduate students to find professors who would push them to publish or would want to publish with them. Professors are busy and have their own publication agenda as well as their established service workload, so they often have no time to devote to publishing with students. In some fields such as sciences, professors work with students because they help them save time with research in the lab. However, even if that sounds very promising there is always the risk that such collaboration could lead to publications where the graduate student does all the work and in the end won't have their name on the paper. In other fields such as the humanities, few professors have research assistants and it is rarer to find professors who would want to take the time to sit down with a student and publish with them, especially the dissertation advisor, as they are probably already busy revising several candidates' dissertations and completing all the other tasks required by their position including their own research. The second issue is as Elliott Shore, executive director of the Association of Research Libraries mentions: professors "might see their own students as rivals" (Flaherty, 2017). Indeed, if publications abound in a certain field, the competition is fiercer, and professors might want to think about their own publication agenda as needed for promotion. To try and avoid all those difficulties, one suggestion for graduate students in need of finding someone who would mentor them and publish with them would be to very carefully choose their dissertation advisor (if the person is too busy there is a great chance they won't have time to publish with you) and to not be afraid to wander outside of their department and seek professors, who would be interested in collaborating with them. Presenting at conferences might also help meeting new researchers who would share their feedback and may be willing to cooperate. Depending on their department's atmosphere, if competition is not an issue, Van Cott (2005) also suggests publishing with another graduate student, as it might be an enriching learning experience.

While conducting this study, the researchers encountered numerous problems and challenges. Those issues ranged from creating the survey to IRB preparation, filing and acceptance, finding participants, training, and managing time. Some problems arose along the way and were often intertwined as one issue led to another.

The first and biggest problem encountered was the lack of time, as the research project was started quite late compared to the deadline for the paper. It is often not easy for researchers to find a call for papers that corresponds exactly to their research and when they do, the research might not be complete. Calls for papers in and of themselves give researchers new ideas that can generate more publications. However, they are also risky because the time frame available might be very short and might leave the researchers with a lot of stress in their efforts to meet deadlines. One also always has the option of coming up with one's own ideas and presenting the research to a journal, which includes a different level of concerns.

One of the reasons the project was started late was the challenges in finding participants. First, the home university no longer had the hybrid classes, then once a class was found, the students were unwilling to participate. This led to sending an appeal to listservs, which did not yield any viable populations. Lastly, only when the friend of the associate professor offered his class were the researchers able to proceed. Finding a class to study took approximately nine months, which wasted a lot of valuable time. In retrospect the researchers should have identified the target population ahead of time to design the study, prepare the IRB, and encourage participation.

Before one can apply for IRB, they first need to train and pass the CITI course about Human Subjects Research. That course includes thirteen modules and one needs to obtain a minimum of 80% to pass. Both researchers already had passed it and were able to provide the IRB office with proof of their successful completion. They were able to fill out an IRB application for their study and sent it to the IRB office along with their exam certificate. There are three different types of IRB applications: exempt (minimal risk to participants), expedited (minimal risk to participants including voice or video recordings, for instance), and full (sensitive population such as children or prisoners). Once the paper is written, there is formatting involved that requires knowledge of Microsoft Word. If the researchers are not savvy with the Microsoft Suite that can make them lose time trying to figure out how to format the document. If the research asks for a survey, then there is also the survey software to master. The Internet is full of resources to help one navigate through such tools. User guides and YouTube videos were particularly helpful. There are many applications that can be used for online surveys (Andrews et al., 2003), but one also has to know how to use them and that takes time. Professors' constant training is thus quite important as they need to be able to wear different hats. However, even if institutions offer some training, faculty often have to seek other sources for help such as online training with professional associations or pay a course out of their own pockets to be able to acquire some new technical knowledge.

At first, when the researchers designed the study, they had quite a big project in mind involving online class activities and a long survey. Of course, the great idea came first, followed by the intense search for a suitable class for them to be able to conduct that study. Unfortunately, the researchers soon realized that their idea was not feasible because of the time constraint and also because of the students' lack of consent to participate. The researchers were only able to secure a class at the last minute, and only thanks to a mutual friend. The researchers were not prepared for the students' lack of involvement. Fortunately, they had the ability to shift the focus of their research. Flexibility and a back-up plan are key to dealing with these challenges. The researchers had to adapt and change their course of action. Fortunately, they were able to restructure their design in a way that was adaptable to the circumstances. This flexibility could be reinforced by planning ahead, for instance, asking a variety of research questions that could each generate their own article would make the researchers save some time in case circumstances change suddenly. In fact, Pultz (2018) encouraged such flexibility, noting how "episodes of unexpected events or experiences of something 'going wrong' have, in hindsight, actually benefited [the] research process…A fixed research design or a deductive approach does not have the flexibility required to take these experiences into consideration" (p. 1). Agee (2009) writes of the importance of reflection while developing quality research questions.

Any study involving human subjects needs to be presented in front of the Institutional Review Board (IRB). The IRB application needs to be very detailed and the whole process can be strenuous. It can be so strenuous that one might consider not taking on any project requiring IRB approval, especially if it involves high school age populations. This fear of the process can be diminished with training and help from IRB members. The researchers did not encounter difficulties with different IRB applications, as only one application had been sent in their own institution. They had to try and go through the process like everyone else and learn on the spot, as every study is different and requires different justifications. Some universities have training options: for example, American University in Washington DC has a web page with useful information about their IRB process (see https://www.american.edu/irb/). According to one of the academics interviewed by Patricia Aufderheide, Professor of communication studies at American University in her article "Does This Have to Go Through the IRB?" (2016), the least prepared IRB application submitters are students. They suggested that faculty train their students better for the IRB process. The same article mentions that what students and faculty also fear is delay. Some easy IRB applications such as exemption cases can be dealt with by the board in a week or so, whereas a full board approval can take weeks. One recommendation would be to keep an eye on the IRB meetings schedule and monitor when the next meeting will take place: make sure to send out your application before that date. Also, contacting the person in charge of IRB for your department might also help. If you are lucky, they will be patient and answer any questions you might have. If the survey is sent to a class, as per compliance regulations, students need to be aware that they can refuse to participate at any time. However, if the survey is made part of their grade somehow, then do they really have a choice? How can the researchers ensure they will have enough participants? According to Padilla-Walker et al. (2005), offering extra credit does not increase participation, it might even limit the generalization of the findings. In a Chronicle of Higher Education Forum from 2011 pertaining to the problem of extra credit for research participation (Chronicle Forums, 2011), an instructor stated that if they were asked to participate in a study and asked to offer extra credit to their students if they participated, they would not accept. To make it fair, if students are offered extra credit for participating in a study, then other students in the same class who decide not to participate in that study also need to have the possibility of doing some extra credit activities. Instructors might not want to put in the extra work to prepare those activities because their students need to concentrate on their course material first instead of being attracted to extra credit. Another instructor in the forum proposed that the researchers offer treats or "a few bucks" to students who participate. However, research guidelines need to be ethical and recruitment cannot be coercive: participants need to be able to opt out at any given time and not feel threatened or influenced in any way. There is a controversy about payments made to participants for taking part in a study: on the one hand, those payments are considered as coercive by IRB members, but on the other hand, studies such as Largent et al.'s (2013) suggest that as coercion is defined as a threat of harm, payments cannot be considered as coercive. As a result, it is suggested that IRB members should change their minds about categorizing payments as coercive. In addition, if coercion is irrelevant for payments, there is still the issue of undue influence, which happens when participants are offered an excessive or inappropriate reward. In order to minimize that risk, payments should be reduced

(Borasky et al., 2019). Kilb and Herzig (2016) found that by organizing a research participation night, with offerings of course credit, pizza, and other prizes, they attracted 95 participants. They recommend it for student researchers struggling to recruit research participants in a short period of time (p. 53).

In the present study, the researchers had used an Amazon gift card of a reasonable amount as incentive, but it did not attract participants. There are three main reasons why students would participate in a study: altruism (that is more the case for women), interest in the topic, and incentives (that is more the case for men) (Singer & Bossarte, 2006). Thus, generating students' interests while giving details on the study (how it can be relevant to their lives) and highlighting how their participation can improve the field can increase the number of students willing to participate (Far, 2018). It seems like curiosity about the topic and possible benefits of the study to students' lives are better motivators than financial incentives (Gul and Ali, 2010).

When looking for a class, the researchers had a hard time finding one, as the hybrid courses offered at their institution were no longer an option. That shows that some work has to be done ahead of time by the researchers—before starting any project, such as inquiring about the courses' status in one's institution. It would be more convenient to be able to use a class close by or at least to have the host instructor close by. One can also use their own class if its description corresponds to the study. However, when using one's own class, some steps need to be taken to avoid bias, such as consent forms distributed by a third party and insurance that all student names are removed from all materials by a third party. Results then also need to be checked to avoid partiality. The class the researchers found at first seemed perfect, but it ended up not working out, as time went by too fast and none of the students expressed the desire to participate in the study. After that, the survey had to be modified and was sent out to listservs, but the researchers did not receive many answers. The problem when using listservs is first to know and decide which listservs to use exactly and second to establish a sampling frame. It will be difficult indeed to know exactly which population is being sampled and one might get different results with a population that is completely heterogenous if the survey is sent to different listservs. There are strategies that can help take control over those uncertainties encountered in the online survey world. According to Wright (2005), choosing online communities that require a membership and that are willing to share statistics with the researchers might help researchers have more insight about the characteristics of the population studied. Another disadvantage of online sampling is that participants might also want to retake the survey especially if there is an incentive offered. In order to prevent that, asking them to contact the researchers to receive a unique activation code or require them to give their email before starting the survey may be a good idea (Wright, 2005). However, it is true that participants might retake the survey while using another email address, which leads to false representation. To make sure the results obtained are not falsified, doing parallel research in similar communities will help having a better picture of the participants' characteristics (Wright, 2005).

Also, from the point of view of the survey creation, the researchers started writing down questions, but except for multiple choice scales answers in Qualtrics, no specific steps were taken to make sure the questions asked were not biased. Not only the wording used in questions, but also the question order in surveys must be carefully chosen, as it can influence the results. It has been shown for example, that in the agree-disagree question format, less educated participants tend to choose "agree" with corresponding statements. That can be called an "acquiescence bias" (Pew Research Center). In addition, when the final survey was ready, there was no test to check how the survey was perceived, except a test done by the researchers themselves. This, again, happened because of a lack of time. If the researchers had had more time, they could have had their survey tested by other colleagues for instance, before distributing it. The researchers had sent an email to the listservs that seemed to correspond to the scope of the study. The email contained the link to the survey, which the researchers asked to share. However, it might have been better to find the person in charge of the listserv and ask them if they could help distributing a first message introducing the study, and then a second message later including the link to the survey (Andrews et al., 2003). In addition, the researchers sent the emails to the listservs with their own email addresses, but Andrews et al. (2003) suggest creating a common email address different from personal email addresses that would be used for survey purposes only. Such a neutral email address would prevent the researchers from being attacked online with abusive comments, for example especially if the survey is published in a forum. Once again, close communication with the person responsible for the forum or the listserv should also help reduce unwanted occurrences.

All these issues with online sampling circle back again to the idea of flexibility, as assumptions about how many participants researchers might find or what the sample should look like will not help the researchers when the time comes to find a decent number of participants or to generalize the findings about the study. For instance, they should not assume a typical college age, but rather anticipate different age groups even if the targeted population is first-year language learners. The Internet is a wonderful tool that can be used to reach so many communities, and one would thus think that it would provide so many more opportunities to find participants, but in the end, that is not an easy task. Even if an incentive is offered, some participants perhaps won't accept to take the survey because they do not see why they would spend time doing this.

It is also worth considering the notion of survey fatigue. That is, since surveys are ubiquitous not only for data collection but for faculty evaluations, improvement of services, and other inquiries, it is not uncommon to receive multiple surveys in a given week that demand one's attention. Or, if participants actually respond to a survey, if it is too long or too cumbersome, they will often abandon it before completion. Lavrakas (2018) defines this phenomenon in the follow way:

Respondent fatigue is a well-documented phenomenon that occurs when survey participants become tired of the survey task and the quality of the data they provide begins to deteriorate. It occurs when survey participants' attention and motivation drop toward later sections of a questionnaire. Tired or bored respondents may more often answer "don't know," engage in "straight-line" responding (i.e. choosing answers down the same column on a page), give more perfunctory answers, or give up answering the questionnaire altogether. Thus, the causes for, and consequences of, respondent fatigue, and possible ways of measuring and controlling for it, should be taken into account when deciding on the length of the questionnaire, question ordering, survey design, and interviewer training (p. 743).

In order to encourage participation in surveys, the researchers suggest using a short and easy to complete questionnaire. It might also be helpful to allow for class time to complete the survey. A good relationship with the host instructor can also serve as a motivator.

Had the researchers had time to get to know the first host instructor better, perhaps even if that instructor had been in the vicinity, the researchers could have met with her to discuss the study more in detail and establish some relationship with her (Hobbs & Kubanyiova, 2008). Maybe the researchers' situation was made more difficult by the fact that everything had to happen online. Establishing a relationship with the host instructor could also have involved Skype calls or phone calls, basically anything that could have helped gain her trust and reassure her that the purpose of the study was solely research and not a critique of her class organization. What other strategies could the researchers have used to motivate the host instructor, as they did not have any contacts with the students? All in all, what could have the host instructor benefited from in this study? Perhaps she could have been mentioned in the paper and maybe more students would have been interested in taking her class? Other than helping researchers finding answers to their study and thus improving the research done in that domain, what other perks could the host instructor have benefitted from?

Hobbs and Kubanyiova (2008) suggest that it is important to be sure to sell the benefits of the study and to make it as easy as possible to participate. This requires knowing the context in which the study will take place and what is workable. They suggest the following steps in order to establish trust early: first, "win the trust of potential participants before requesting participation, familiarize yourself with the context and adopt an unassuming approach, and be sensitive to participants' anxieties and adopt a flexible approach" (pp. 501-502). They also emphasize the importance of checking in periodically, showing appreciation, and adopting a stance of contributor rather than evaluator.

It is well known that professors work long hours and they make and manage their own schedule, which should be kept very strict especially during a research project, but that schedule is also dictated by other tasks than research. The results of a survey mentioned in an Inside HigherEd article show that faculty spent most of their time preparing classes, answering emails, and attending meetings (Flaherty, 2014). Among all those tasks, it is difficult to maintain a strict schedule for research, especially also because professors work on their own and are on their own most of the time. The same article reports that faculty work on their own "57 percent of the time." Unlike in the corporate world, faculty do not have anyone holding them accountable for their work on a daily basis and there is no program in place to help them organize their time. It is also worth noting that teaching in and of itself deserves to be acknowledged and valued in the same way as research (Cole, 2000). O'Connor et al. (2011) believe it is instrumental for research institutions to find ways to help untenured faculty find balance in their equally demanding roles of instructor and researcher. They suggest that universities should strive to "(a) make supporting untenured faculty members an institutional priority, (b) have a "point person" who would inform untenured faculty members about deadlines, tenure procedures, etc., (c) assign a research, publication, grant and/or tenure mentor, (d) find ways to reduce high teaching loads if there are also high research expectations, and (e) receive constructive feedback about research" (p. 3).

In the midst of such challenges, it is important to have support throughout the process. More than once, the researchers wondered whether they should continue with the study in light of the persistent issues that arose along the way. To this end, they suggest that researchers have someone who can help to keep them motivated and on course, especially when things get discouraging. Driscoll et al. (2009) found that a peer mentoring collaborative group "revealed experiential, emotional and developmental commonalities that guided the women's navigation of the tenure track at a large public research university via understanding of self, others and the environment" (p.5). The researchers felt fortunate to have each other to talk to as well as the friendship of the one who helped make the study happen. Pegg et al. (2014) talked about how they drew on previous connections and built new ones with academics at other universities as a means to find the support they did not experience at their home institutions. Hellsten et al. (2011) also found that a self-created peer support group was instrumental for pre-tenure faculty to find encouragement and insight with each other (p. 275). Hobbs and Kubanyiova (2008) suggest the following as a way to maintain sanity: schedule a specific time each day to relax and collect your thoughts, prepare additional data collection methods to employ should holes in the data emerge after the primary data collection has been completed, and keep a research journal. They offer some perspective with the reminder that in the end "it's just a project" (p. 509).

CONCLUSION

In this particular instance, the researchers had to deal with many obstacles that affected their ability to conduct the type of research they wanted to pursue. Challenges included choosing an appropriate setting, enlisting a willing host instructor and class participants, dealing with time, curriculum, and content restraints, developing appropriate research questions and methods of analysis, etc. As a result, the researchers found themselves having to constantly and consistently adjust and regroup during a two-year period. At times, the study proved to be demoralizing to the researchers. Consequently, they now believe they are in a unique position to caution future researchers of potential pitfalls and offer realistic solutions to either avoid and/or resolve these issues. Future researchers should be advised of the following.

- Organize one's time by defining a plan of action
- Approach the research design with multiple possibilities and flexibility
- Expand knowledge about Word processing and survey software
- Identify multiple data sources
- Familiarize oneself with the Internal Review Board process at one's institution and any constraints
- Monitor the IRB schedule and send out the application before the board meets
- Strengthen relationships with host instructors and with participants where possible
- Incentivize participants
- Make the data collection process as straightforward and efficient as possible
- Remain objective and optimistic during the research process
- Seek support as needed

In sum, the researchers offer the following suggestions. First, effective planning is paramount. This involves some preliminary work in identifying potential target populations, creating a user-friendly yet effective survey, and expediting the IRB process. Time can be saved by becoming familiar with technological issues such as using online survey forums as well as formatting. It is also worth having a research design that is flexible and adaptable and can take the study in several different directions. Relationships and trust are vital to ensure participation on the part of the host instructor as well as the subjects. They must be able to see the benefits of the study and the importance of their contribution; time in class could be designated for survey completion. In addition, it is also important for the researchers to maintain balance during the process and to find a support system to keep them motivated and on-track. If future researchers tend to this advice, they, too, will find that it all works out in the end.

REFERENCES

Agee, J. (2009). Developing qualitative research questions: A reflective process. *International Journal of Qualitative Studies in Education: QSE, 22*(4), 431–447. doi:10.1080/09518390902736512

Alshenqeeti, H. (2014). Interviewing as a data collection method: A critical review. *English Linguistics Research, 3*(1), 39–45. doi:10.5430/elr.v3n1p39

American Council on the Teaching of Foreign Languages. (2012). *The ACTFL performance descriptors for language learners 2012*. Retrieved June 4, 2018 from https://www.actfl.org/sites/default/files/pdfs/PerformanceDescriptorsLanguageLearners.pdf

American University. (2020, May 24). *Human subjects research IRB*. https://www.american.edu/irb/

Anand, P., Chung, S., & Wagers, M. (2010). Widening the net: Challenges for gathering linguistic data in the digital age. *NSF SBE 2020: Future Research in the Social, Behavioral & Economic Sciences,* 1-5.

Andrews, D., Nonnecke, B., & Preece, J. (2003). Electronic survey methodology: A case study in reaching hard-to-involve internet users. *International Journal of Human-Computer Interaction, 16*(2), 185–210. doi:10.1207/S15327590IJHC1602_04

Aufderheide, P. (2016, August 17). Does this have to go through the IRB? *The Chronicle of Higher Education.* https://www.chronicle.com/article/Does-This-Have-to-Go/237476

Baker, A., & Lee, J. (2011). Mind the gap: Unexpected pitfalls in doing classroom research. *Qualitative Report, 16*(5), 1435–1447.

Ballamingie, P. C., & Johnson, S. (2011). The vulnerable researcher: Some unanticipated challenges of doctoral fieldwork. *Qualitative Report, 16*(3), 711–729.

Benson, P., Chik, A., Gao, X., Huang, J., & Wang, W. (2009). Qualitative research in language teaching and learning journals, 1997–2006. *Modern Language Journal, 93*(1), 79–90. doi:10.1111/j.1540-4781.2009.00829.x

Bernhardt, E., & Hammadou, J. (1987). A decade of research in foreign language teacher education. *Modern Language Journal, 71*(3), 289–299. doi:10.1111/j.1540-4781.1987.tb00369.x

Borasky, D., Cooper, J., & FitzGerald, K. (2019). Paying subjects to take part in research: A new perspective on coercion and undue influence. *Clinical Research, 33*(3). https://acrpnet.org/2019/03/12/paying-subjects-to-take-part-in-research-a-new-perspective-on-coercion-and-undue-influence/

Brockington, D., & Sullivan, S. (2003). Qualitative research. In R. Scheyvens & D. Storey (Eds.), *Development Fieldwork: A Practical Guide* (pp. 57–74). Sage. doi:10.4135/9781849208864.n4

Buchmann, M. (1984). The use of research knowledge in teacher education and teaching. *American Journal of Education, 92*(4), 421–439. doi:10.1086/443768

Chism, R., & Graff, C. (2020). The impact of online lower-level courses on world language learners' self-perceptions, mindset and willingness to communicate. In M. Freiermuth & N. Zarrinabadi (Eds.), *Technology and the Psychology of Second Language Learners and Users: New Language Learning and Teaching Environments* (pp. 543–569). Palgrave Macmillan. doi:10.1007/978-3-030-34212-8_21

Chronicle Forums. (2011, September 2). *The Chronicle of Higher Education.* Retrieved May 20, 2020 from https://www.chronicle.com/forums/index.php?topic=81783.0

Cole, A. (2000). Academic freedom and the publish or perish paradox in schools of education. *Teacher Education Quarterly, 27*(2), 33–48.

Douglas, J. D. (1976). *Investigative social research: Individual and team field research.* Sage.

Driscoll, L., Parkes, K., Tilley-Lubbs, G., Brill, J., & Bannister, V. (2009). Navigating the lonely sea: Peer mentoring and collaboration among aspiring women scholars. *Mentoring & Tutoring, 17*(1), 5–21. doi:10.1080/13611260802699532

Far, P. K. (2018). Challenges of recruitment and retention of university students as research participants: Lessons learned from a pilot study. *Journal of the Australian Library and Information Association, 67*(3), 278–292. doi:10.1080/24750158.2018.1500436

Flaherty, C. (2014, April 9). So much to do, so little time. *Inside Higher Ed.* Retrieved June 26, 2020, from https://www.insidehighered.com/news/2014/04/09/research-shows-professors-work-long-hours-and-spend-much-day-meetings

Flaherty, C. (2017). Should grad students publish? *Inside Higher Ed.* Retrieved June 26, 2020, from https://www.insidehighered.com/news/2017/08/23/renewed-debate-over-whether-graduate-students-should-publish

Gul, R. B., & Ali, P. A. (2010). Clinical trials: The challenge of recruitment and retention of participants. *Journal of Clinical Nursing, 19*(1–2), 227–233. doi:10.1111/j.1365-2702.2009.03041.x PMID:20500260

Hafernik, J. J., Messerschmitt, D. S., & Vandrick, S. (1997). Research news and comment: Collaborative research: Why and how? *Educational Researcher, 26*(9), 31–35.

Hellsten, L., Martin, S., McIntyre, L., & Kinzel, A. (2011). Women on the academic tenure track: An autoethnographic inquiry. *International Journal for Cross-Disciplinary Subjects in Education*, *2*(1), 271–275. doi:10.20533/ijcdse.2042.6364.2011.0039

Hobbs, V., & Kubanyiova, M. (2008). The challenges of researching language teachers: What research manuals don't tell us. *Language Teaching Research, 12*(4), 495–513. doi:10.1177/1362168808097162

Kilb, A., & Herzig, K. (2016). Increasing the efficiency of data collection with a research participation night. *Teaching of Psychology*, *43*(1), 53–58. doi:10.1177/0098628315620881

Kubanyiova, M. (2008). Rethinking research ethics in contemporary applied linguistics: The tension between macroethical and microethical perspectives in situated research. *Modern Language Journal*, *92*(4), 503–518. doi:10.1111/j.1540-4781.2008.00784.x

Largent, E., Grady, C., Miller, F. G., & Wertheimer, A. (2013). Misconceptions about coercion and undue influence: Reflections on the views of IRB members. *Bioethics*, *27*(9), 500–507. http://healthpolicy.ucla.edu/chis/tac2015/Documents/SDSM%20TAC/Langert_etal_2013.pdf. doi:10.1111/j.1467-8519.2012.01972.x PMID:22493972

Lavrakas, P. (Ed.). (2008). *Encyclopedia of survey research methods*. Sage Publications. doi:10.4135/9781412963947

Mehra, B. (2002). Bias in qualitative research: Voices from an online classroom. *Qualitative Report*, *7*(1), 1–19. https://nsuworks.nova.edu/tqr/vol7/iss1/2

Mligo, E. S. (2013). *Doing effective fieldwork: A textbook for students of qualitative field research in higher learning institutions*. Resource Publications.

Norris, N. (1997). Error, bias and validity in qualitative research. *Educational Action Research*, *5*(1), 172–176. doi:10.1080/09650799700200020

O'Connor, K., Greene, H. C., Good, A., & Zhang, G. (n.d.). Finding balance: A challenge for untenured faculty. *International Education Studies*, *4*(4), 3–12.

Opdenakker, R. (2006). Advantages and disadvantages of four interview techniques in qualitative research. *Forum Qualitative Social Research*, *7*(4), 1–13.

Padilla-Walker, L. M., Thompson, R. A., Zamboanga, B. L., & Schmersal, L. A. (2005). Extra credit as incentive for voluntary research participation. *Sage Journals Teaching of Philosophy*, *32*(3), 150–153. doi:10.1207/ s15328023top3203_2

Partington, G. (2001). Qualitative research interviews: Identifying problems in technique. *Issues in Educational Research*, *11*(2), 32–44.

Pegg, J., Adams, A., Risser, H., Bottoms, S., Kern, A., & Wu, K. (2014). Finding Friends: Creating a community of support for early career academics. *Brock Education*, *24*(1), 47–54.

Pew Research Center. (n.d.). *Pew Research Center Methods. Questionnaire Design.* Pew Research Center. https://www.pewresearch.org/methods/u-s-survey-research/questionnaire-design/

Pole, C., & Hillyard, S. (2016). *Doing Fieldwork*. Sage. doi:10.4135/9781473966383

Pultz, S. (2018). *Flexibility in research design: How unexpected events can improve learning and research*. SAGE Research Methods Cases., doi:10.4135/9781526435217

Rahman, M. S. (2017). The advantages and disadvantages of using qualitative and quantitative approaches and methods in language "testing and assessment" research: A literature review. *Journal of Education and Learning*, *6*(1), 102–112. doi:10.5539/jel.v6n1p102

Richards, K. (2009). Trends in qualitative research in language teaching since 2000. *Language Teaching*, *42*(2), 147–180. doi:10.1017/S0261444808005612

Rodham, K., & Gavin, J. (2006). The ethics of using the internet to collect qualitative research data. *Research Ethics Review*, *2*(3), 92–97. doi:10.1177/174701610600200303

Roulston, K., & Shelton, S. (2015). Reconceptualizing bias in teaching qualitative research methods. *Qualitative Inquiry*, *21*(4), 332–342. doi:10.1177/1077800414563803

Saba, F. (2000). Research in distance education: A status report. *International Review of Research in Open and Distance Learning*, *1*(1), 1–9. doi:10.19173/irrodl.v1i1.4

Sim, J., & Waterfeld, J. (2019). Focus group methodology: Some ethical challenges. *Quality & Quantity*, *53*(6), 3003–3022. doi:10.100711135-019-00914-5

Singer, E., & Bossarte, R. M. (2006). Incentives for survey participation: When are they 'coercive'? *American Journal of Preventive Medicine*, *31*(5), 411–418. doi:10.1016/j.amepre.2006.07.013 PMID:17046413

Van Cott, D. L. (2005). A graduate student's guide to publishing scholarly journal articles. *PS, Political Science & Politics*, *38*(4), 741–743. http://www.jstor.com/stable/30044359. doi:10.1017/S1049096505050237

Wright, K. B. (2017). Researching internet-based populations: Advantages and disadvantages of online survey research, online questionnaire authoring software packages, and web survey services. *Journal of Computer-Mediated Communication*, *10*(3), 00. Advance online publication. doi:10.1111/j.1083-6101.2005.tb00259.x

Zarrinabadi, N. (2017, June 1). *Call for chapters: Technology and language learner psychology* [Msg 3]. Message posted to https://listserv.dartmouth.edu/scripts/wa.exe?A1=ind1706A&L=LLTI

KEY TERMS AND DEFINITIONS

ACTFL Performance Descriptors: The articulation of levels of ability (novice, intermediate, advanced) in usage of the target language in the interpretive, interpersonal, and presentational modes of communication.

Computer-Mediated Communication (CMC): The use of online networking forums to engage in the construction of mutual understanding and information exchange.

Growth Mindset: A set of beliefs held by an individual that supports the expansion and growth of one's abilities and knowledge.

Information-Gap Activity: A task or assignment where participants lack information that requires communication with another to complete.

Integrated Performance Assessment (IPA): An instrument that encompasses the three modes of communication centered around an authentic cultural theme that allows learners to demonstrate their proficiency in the target language.

Internal Review Board (IRB): An administrative body created to protect the rights of human research subjects.

Learning Management System (LMS): Software that helps educational institutions to create and manage courses.

Willingness to Communication (WTC): A gauge of one's interest and initiative in engaging in conversation with others.

Section 3

Fieldwork Partnerships and Researching Corporate Challenges

Chapter 5
Meaning Making Change:
Interpretive Phenomenological Analysis and an Epistemology of Power

Denise Michelle Brend
https://orcid.org/0000-0003-3473-6123
Concordia University, Canada

ABSTRACT

This chapter describes how an interpretive phenomenological analysis (IPA) study generated perceived risk for stakeholders and for participants. Here, perceived risk was interpreted through discourses and practices specific to intimate partner violence contexts that influenced intimate partner violence professionals' subjective experiences. These risk-responses were a fundamental threat to the purpose of the research: to contribute to meaningful change for the participants in their contexts. The clash between the research aim and the risk-responses opened a theoretical space for reflection about power and knowledge relationships in lived experience and meaning-making in IPA research. Specifically, this chapter addresses the question of whether the current epistemological stance grounding IPA research leads to meaning-making that reproduces knowledge in a form that overlooks the omnipresent influence of power and knowledge dynamics. Butler's philosophies of power, knowledge, subjectivity, and performativity are explored as means of expanding the epistemological foundation of IPA.

DOI: 10.4018/978-1-7998-5826-3.ch005

INTRODUCTION

This chapter presents several unexpected issues encountered in the field during my doctoral research. Notably, these issues revealed that my methodology had the potential to obscure dynamics of power and knowledge in my data, and they therefore led me to reflect on how to improve future interpretive phenomenological inquiry. This chapter first provides context for the situation through a literature review that introduces the reader to my substantive area and establishes the reasons for this research project. I then describe how I designed and carried out my research protocol and discuss how I responded and adapted to the unexpected issues that occurred at specific steps in my protocol. Finally, drawing on my experience of how research methodologies can conceal power and knowledge dynamics, I propose a mechanism for improving future interpretive phenomenological inquiry.

My study used the Interpretive Phenomenological Analysis (IPA) approach (Smith, Flowers, & Larkin, 2009) to examine workplace social support (WSS) in intimate partner violence (IPV) practice. As my research shows, social workers' disclosing of WSS experiences also involved disclosing events that prompted the need for WSS and therefore the revealing of difficult experiences. To the participants in my study, revealing these difficult experiences felt risky, and by examining their sense of risk I became aware that my study might have benefitted from an expanded epistemological grounding. This chapter uses my experience to guide researchers designing similar studies. To that end, I first present the unanticipated risks perceived by the participants when they were sharing their accounts of WSS and describe how I responded. I then theorize that these risk-reactions are uses of power in response to knowledge about the potential risks associated with exposing existing challenges within IPV practice. I describe how the unexpected issues I encountered throughout my sampling, recruitment, and interviewing drew my focus to a limitation in the methodology I employed. Finally, I advance a novel epistemological approach through which to consider dynamics of power and knowledge in IPA research. Ultimately, this chapter describes how unexpected reactions and perceptions among participants and gatekeepers led me to a reappraisal of the epistemology that grounded my study. Other researchers may benefit from incorporating the lessons I learned in my research project into the designs of their IPA research projects, potentially improving their methods and enriching their findings.

BACKGROUND

The aim of my doctoral research was to explore the phenomenon of WSS among IPV social workers—both because IPV social workers are at elevated risk of developing work-related post-traumatic stress and because WSS is a promising factor in moderating that risk of post-traumatic stress (Bride, Jones, & MacMaster, 2007; Regehr, Hemsworth, Leslie, Howe, & Chau, 2004; Slattery & Goodman, 2009). However, the meaning of WSS for IPV social workers had yet to be inductively described. My study therefore aimed to clarify the phenomenon of IPV social worker WSS in order to support future efforts to engage this promising protective factor. The method I used—IPA—necessitates in-depth interviews with a small homogenous sample of participants. However, despite a vigorous recruitment effort, only few IPV social workers came forward to be interviewed. Throughout the process of recruitment, it also became clear that this research was considered threatening both to the potential participants and to the administrators of IPV services. This issue was particularly important since these administrators served as gatekeepers through which I needed to pass to gain access to participants. The participants I interviewed, expressed fear that they could damage their professional relationships, lose their jobs, or put their workplaces in jeopardy by sharing their experiences. However, it was less clear what the administrators feared, though one shared the concern that operational funding could be lost if staff shared information that cast a negative light on the organization. In the following section, I review the literature demonstrating the importance of this research project in the field of IPV social work practice. This literature also helps make clear the inherently risky nature of IPV practice and the imperative to protect IPV professionals.

LITERATURE REVIEW

Social workers responding to IPV are at an elevated risk of developing posttraumatic stress disorder (PTSD), which may result in symptoms including frightening psychological re-experiencing of traumatic events, dissociation, or the inability to experience pleasure (APA, 2013). Being at risk of PTSD is concerning since the associated symptoms include physical disorders, disability, and suicidality (Pagura et al., 2010; Sareen et al., 2007). Exposure to traumatic events, including the aversive details of traumatic event(s) in the course of professional duties, is a criterion that can result in professionals developing PTSD (APA, 2013). Thus, routine exposure to the aversive details of IPV through human service work is a risk factor for PTSD. Prior to 2013, when the American Psychiatric Association (APA) included work-related exposure to the aversive details of traumatic events in the diagnostic criteria for PTSD, the predominant concepts employed to investigate this phenomenon were vicarious trauma (McCann & Pearlman, 1990; Pearlman & MacIan, 1995), compassion fatigue, and secondary traumatic stress (Figley, 1995; Stamm, 1995). For clarity, work-related PTSD is referred to here as secondary traumatic stress

(STS). IPV practice routinely exposes professionals to aversive details of traumatic experiences. The World Health Organization (WHO) estimated that one in three women globally will experience IPV (2013), suggesting that this phenomenon impacts many people worldwide. Therefore, the risk of STS for IPV professionals is also arguably far-reaching.

Both social workers in general and IPV human service professionals in specific have been shown to be at an elevated risk of STS. The projected lifetime risk of PTSD in the United States is 8.7% (Kessler, Chiu, Demler, & Walters, 2005). This rate was found to be elevated among the following types of workers: social workers at a rate of 15% (Bride, 2007); social workers practicing in the fields of IPV or sexual abuse at a rate of 29% (Choi, 2011b); and domestic violence advocates at a rate of 43.7% (Slattery & Goodman, 2009). Despite the formalized acknowledgement by the APA of the inherent risks posed to professionals who may experience such exposure, evidence suggests that IPV workplaces have yet to acknowledge and effectively manage this workplace hazard (Hogan, Hegarty, Ward, & Dodd, 2012; Iliffe & Steed, 2000; Morran, 2008). As a result, professionals may continue to engage in IPV practice in spite of the potential for extreme stress to impair their professional capacities (Bell, 2003). Both their suffering and their potential to underperform in their roles helping vulnerable and volatile clientele present a clear and pressing need to mitigate the potentially harmful impacts of IPV practice.

Recent scholarship suggests that WSS may play an important role in moderating the risk of STS among IPV human service professionals (Brend, Krane, & Saunders, 2019). Diverse types of WSS (e.g., clinical supervision, group supervision, peer support) showed mixed results in reducing the risks associated with exposure to the potentially upsetting details of IPV. Choi (2011a) found that "social workers who received more support from their coworkers, supervisors, and work teams demonstrated lower levels of STS" (n = 154) (p. 252). Slattery and Goodman (2009) confirmed their hypothesis that the greater the WSS between domestic violence advocates is, the lower their levels of STS are (n=148). These findings concur with the findings of two meta-analyses investigating PTSD more generally. Brewin, Andrews, and Valentine (2000) identified the three factors carrying the strongest risk for PTSD as greater trauma severity, lack of social support, and more subsequent life stress. A second meta-analysis found that the strongest predictor of PTSD was dissociation occurring around the time of the trauma exposure (ES=0.35) and that the second strongest predictor was a lack of social support (ES= 0.29) (Ozer, Best, Lipsey, & Weiss, 2003). Ozer et al. (2003) speculated that social support had a cumulative effect over time and that the most valuable type of social support appeared to be that which assisted people in processing the meaning of what they had experienced and in managing their resulting psychological distress. Overall, there is an agreement that WSS can promote the well-being of professionals who work in the IPV field

(Morran, 2008). However, the factors that make WSS effective in reducing STS remain unclear (Brend et al., 2019). Indeed, there is a lack of clarity in the larger body of literature that examines social support. To address this issue, Williams, Barclay, and Schmied (2004) undertook a critical appraisal of the definitions of social support in the academic literature between 1996-2001 “to clarify the way in which social support has been developed and defined, and to assess the impact this has on the utility of the concept for research, intervention, and clinical practice” (p. 945). They identified 30 different definitions of social support and concluded “that any researchers attempting to measure social support or intervene at the level of social support [must] first employ qualitative methods to understand the meaning of social support to the people they wish to study” (p. 958).

Aim

I designed the study discussed in this chapter in order to identify what WSS meant for a professional group at risk of PTSD—given both that social support showed promise as a moderating factor of traumatization and that WSS specifically showed promise in reducing work-related stress among IPV professionals. In keeping with the recommendations for research investigating social support, I selected a methodology appropriate for eliciting descriptions of WSS from IPV social workers and for exploring what WSS meant to this group of professionals specifically. However, over the course of this study, an unexpected reaction emerged during recruitment and data collection. Specifically, several administrators, who were also the gatekeepers for access to the IPV social workers working in their organizations, chose not to share the recruitment materials with their staff. As well, the participants expressed fear during the interview process that their participation could have diverse negative ramifications. This chapter describes how these unexpected issues emerged during the research protocol and how they shed light on participant experiences, influenced my methods, and illuminated an epistemological gap in my chosen methodology.

METHODOLOGY AND ANTICIPATED RISKS

Methodology

I selected the IPA approach, as detailed by Smith et al. (2009), to elicit detailed accounts from IPV social workers about their WSS experiences and about how they made sense of those experiences. IPA is a qualitative research approach grounded in the phenomenological and hermeneutic traditions for use within the practices of psychology, human health, and social sciences. I chose IPA because the aim of researchers using IPA is to collect data that describes experiences and understandings of particular phenomena from the people who experience them (Smith et al., 2009). Research questions appropriate for IPA "focus on personal meaning and sense making in a particular context, for people who share a particular experience" (Smith et al., 2009, p. 45). To this end, I generated three interrelated research questions: How do IPV social workers describe experiences related to their workplaces that they perceive as socially supportive? How do experiences that are perceived as socially supportive or unsupportive impact these workers? And, how do IPV social workers make sense of their WSS-related experiences? When seeking research ethics approval at my university, I also verified whether the sites where I planned to recruit had independent research ethics protocol in place. This project was ultimately approved by the Research Ethics Board of McGill University and by the independent ethics protocol in place at one of the recruitment sites.

Anticipated Risks and Risk Management

I did not anticipate much risk for those taking part in my study. Although sensitive information and experiences about moments that required WSS were to be discussed, the setting was confidential. I did, however, assume it was possible that participants could become upset or distressed because of the discussion. Thus, I implemented a plan to support any participants who became distressed or showed a clear and present danger to themselves or others. This plan included a list of available resources in local services. A detailed informed consent process was carried out prior to any data collection in order to describe the aims of the project and the steps taken to protect confidentiality. In the following section, I describe the methods I employed within my research protocol that elicited unexpected risk-responses.

METHODS AND UNANTICIPATED RISKS

Recruitment and Sample Size

I employed purposive sampling to access participants who had the necessary professional experience to respond to the research questions (Campbell et al., 2020). In this case participants were considered eligible for inclusion if they met the following criteria: they held a Bachelor of Social Work degree or higher; they had carried half a caseload or more focussed on IPV practice during the previous two years; and they had a minimum of six months of experience providing face-to-face psychosocial services to people impacted by IPV. Recruitment was carried out through agencies, organizations, and educational institutions that provided psychosocial services or training for IPV practice. Sites were initially contacted by phone, and then—with permission from the appropriate individual or gatekeeper at each site—an email invitation and a recruitment advertisement were sent or posted in public areas at the site.

Unanticipated Risks and Risk Management During Recruitment

In total, 24 sites were contacted. Some of these sites also managed newsletters with expansive mailing lists and offered to include my recruitment advertisement—meaning that it would potentially reach the majority of IPV services in Central Canada. During the initial contacts by telephone, I either spoke with human service professionals or administrators, or I left a voicemail message. On several occasions the professionals providing psychosocial services who answered the phones expressed enthusiasm about the value of this research. In each such instance the professional then directed my call to an administrator. In some cases, an administrator answered the phone. Since IPV services may be confronted with crisis situations or be under-resourced, I anticipated that recruitment would be challenging, and indeed voicemail messages were rarely returned. In three instances, interactions with administrators during recruitment were notable.

In one such case, the administrator became agitated as the project was briefly described. In a raised voice, this administrator said not to call again or in any other way attempt to contact the staff there and abruptly hung up. Another administrator stated that the recruitment materials I had emailed would not be distributed since no one on their staff had a social work degree. It later emerged that this statement was untrue when an IPV social worker who worked at that site responded independently to a recruitment poster they had seen in the hallway of another institution. On the third occasion, a research coordinator employed at a recruitment site called me back after receiving my message. This person explained that while they perceived the project to be of value, they were declining to inform their staff. My project had been discussed at an administrative meeting and had been deemed a potential threat to their organization. I asked this coordinator how my research might be a threat and I was told that they feared the organization could be identified due to the small sampling frame and, further, that if the findings cast a negative light on their organization, they believed their funding could be jeopardized.

The administrators were the main gatekeepers to accessing the participants sought for this study. Although only three administrators made their refusal to participate known, the interviews confirmed that even my participants' administrators were not always forthcoming with the recruitment materials. For example, in one of the interviews a participant shared that they had only found out about the study because they had the task of monitoring the email address to which the recruitment materials were sent. Their administrator had received the recruitment materials yet had not circulated them to staff. There was little to be done to manage this recruitment issue differently. The administrators could not have been circumvented without deception, a practice that would not have been in keeping with the ethical approval for this study.

Unanticipated Risks and Risk Management During Sampling

I was engaged in an ongoing recruitment when my first interviews began. From my first interview, participants shared a perception that their involvement in this study was threatening. Participating caused feelings of vulnerability in their jobs, discomfort in disclosing information about their colleagues, or fear that they might jeopardize public support or funding for their organizations. These accounts raised a concern related to the core principal of concern for welfare as set out by the Canadian Institutes of Health Research and and Social Sciences and Humanities Research Council of Canada (December 2014). Welfare is broadly defined by Canadian Research Funders to include mental health, economic and social circumstances, employment, security, community membership, and social participation, among other aspects of life (Canadian Institutes of Health Research & and Social Sciences and Humanities Research Council of Canada, December 2014, p. 7). Ultimately, the discomfort that participants expressed prompted me to re-evaluate my intended sample size of 8-12 participants.

Guiding factors in my re-evaluation, as suggested by Smith et al. (2009), were the richness of the data being collected and the general recommendations for sample sizes in IPA studies. Since small sample sizes will be challenged by dissertation examiners in oral defenses and by peer reviewers for ensuing publications, it was important that my choice to end recruitment with a small sample size was defensible in my data and in the extant IPA literature. I was satisfied that the accounts being collected were rich and detailed, and I could demonstrate this with extracts of data as well as the number and lengths of interviews. To further contribute to the rigorous re-evaluation of my sample size, I also performed an extensive examination of the number and lengths of interviews reported in recent IPA literature. This review confirmed that a sample of five participants with more than one interview each and with interviews averaging an hour to an hour-and-a-half was consistent with both the recommendations of IPA theory and the sample sizes reported in the extant literature. Thus, once five participants had completed more than one interview, I deemed further recruitment unwarranted. This decision was based on participants' accounts of discomfort for their personal welfare, the richness of the accounts collected, and my determination that the sample size and dataset were appropriate for the aims of this particular study and for an IPA study more generally.

Interviewing, Socio-Demographic Survey Data, and Experiential Data

The ethical responsibility of participants to protect client confidentiality made their sharing of experiences related to practice a sensitive matter. As recommended by Pearlman and Saakvitne (1995), I selected one-on-one interviews to help establish a trusting relationship with each participant and to make them comfortable enough to speak freely. I developed an interview guide in consultation with my doctoral supervisor, and I pre-tested my data collection methods through a pilot study to ensure rigour (Elo et al., 2014). At the time of the interviews I had over a decade of experience as a social worker and psychotherapist engaging with individuals about potentially sensitive topics. As such, I felt capable of managing challenging emotional reactions should they arise in these one-on-one interactions. Based on this experience, I recommend that researchers prepare themselves before attempting interviews or focus groups, particularly with vulnerable participants or on potentially sensitive topics. This preparation can be done through a review of research methods literature describing approaches that center individual, group, and community well-being, such as community-based participatory research, Black participatory research, and Indigenous research methodologies (Coughlin, Smith, & Fernández, 2017; Drame & Irby, 2016; Windchief & San Pedro, 2019). Becoming familiar with the general guidelines for promoting respectful and sensitive interactions can help researchers avoid creating conflict or unintentional emotional upset during data collection (Tilley, 2016).

In keeping with Smith et al. (2009), I also collected additional data to assist me in contextualizing the interview data. Using an orally administered self-report questionnaire, I collected demographic information such as the ages and genders of the participants, as well as their employment and educational histories. This information assisted me in understanding participants' work histories and their current work environments, such as who they worked with and the structure of the organization. Establishing both a description of the current workplace and a work history helped me to understand how different participants interpreted their WSS experiences. Mapping out the staff structure and identifying the individual colleagues were instrumental actions in understanding the idiographic accounts provided by the participants. Eliciting these descriptions allowed me to make visible participants' social locations, their roles at work, and the roles of their colleagues in relation to themselves and in the workplaces more generally.

All interviews were audiotaped. I also used a Livescribe pen to make notations during the interviews in order to synchronize what I wrote during interviews with the corresponding time point in the audio recording. This technology was also used to note the embodied reactions I experienced and those I noticed participants experiencing. Notations were created for reactions including blushing, fidgeting, or the furrowing of a brow. Also noted were my own impressions, such as my sense that the participant was reticent to answer a question or discuss a topic. In some instances, strong emotional reactions on my part were also noted. The audiotapes of the interviews were transcribed verbatim, including all sounds, pauses, and expressions during interviews. The transcription of these non-verbal expressions acknowledges the importance of how meaning may be embodied and considers the potential for context to influence meaning-making (Smith et al., 2009). These notes became very important during my analysis, as they brought me back to the experience of conducting the interviews and helped me to recall how I had felt and how participants had appeared at specific time points in the transcriptions.

Unanticipated Risks and Risk Management Related to Data

Participants were willing to share the socio-demographic information I requested, but it was immediately clear that my using any of this contextual information required the transcripts to be carefully screened for additional potentially identifiable details. The context of IPV produces many opportunities for research participants to be identified by insiders. First, IPV services often occur in unique community settings rather than in generic institutional settings, and some IPV organizations provide services to specific sub-populations of people based on religion, culture, ethnicity, gender expression, etc. Second, in many geographic locations there is only one IPV service setting. Third, depending on the subspecialty of the setting and resulting positions available, there can be characteristic patterns of gender among the staff in specific IPV organizations. Finally, the religion, culture, ethnicity, physical characteristics, and gender expression of the participants and their colleagues are all potentially identifying information. Ultimately, details related to physical spaces, type of IPV setting, services offered, geographic locations, and characteristics of clients, IPV social workers, and their colleagues were all factors that could render the participants in my study, their workplaces, or their colleagues recognizable to insiders. Thus, these details were stripped from the transcripts prior to my analysis of the data. Most notably, the participants were given gender-neutral names and no gendered pronouns were used in the doctoral thesis or in resulting manuscripts. I also decided to expand the sampling frame to include sites throughout Central Canada. My intention here was both to attract more participants as needed and to further de-identify existing participant accounts.

The interview transcription process included the notation of pauses, non-verbal sounds, and descriptions of vocalized sounds (a cracking voice, vocal fry, etc.). I then added my observations of participants' body language and my own reactions and thoughts at the associated time points in the interviews. This level of detail allowed me to make sense of more than words during my analysis of the interview transcripts. The resulting collection of emotional expressions embedded within the dataset facilitated the creation of data groupings, which could also be analyzed. In some cases, distinct themes resulted from that analysis. The following datum demonstrates how this level of detail enabled emotional expression and emotional intensity to be captured and analyzed. The extract below is an example of data that I annotated with multiple notes related to experiential observations. Such highly emotive data helped establish a theme representing participants' sense that sharing information about unsupportive experiences felt like a moral transgression. Across this sample, each participant expressed a strong desire to be perceived as supportive of their colleagues and workplaces despite any struggles they might have experienced. Each participant also expressed strong commitments to people impacted by IPV as well as a fear of taking any action that could undermine any aspect of IPV service delivery. In the datum below, the participant was speaking about why sharing information about their WSS experiences felt "delicate." In order to highlight the experiential data, I have underlined language where the participant made a facial grimace and bolded language where the participant placed emphasis on a word.

What makes it delicate? Because it is definitely uncomfortable. It's uncomfortable to **talk** *about…when you start to talking about the fact that in reality there isn't that much support, and it feels* as though you're- you're talking behind peoples' backs, or being disloyal to the organization if you start talking *on the outside about the reality of what's going on. Because it's a closed world, and only people that really get in there are the people that are on the inside. So as soon as you start to talk to people on the outside- especially because the image can be (pauses and resumes in a lower voice) 'Oh you know, there's a [workplace] and it's for [people who have experienced IPV], it's such a great job that people are doing' (resumes normal tone) and then when you dig beneath the surface and you realize the people that are doing the brunt of the hard work are (pause) the* **front** *line workers (pause) working in- in- in really difficult circumstances and not getting the support they need. And then the people who* **do** *get a lot of the praise and the glory for the great work that's going on in the [workplace], are actually quite disconnected from the reality of what's going on in the day to day. So,* **that** *(chuckles) all of that makes the whole thing quite delicate, I would say. And there's definitely a part of me saying, 'Oh, should I be saying all that?' (laughs while saying this last statement)*

Since a significant amount of other data was similar to this example, I established a theme called "sharing feels wrong." Within this theme, the participants expressed strong discomfort that what they were sharing could put themselves, their organizations, colleagues, or the IPV field at risk.

This discomfort resulted in some participants expressing ambivalence about sharing with me workplace experiences that were difficult or uncomfortable. However, the goal of my data collection was to elicit detailed accounts of experiences related to WSS and not to selectively learn about instances in which WSS was effective. As such, I chose to address participants' fears that their words could reflect poorly on them, their colleagues, their workplaces, or the broader domain of IPV practice. To address these fears, in each interview I reiterated to the participant that the ultimate aim of this research was to identify what type of WSS IPV social workers needed and potentially identify better ways of offering WSS to IPV social workers. This emphasis helped re-focus the participant on the goals of the interview—rather than on the concerns about secrecy and protection conventions in IPV practice. It also helped make clear that my casting aspersions on any aspect of IPV service delivery would not be helpful in meeting the goal of the study. As well, in order to reassure participants that IPV social worker accounts of ineffective WSS were normal, I underscored the lack of agreed-upon conceptual understanding of WSS in IPV practice. Indeed, as I let participants know, it would be remarkable if *any* WSS experiences were uniformly effective in any given IPV context.

CHALLENGES

The aim of this research project was to clarify the phenomenon of WSS among IPV social workers in order to identify strategies more effective in promoting their wellbeing. Yet, from the moment recruitment began, two issues emerged: first, the perception that this project created risks for IPV social workers, their colleagues, their workplaces, or IPV practice in general and second, the appearance that some administrators used their power as gatekeepers to stop the IPV social workers in their employ from participating. While less can be understood about what motivated administrator choices, all participants, each from a different setting and without knowledge of one another, perceived that they were doing something wrong by sharing their knowledge about IPV practice with an outsider (me). They expressed fear they could damage their collegial relationships, lose their jobs, undermine public support for IPV work, or even cause operational funding to be revoked from their workplaces. Despite their fears, they agreed to participate and trusted me, to some degree, not to cause harm. My impression was that they were motivated by a desire to contribute to the improvement of IPV social worker working conditions. I offered no incentives for their participation other than a cup of tea or coffee and, if needed, bus fare to travel to the interview.

Protecting participant confidentiality was one way I protected participants from harm. Managing the risks associated with protecting confidentiality, however, undermined the richness of what I could report, since many details that contributed to the unique experiences of these participants had to be stripped and hidden from view following the analysis. Baez (2002) struggled with this same inability to report details that offered core information in his study examining the experiences of racial minority faculty members in higher education. The author went so far as to suggest that the commitment to confidentiality in qualitative research itself needed to be reconsidered: "Secrets thwart the politically transformative goals of qualitative research. Transformative research, if it is to be possible, must expose and resist oppression, which is often hidden…To do so, it requires openness and risk-taking, which are undermined by secrecy" (Baez, 2002, p. 46). Indeed, if any of the IPV social workers in my study had experienced any form of discrimination in their workplaces, it would not have been possible to address those experiences in disseminated materials because details of discrimination necessarily involve characteristics of the people involved that could lead to their identification (Baez, 2002). However, experiences of workplace discrimination would be germane to describing and understanding effective versus ineffective WSS. In my study, as in Baez's, the process of applying the methods and conducting ethically sound research resulted in blind spots in the reported findings since crucial information representing the phenomenon of interest had to be obscured. Reducing the ethical constraints around confidentiality, however, would have put the vulnerable participants at risk of retaliation—and using participants for the greater good is indeed not the goal of transformative qualitative research.

Unlike those in Baez's research, the participants in my study were purposely cloistered by their organizations from outsiders. Barriers between outsiders and insiders in IPV settings were viewed as essential for the safety of both the people served and the professionals serving them (for example, from potentially violent ex-spouses). These barriers were enforced by keeping the names of staff in IPV settings off of organizational websites, keeping addresses of shelters confidential, keeping personal phone numbers unlisted, and using pseudonyms for social media accounts. My research project uncovered how these barriers, essential for safety in the IPV context, also served to stop potentially valuable information from getting to IPV social workers. Outsiders were purposely barred from direct contact with the staff, a fact which, paradoxically, could also stop staff from accessing outside support or information about how to protect themselves from the risks they faced a work.

Beyond the strategies used to hide the identity of these professionals, a sense of risk also influenced their actions. Participants who did learn about this project felt threatened by the potential to identified as the sources of information about deficits in their workplaces. Even after consenting to participate, might these IPV social workers have made further conscious or unconscious choices about how they presented themselves and their experiences to minimize perceived risks of harm to themselves, their colleagues, their workplaces, or the entire profession? Might they still have hidden significant details, and shown only those details that minimized the power of the researcher/research project to cause any harm? Given these possibilities, we must recognize that critical aspects of a studied phenomenon may remain invisible because we lack an epistemological ground from which to consider the potential influences of power. In my study, I see this lack as having influenced my recruitment and sampling strategies. As well, my resulting analysis did not include these performances as data, although I made meaning out of how gatekeepers, potential recruits, and participants performed their roles related to my research. As this experience makes clear, the methodological infusion of an epistemology of power and knowledge would assist researchers in identifying aspects of human experience that are often unconscious, hidden, or otherwise unshown.

Reconsidering Power and Knowledge Relationships

I was surprised to find that I was shut out of contact with IPV social workers by some administrators. My research aim did not seem controversial and I was not asking administrators to commit any resources, such as paid time off for their employees. Yet, by getting responses to my questions, I was perceived as having power that could be used to cause some type of harm. To counteract this perceived risk, some administrators used their power to shield their participants from the recruitment effort. As well, IPV social workers seemed to share the understanding that they ought not to reveal information about their working realities with an outsider. This power (to stop the access of information) and knowledge (that sharing information with an outsider about WSS was dangerous) informed IPV social workers and potentially their administrators about what they could or would share. As a result, I was forced to modify my recruitment strategy to minimize the potential for risk and address the impacts of this power and knowledge relationship in my interviews. Therefore, relationships between the power and the knowledge intrinsic to IPV insiders influenced the methods I used and the results of my study.

Participants spoke about WSS experiences occurring in their workplace contexts, yet they did not talk about power and knowledge as central to how they made sense of their WSS experiences. Thus, while an analysis of power might be an interesting secondary analysis of my data—and would undoubtedly enrich how IPV WSS is understood—power was not central to these IPV social workers' descriptions of effective WSS. As a consequence, power was not a central element identified and addressed by my analysis. Nonetheless, the clear influence of power relations evident in this process suggests that the power dynamics pervasive throughout both the research protocol and participant contexts needed somehow to be conceptualized.

Looking to the philosophical underpinnings of IPA may offer a novel approach to the challenges presented by the power and knowledge relationships made evident by this research project. In their description of the philosophical foundations of IPA, Smith et al. (2009) detailed how Heidegger theorized about meaning-making as a social process through the phenomenological concept of intersubjectivity. Intersubjectivity involved Heidegger's complex view of people as "being with [other people] in the world" (Heidegger, Macquarrie, & Robinson, 1962, pp. 157-157), which described human experience as constructed through interpersonal interactions. Smith (2009) defined intersubjectivity as "the shared, overlapping and relational nature of our engagement in the world" (p.17). Heidegger proposed that people understand experiences *through shared being* with others and thus that human understanding results from experiencing the world intersubjectively, rendering the interpretation of people's meaning-making unavoidable. As Smith et al. (2009) summarized, "Our being-in-the-world is always perspectival, always temporal and always 'in-relation-to' something…" (Smith et al., 2009, p. 18). This view of phenomena moved phenomenology away from Husserl's original theorizing about the existence of essential reality and toward an understanding of reality as grounded in intersubjective meaning-making.

Echoing hermeneutics, or the importance of context in interpreting text (Schleiermacher, 1998), Heidegger wrote about phenomenology as a practice of being exposed to a phenomenon as it showed itself and then engaging in an analytical process to "facilitate, and grasp, this showing" (Smith et al., 2009, p. 24). The showing referred both to what was explicitly shown and what was hidden or concealed. Therefore, what an individual knew and believed in advance of seeing a phenomenon was a complex and dynamic aspect of how they made meaning. While Husserl attempted to bracket out, or set aside, fore-understanding, Heidegger discussed its central role in meaning making. To summarize Smith et al. (2009), what a person knows impacts how they interpret things and what a person learns impacts what they know. My research project raises questions about the role of power in this intersubjective experience of learning and knowledge-shaping and suggests that a philosophical base that includes an epistemology of power might enhance phenomenological hermeneutic inquiry.

Smith briefly addresses power by introducing methodologies drawing on Foucault, explaining that these approaches have been applied to research expressly interested in power. In IPA, the topic is the experience (such as WSS) and the person and their meanings are the unit of analysis. While power might easily fit as an explicit topic of study, it is not described as a fundamental aspect of human experience in the theoretical foundations of IPA. However, Butler offers a way of conceptualizing power as intrinsic to the development of meaning. Butler extends Heidegger's notion of understanding through shared, intersubjective being-with-others-in-the-world. She introduces the idea that subjectivities are recognizable because they have been subjected to a collection of discourses suffused by power and knowledge that legitimize certain discourses over others (Nagington, 2016). These unconscious performances help us to recognize each other and to find ways of belonging together (Nagington, 2016). In doing so, we are subjecting ourselves to discourses suffused by power and knowledge about who we are, what we are doing, and why we are doing what we do (Salih, 2002). Indeed, the research interview could be considered from multiple subjectivities, each invoking their own conditions of power and knowledge relations. The fact that IPV social workers participated in my study—and did so despite the sense of risk created by the research activity—suggests they were enacting the power to reform or deform their understandings of themselves or of WSS. By participating in the research project, they were claiming the power to engage as knowledge shapers.

If we adopt a Butlerian view of the roles of power and knowledge in discourses that shape subjectivities and make them recognizable to others, we see that meanings are fluid and dependant on "a complex and constantly maintained set of discourses that produce meaning through extant power and knowledge" (Nagington, 2016, p. 311). Butler's view of power is not deterministic; rather, she views our performances as animated and shaped by the complex and constantly maintained set of discourses to which a given individual is subjected (Butler, 1997). How these performances are then interpreted depends largely on the discourses adopted by the counterpart in intersubjective exchange. For Butler, as for Heidegger, meaning cannot exist outside of social relations. In contrast to Heidegger, however, for Butler social relations are necessarily influenced by discourses and practices that have been established through power and knowledge relationships. Perhaps most pertinent to Butler's stance is that *changes* in power and knowledge relations occur through the adaptations made within intersubjective exchanges. Subjectivities are promulgated through ongoing social relations and inevitably differences are produced through the variety of performances enacted. Knowledge is thus advanced through repeated types of intersubjective exchange over time, and it gets adjusted or rejected through new performances when sufficient power has gathered in the ways that the performances have been altered. In short, knowledge alters as the counterparts in an intersubjective exchange recognize the plausibility or efficacy of new performances.

How would my research project have differed had it been infused with Butler's philosophies of power, knowledge, subjectivity, and performativity? Perhaps by approaching participant actions as performances communicating meaning, I might have made visible the ways in which gatekeepers' and participants' performances strengthened accepted power and knowledge or altered it by performing other ways of knowing or being. One example is the participant who stated that IPV practice looks different from the outside than on the inside. If we consider Butler, this participant's sharing of insider knowledge can be interpreted as a powerful act of deviation from the norm with the potential to reshape accepted intersubjective meaning-making around WSS and IPV social worker subjectivity. Without Butler, the fundamental power enacted in this performance might remain invisible, and the researcher might underestimate the potential for change supported by the intersubjective exchange occurring in the research interview. When I read the data through a Butlerian lens, I feel no impetus to break confidentiality to allow the ugly truths of participant experiences to be seen—as Baez wonders if we must. Thus, the potential for a research project to enact change shifts from the *content* of the interview—and what is done with that content—to the full *performance* of the research project. In this sense, the interview serves as one of several performances within which subjectivities can be reoriented and power and knowledge dynamics can be altered.

SOLUTIONS AND RECOMMENDATIONS

My observations during the recruitment process suggest that administrators in the IPV field can exercise their power to cloister staff from outside danger, influence, or assistance. While I cannot generalize or suggest that such administrative action is the norm, researchers investigating this population would do well to design their research in a way that considers IPV social workers a vulnerable and hidden population. In the case of my project, my lack of awareness of the power and knowledge relations to which my potential participants were subjected may have resulted in many of them being denied access to my recruitment materials. Had I known this, I might have changed my design to actively seek previous IPV social workers or to engage in more sustained strategies to reach active IPV social workers.

Although oppressive acts are often hidden, the power that makes them possible is hiding in plain view. IPA might, therefore, be better adapted to transformative ends if it were to claim a philosophical stance about power and develop approaches that are attuned to the fundamental presence of power and knowledge relationships in intersubjective exchange. I do not mean to suggest that IPA adopt power as a main research focus, as in Foucauldian Discourse Analysis. Rather, I suggest that power and knowledge relationships be acknowledged as influencing the discourses and practices that shape subjectivity through intersubjective exchange. I suggest that they do so unavoidably and potentially at all times. Based on the benefits I have identified in considering how power and knowledge can influence intersubjective meaning making, I recommend that researchers interested in applying the IPA approach expand the philosophical foundations that guide IPA research. Such an expansion will help them attend to the potential influences of power on their fore-understandings and to how the research relationship—including the contexts within which and about which the research occurs—are influenced by power and knowledge relationships. Here I introduce Butler's philosophy in order to expose the gaps upheld by research practices using a philosophical framework risking to reify phenomena while overlooking the pervasive influence of power and knowledge dynamics. Overall, the analytic frame of an IPA study might be expanded through Butler's view of how power and knowledge relations operate in intersubjective meaning making. A Butlerian lens might also sensitize the researcher to how power-knowledge relationships can divert attention, amplify phenomena, and influence meaning within participant discourses—potentially adding depth or changing the researcher's own interpretations and meaning-making about the phenomenon under study. Whether and if Butler's work is the ideal choice for future IPA research remains to be explored.

Researchers have already begun expanding the interpretive process of IPA to incorporate a "a sensitivity to and an analysis of historical and current forms of oppression as part of its interpretive process" (Aparicio, Birmingham, Rodrigues, & Houser, 2018, p. 332). Much discussion and inquiry remain in order to determine whether and how IPA could benefit from an epistemological approach that acknowledges power. It is time to break the relative silence in IPA discourse about Heidegger having been a Nazi (Sheehan, 1988). This unsettling fact prompted me to use an ongoing reflexive process to stay vigilant to the potential impacts of adopting an epistemology that did not preclude Heidegger himself adopting Nazism in his personal ideology, actions, and intersubjective exchanges. This process involved drawing on Heidegger with suspicion and carefully weighing the potential impacts of the methods I intended to use on the participants, the larger IPV community, and myself. Ultimately, adopting an epistemology of power when invoking Heidegger may be a necessary ethical act.

CONCLUSION

My small-sample IPA study revealed large-scale issues faced both by participants and by a researcher examining a phenomenon from insider perspectives. The roles of power and the risk-responses generated by my research suggest that the epistemological foundations of IPA might be enhanced by including a philosophical lens through which to conceptualize power and knowledge relationships. Such an expansion could increase the effectiveness of IPA methods by systematically broadening researchers' horizon of understanding. In order to grasp a phenomenon, we must be able to see the power and knowledge relationships embedded within participant and researcher fore-understandings and operating in our unconscious intersubjective performances. In fact, researchers who fail to acknowledge the roles of power embedded within intersubjective exchange may unknowingly promote situations in which the powerful can separate others from their capacity to act for change. Thus, as qualitative researchers our fundamental ethical questions may not be "'What *must* I do?' (which is the question of *morality*) but rather 'What *can* I do?...Given my degree of power, what are my capabilities and capacities? How can I come into active possession of my power? How can I go to the limit of what I "can do"[']" (Smith, 2012, p. 285). Not every IPA study will be aimed at analyzing power, but every IPA study may be enriched by systematic attention to how power and knowledge relationships shape the research effort and how these relationships embed themselves within the intersubjective exchanges through which we make meaning.

ACKNOWLEDGMENT

I would like to thank Dr. Gail Boldt for her wisdom, generosity, and guidance concerning the work of Judith Butler and my doctoral supervisor, Dr. Julia Krane, who tirelessly read through every draft of my dissertation. This research was funded by MITACS & Boscoville through the Accélération postdoctoral fellowship program (FR41949) and the Fonds de Recherche du Québec - Société et Culture (FQRSC) bourse de doctorat en recherche - B2.

REFERENCES

APA. (2013). Posttraumatic Stress Disorder. In *Diagnostic and statistical manual of mental disorders: DSM-5*. American Psychiatric Association.

Aparicio, E. M., Birmingham, A., Rodrigues, E. N., & Houser, C. (2018). Dual experiences of teenage parenting and homelessness among Native Hawaiian youth: A critical interpretative phenomenological analysis. *Child & Family Social Work*. Advance online publication. doi:10.1111/cfs.12618

Baez, B. (2002). Confidentiality in qualitative research: Reflections on secrets, power and agency. *Qualitative Research*, *2*(1), 35–58. doi:10.1177/1468794102002001638

Bell, H. (2003). Strengths and Secondary Trauma in Family Violence Work. *Social Work*, *48*(4), 513–522. doi:10.1093w/48.4.513 PMID:14620108

Brend, D., Krane, J., & Saunders, S. (2019). *Exposure to trauma in intimate partner violence human service work: A scoping review. Traumatology*. doi:10.1037/trm0000199

Brewin, C., Andrews, B., & Valentine, J. (2000). Meta-analysis of risk factors for posttraumatic stress disorder in trauma-exposed adults. *Journal of Consulting and Clinical Psychology*, *68*(5), 748–766. doi:10.1037/0022-006X.68.5.748 PMID:11068961

Bride, B. (2007). Prevalence of Secondary Traumatic Stress among Social Workers. *Social Work*, *52*(1), 63–70. doi:10.1093w/52.1.63 PMID:17388084

Bride, B. E., Jones, J. L., & MacMaster, S. A. (2007). Correlates of secondary traumatic stress in child protective services workers. *Journal of Evidence-Based Social Work*, *4*(3-4), 69–80. doi:10.1300/J394v04n03_05

Butler, J. (1997). *The psychic life of power: theories in subjection*. Stanford University Press.

Campbell, S., Greenwood, M., Prior, S., Shearer, T., Walkem, K., Young, S., ... Walker, K. (2020). Purposive sampling: Complex or simple? Research case examples. *Journal of Research in Nursing, 174498712092720*. Advance online publication. doi:10.1177/1744987120927206

Canadian Institutes of Health Research, N. S. a. E. R. C. o. C., & and Social Sciences and Humanities Research Council of Canada. (2014, December). Tri-Council Policy Statement: Ethical Conduct for Research Involving Humans. Author.

Choi, G. (2011a). Organizational Impacts on the Secondary Traumatic Stress of Social Workers Assisting Family Violence or Sexual Assault Survivors. *Administration in Social Work*, *35*(3), 225–242. doi:10.1080/03643107.2011.575333

Choi, G. (2011b). Secondary Traumatic Stress of Service Providers who Practice with Survivors of Family or Sexual Violence: A National Survey of Social Workers. *Smith College Studies in Social Work*, *81*(1), 101–119. doi:10.1080/00377317.2011.543044

Coughlin, S., Smith, S., & Fernández, M. (2017). *Handbook of community-based participatory research*. doi:10.1093/acprof:oso/9780190652234.001.0001

Drame, E. R., & Irby, D. J. (2016). *Black participatory research: power, identity, and the struggle for justice in education*. Academic Press.

Elo, S., Kääriäinen, M., Kanste, O., Pölkki, T., Utriainen, K., & Kyngäs, H. (2014). Qualitative Content Analysis: A Focus on Trustworthiness. *SAGE Open*, *4*(1). doi:10.1177/2158244014522633

Figley, C. (1995). *Compassion fatigue: Coping with secondary traumatic stress disorder in those who treat the traumatized*. Bruner/ Mazel.

Heidegger, M., Macquarrie, J., & Robinson, E. (1962). *Being and time*. Blackwell.

Hogan, K., Hegarty, J., Ward, T., & Dodd, L. (2012). Counsellors' experiences of working with male victims of female-perpetrated domestic abuse. *Counselling & Psychotherapy Research, 12*(1), 44–52. doi:10.1080/14733145.2011.630479

Iliffe, G., & Steed, L. (2000). Exploring the Counselor's Experience of Working With Perpetrators and Survivors of Domestic Violence. *Journal of Interpersonal Violence, 15*(4), 393–412. doi:10.1177/088626000015004004

Kessler, R. C., Chiu, W., Demler, O., & Walters, E. E. (2005). Prevalence, severity, and comorbidity of 12-month dsm-iv disorders in the national comorbidity survey replication. *Archives of General Psychiatry, 62*(6), 617–627. doi:10.1001/archpsyc.62.6.617 PMID:15939839

McCann, I., & Pearlman, L. (1990). Vicarious Traumatization: A framework for understanding the psychological effects of working with victims. *Journal of Traumatic Stress, 3*(1), 131–149. doi:10.1007/BF00975140

Morran, D. (2008). Firing up and burning out: The personal and professional impact of working in domestic violence offender programmes. *Probation Journal, 55*(2), 139–152. doi:10.1177/0264550508090272

Nagington, M. (2016). Judith Butler's theories: Reflections for nursing research and practice. *Nursing Philosophy, 17*(4), 307–316. doi:10.1111/nup.12134 PMID:27321751

Ozer, E., Best, S., Lipsey, T., & Weiss, D. (2003). Predictors of posttraumatic stress disorder and symptoms in adults: A meta-analysis. *Psychological Bulletin, 129*(1), 52–73. doi:10.1037/0033-2909.129.1.52 PMID:12555794

Pagura, J., Stein, M. B., Bolton, J. M., Cox, B. J., Grant, B., & Sareen, J. (2010). Comorbidity of borderline personality disorder and posttraumatic stress disorder in the U.S. population. *Journal of Psychiatric Research*, 44. PMID:20537660

Pearlman, L., & MacIan, P. (1995). Vicarious Traumatization: An empirical study on the effects of trauma work on trauma therapists. *Professional Psychology, Research and Practice, 26*(6), 558–565. doi:10.1037/0735-7028.26.6.558

Pearlman, L., & Saakvitne, K. (1995). *Trauma and the therapist: countertransference and vicarious traumatization in psychotherapy with incest survivors*. W.W. Norton.

Regehr, C., Hemsworth, D., Leslie, B., Howe, P., & Chau, S. (2004). Predictors of post-traumatic distress in child welfare workers: A linear structural equation model. *Children and Youth Services Review*, *26*(4), 331–346. doi:10.1016/j.childyouth.2004.02.003

Salih, S. (2002). *Judith Butler*. Academic Press.

Sareen, J., Cox, B., Stein, M., Afifi, T., Fleet, C., & Asmundson, G. (2007). Physical and mental comorbidity, disability, and suicidal behavior associated with posttraumatic stress disorder in a large community sample. *Psychosomatic Medicine*, *69*(3), 242–248. doi:10.1097/PSY.0b013e31803146d8 PMID:17401056

Schleiermacher, F. (1998). *Hermeneutics and criticism and other writings* (A. Bowie, Trans.). CUP.

Sheehan, T. (1988). Heideggar and the Nazis. *The New York Review of Books*, *35*(10), 1–12.

Slattery, S., & Goodman, L. (2009). Secondary traumatic stress among domestic violence advocates: Workplace risk and protective factors. *Violence Against Women*, *15*(11), 1358–1379. doi:10.1177/1077801209347469 PMID:19809098

Smith, D. (2012). *Essays on Deleuze*. Academic Press.

Smith, J., Flowers, P., & Larkin, M. (2009). *Interpretative phenomenological analysis: theory, method and research*. SAGE.

Stamm, B. (1995). *Secondary Traumatic Stress: Self-care issues for clinicians, researchers and educators*. Sidran Press.

Tilley, S. (2016). *Doing respectful research: power, privilege and passion*. Fernwood Publishing.

WHO Department of Reproductive Health and Research. London School of Hygiene and Tropical Medicine, & South African Medical Research Council. (2013). Global and regional estimates of violence against women: Prevalence and health effects of intimate partner violence and non-partner sexual violence. WHO.

Williams, P., Barclay, L., & Schmied, V. (2004). Defining Social Support in Context: A Necessary Step in Improving Research, Intervention, and Practice. *Qualitative Health Research*, *14*(7), 942–960. doi:10.1177/1049732304266997

Windchief, S., & San Pedro, T. (2019). *Applying indigenous research methods: storying with peoples and communities*. doi:10.4324/9781315169811

Chapter 6
Data Collection as a Journey:
A Multi-Step Approach to Navigating Access to an Organization

Géraldine Bengsch
King's College London, UK

ABSTRACT

The journey approach for data collection may enable the novice researcher to reflect on their data collection processes, and it aims to point to potential creative solutions that can help create a coherent study deeply rooted in its social environment. Fieldwork becomes a part of the project, rather than an isolated element that needs to be done. Through this approach, even a novice researcher can demonstrate connected insights that are not only relevant to the study itself but also the subject of study. A creative, multi-layered approach to resources and created opportunities may help increase the feasibility of a study by reacting to and interacting with the social context of the field. This chapter reflects on the author's data collection progress during her PhD program and invites readers to discover actionable steps that can be used to overcome the inertia due to inability to secure access to a field site.

INTRODUCTION

In this chapter, the author draws on her PhD data collection journey to provide pointers and potential solutions for novice researchers attempting to interact and collaborate with an organization in the real world to generate data for research projects. The success of a research project is dependent on the data it is based on, and many projects start with a grand idea of what would constitute not only a perfect

DOI: 10.4018/978-1-7998-5826-3.ch006

data site, but also the perfect type of data. However, the author quickly discovered that data collection is not merely an end to a means on the route to a PhD, but all of data collection's jumps and hurdles can significantly enrich the researcher's understanding of a study and can help researchers grow academically, professionally and personally. This chapter draws on the researcher's experience in the field and discusses how the author made sense of the struggles she faced to gain access to the field and related endeavors in the field.

Establishing contact, working in the field and then leaving that field is a process that needs to be carefully orchestrated in order to be successful. As a concept, fieldwork has been described as different to data collection in the sciences (Clarke, 1975; Lofland, Snow, Anderson, & Lofland, 1971). The data collection process is compared here to a journey that a researcher undertakes. Much like a physical journey, the success is dependent on the itinerary of the research project, but also on the reflexivity of the researcher throughout the collection on the other. The idea of data collection as a process, rather than a single step, is nothing new. Creswell (2007) describes data collection as a circular process with locating and gaining access to an organization as only two briefly mentioned sub-points. Those aspects take center stage in this chapter. However, it should be acknowledged that in order to successfully collect data of high quality, all aspects of the collection process are very important.

BACKGROUND

This study took place in the tourism sector at four hotels in three countries (England, Germany and Spain) as an interdisciplinary project in communication studies. The study used approaches from pragmatics, sociology, and communication sciences. The data needed for the research were video recordings of naturally occurring service encounters between hotel receptionists and guests. The data was analyzed using a conversation analytic framework. Tourism has been described as a natural laboratory for analyzing particular sets of behaviors (Berno & Ward, 2005; Hall, 2010) – a notion that resembles Bourdieu's sentiment of seeing the field as a natural laboratory (Wacquant, 2004). Interactions or service encounters at the hotel front desk has been deemed a crucial 'moment of truth' (Grönroos, 2001) for an organization in which insiders and outsiders interact, which then influences the perception a guest has of their stay at a hotel and thus the organization itself. The author was interested in interpersonal, intercultural, and asymmetric elements in conversation. The aim of the study was to study interaction patterns within these parameters, so the international hotel front desk was deemed a suitable site for data collection.

The author and her supervisors originally believed that accessing data would be straightforward, given the seemingly public nature of such interactions. However, recent literature has begun to challenge the notion of 'simplicity' in gaining access to an organization where rejection is a constant risk (Peticca-Harris, deGama, & Elias, 2016). Front-line conversations add an extra challenge to requirements of access, because the customer provides an external element to the site. Further, literature on service encounters distinguishes between public, semi-private, and private interactions (Félix-Brasdefer, 2015; Kerbrat-Orecchioni, 2006). For instance, according to the results from the PhD study reported on here, conversations at the hotel front desk are semi-private: bystanders actively disengage from interactions underway between the hotel representative and the guest. This understanding only arose after the data had been analyzed. During the data collection, much of the underlying values and implicit understandings that govern behavior in such organizations was slowly revealed. While the author had a strong understanding of the epistemological viewpoints governing the research project, it became apparent that this was not a correct assumption for the organization in the real social world. Academia makes epistemic notions explicit, while organizations also hold their own belief systems, even if they are unlikely to contemplate them in the same way that academic discourse does (Robson, 2002).

The author had worked in tourism and had also conducted her master's level research in the same sector. As a result, she was able to construct a valid research project, containing relevant and applicable research questions, but not nearly enough to anticipate strategic considerations. The "real world" is often considerably different compared to how it is described in academic discourse (Robson, 2002). This led to two challenges for researchers and research: 1) false assumptions can exist about the industry in question, and 2) research can remain rather superficial by the omission of sector specific norms and beliefs. This may be more pronounced in inter- or cross-disciplinary research because the knowledge of the principal author tends to be restricted to one field of expertise, even in such studies. In a cross-disciplinary research, the boundaries of each area are crossed, but not their methodologies. By contrast, interdisciplinary research combines beliefs and methods from the area it draws from ("Cross-Disciplinary Learning," 2012). The author found that it was beneficial to use literature from a business perspective to understand how economic notions are constructed, without drawing on the methods used in the field. This cross-disciplinary element strongly informed the background and the approach to data collection. It also highlighted the sometimes naïve-sounding approaches that her own discipline conventionally takes to researching organizations; often, a look at the literature in which the target sector is located provides the necessary information for understanding how the industry operates.

LITERATURE REVIEW

While different sectors, industries, and fields may present what seems to be unique challenges, considerable barriers to access often exist for researchers who have no pre-established ties to the organizations they are proposing to collect data in. Fieldwork is a tactical endeavor, requiring the researcher to become part of the community – in this case, the target organization (Hammersley & Atkinson, 2007; Van Maanen, 1979). In addition, it needs to establish a convincing framework for the study (Bryman, 2001). Fieldwork here is understood to encompass data collection, both in a researcher's 'backyard' and endeavors further away as a manner to address "real people, real problems, real organizations" (Edmondson & McManus, 2007, p. 1155) in a naturalistic setting (Kaler & Beres, 2016; Paluck & Cialdini, 2014; Robson, 2002). The focus in this chapter is placed on organizations as a site for research. An organization is understood here as "a container of a particular set of social relations" (Falzon, 2009, p. 1). It may be described as a "servicescape", the physical home environment of a service encounter (Touchstone, Koslow, Shamdasani, & D'Alessandro, 2017, p. 147). Having access to an organization is a vital requirement for field research that is to take place within the realm of a company, but it has been said to have been poorly conceptualized in the literature (Horwood & Moon, 2003; Johl & Renganathan, 2010; Peticca-Harris et al., 2016; Riese, 2019). Yet, obtaining access is often difficult as it is subject to navigating complex social situations.

To conquer considerations regarding the social situations in a given field site, reflexivity in data collection has become a recurring theme in various parts of the planning and execution process. Self-reflection is increasingly addressed in the literature (see Reeves, 2010). Crossley (2019) calls for researchers to not only employ it when considering entering a field, but also to critically inform the research. Crossley notes that while reflexivity is generally encouraged, social scientists are prone to self-censor their experiences in the field. She urges researchers to "confront uncomfortable moments of uncertainty, doubt, and failure" (p. 5). Still, there is no one-fits-all approach. For instance, Cunliffe and Alcadipani (2016) suggest three reflexive perspectives that affect trust and outcomes of the project: instrumental (short-term relationship; researcher as privileged member; organization members viewed as 'subjects'), transactional (reciprocal relationship, based on informational exchange; organization members viewed as 'informants'), and relational (fluid and engaged, medium to long-term relationship; organization members viewed as 'research participants'). Given the general aim of qualitative research to understand the social reality of a population, the relational process to access can enhance the notion of co-creating research between researcher and researched (Riese, 2019). As such, it is the approach utilized in the present study.

In addition, fieldwork has been described not only as a form of data collection, but an emotional endeavor. As such, it also invites reflexivity on part of the researcher to navigate the time in the field (Caretta & Jokinen, 2017). It is an opportunistic undertaking that revolves around circumstances available that is becoming increasingly difficult (Buchanan, Boddy, & McCalman, 1986). Organizations are dynamic and evolving social places that do not readily welcome outsiders, such as academic researchers, but are closed to them by default (Homan, 2001; Okumus, Altinay, & Roper, 2007). Deliberating access is not only subject to preparation of formal methods, but it is also preparation for working through emotional tensions in the field, e.g., when the pre-designed methods need to be adjusted while in the field during data collection (Kalir, 2006).

To accommodate the social dynamics in an organization together with the researcher's persona, researchers have begun to address the individual experiences and challenges of fieldwork, but there is still much to explore (McFarlane-Morris, 2020; Peticca-Harris et al., 2016). The present chapter aims to contribute to the current discussion and follows researchers such as Ivanova, Buda, & Burrai (2020) who question the standard, traditional practices in their disciplines and suggest creative and disruptive ways to access knowing and doing in the field. However, ethical requirements and academic scrutiny often create premises and restrains for methods to be applied in the field, which may ultimately hinder the lived field experience. Concerning this, Bell (2020) speaks of "small-time fieldwork delinquency" where a researcher has their "antennae out, looking for any little fragments to enlarge insights" (p. 2), even if this contradicts established practices in a field. Again, reflexivity is called upon to navigate what an ethics committee has deemed a safe project and the real world a researcher finds in the field (Bell, 2020). A researcher needs to be able to adapt to the conditions they encounter. Accessing and navigating the field is complex and far from a neutral experience, and researchers need to be explicitly trained to be successful (Okumus et al., 2007; Peticca-Harris et al., 2016). Access to the field within the social sciences is based on value judgment, motives and interests by the researcher as much as by the organizational representatives (Clark, 2011; Homan, 2001). The literature shows that fieldwork is not a static undertaking, but thoroughly intertwined with the researcher and their chosen site.

In this chapter, access to the field and data collection therein is compared to a journey, an iterative process that is prone to detours; other researchers have described it as a game of hopscotch (Peticca-Harris et al., 2016), a long hallway (Feldman, Bell, & Berger, 2003) or the art of the possible (Buchanan et al., 1986). However, whatever metaphor is used, gaining access to an organization is only a first stop in the journey; access to the field is subject to continued cooperation in all areas (Reeves, 2010; Wanat, 2008). This chapter walks the readers through the full journey and includes strategies that worked for the author in order to offer ideas for the readers' projects.

THE START OF THE JOURNEY: PREPARATION AND PROSPECTING ORGANIZATIONS

Without prior ties to an organization, a researcher needs to develop strategies to attain access before embarking on a data collection journey into an organization. This requires preparation, or as the sales literature notes: "prospecting" or searching for potential sites and "preapproaching," or in academic terms, researching them (Moncrief & Marshall, 2005, p. 15). However, documentation of how researchers manage this process is often sparse in published studies. Research methods textbooks also commonly gloss over the difficulties that exist in gaining access to organizations, or they keep advices brief and nonspecific (e.g., Finn, Elliott-White, & Walton, 2000; Gray, 2014; Silverman, 2013; Veal, 1997). The issue of securing access to data prevails even for experienced and well-established researchers (Holmes, 2014). Here the aim for the researcher is to be prepared in order to understand implicit and explicit behavioral rules in the chosen field (Peticca-Harris et al., 2016).

The lack of guidance led the author of this chapter to seek advice and guidance in exploring a wide range of literature to find explicit advice. This included not only books on methodology in general and subject specific, but also literature on ethnography – despite the current study not using this method for data analysis. However, ethnography is often associated with the notion of 'fieldwork' and at its core produces a narrative on how it unfolds. For novice researchers, such content provides rich insights into what life in the field looks like. In the author's experience, this theoretical understanding and expertise grounded in ethnography can help in negotiations with the potential participating organizations. In addition, an extensive search can also reveal that certain protocols exist on recruiting participants, depending on the industry (Mack, Woodson, MacQueen, Guest, & Namey, 2005). This realization was highly important for the author of this chapter, as the study was situated within, but not on the tourism industry as an academic subject.

This section discusses actionable points on how to prepare academic research material to make it not only attractive, but also accessible and relevant to the target organization. The literature contains a selection of works that provided an understanding for the author on aspects to field access that might help overcome issues in securing participating organizations. This includes:

- Researching relevant organizations
- Rewriting academic proposals into a 'sales pitch' (Brady, 2009)
- Demonstrating knowledge about not only the organization, but also the industry
- Considerations for 'cold canvassing' (Collis & Hussey, 2003) potential participating organizations

- Considering different identities (observer vs. observed) and their impact on the study (Laws, Prideaux, & Moscardo, 2006)

RESEARCHING RELEVANT ORGANIZATIONS

Once the author had identified the subject of the study, she considered what organizations would make suitable sites for fieldwork, both through personal experience with them and online research about them (Paluck & Cialdini, 2014). Research involving fieldwork is commonly interested in identifying patterns, learning about processes, or understanding phenomena. Thus, by nature, it is a rather unthreatening undertaking, which is likely to identify best practices and positive behaviors. Unfortunately, we live in a world where the act of 'uncovering' practices is often linked to sensationalism and negativity. In the author's experience, a sector unfamiliar with academic research can easily understand a research proposal as a threat to how an organization works or to how employees perform in their work.

The author found that in conversations with receptionists (the prospective participants), they responded positively to the idea of the research proposal and viewed it as an opportunity to be represented, since they felt that they were often overlooked. They were confident in their work and happy to participate. However, the decision makers, managers removed from the day-to-day frontline activities, were hesitant. They cited various reasons why they did not want to participate: (1) staff overwhelm through a researcher on site; (2) unsuitability of their guest population for research in general; (3) or the notion that they were doing fine and did not need to be researched. The author learned that organizations need to learn of potential benefits and reciprocity of their participation in a language that they understand (Creswell, 2007). An additional benefit of researching an 'ideal' organization lies in the ability to transfer the knowledge gained to other potential participating organizations.

Step 1: Thinking of organizations, the author envisioned the properties that a hotel should have to be suitable for the type of data required for the study. Some key components included the following: both national and international clientele, both business and leisure guests, and a large size to allow for more data to be collected, possibly a chain to multiply available data sites. She had previously worked in such an organization and used it as a starting point.

Step 2: The author began to research information available on the organization. The targeted hotel chain has a website, so the author read through the material available to find out if and how their values align to hers and the project. As a multi-national, large organization, they also provide additional information in the form of press kits that can give valuable insights. For example, they mention other companies they work with that potentially provide a point a contact – in this case, the name of the communication agency they use. The author also noted the company structure and how this might help or hinder research. The large size of the company meant that they had all of the resources the author was looking for, but a considerable larger number of gatekeepers.

Step 3: To maximize understanding and potential shortcuts to bypass gatekeepers, the author also considered other avenues of information, such as the social media presence of the company (including LinkedIn), reviews online, and companies they work with.

Step 4: The author proceeded to write up the findings in a detailed report for herself. She explicitly connected the information found with the aims of the project. In this case, the report was 16 pages long to provide a solid backbone for considerations about recruitment in and your expectations of the field.

The author found that the knowledge gained through this exercise helped her to formulate the proposal sent to prospective participating organization. When her first choice fell through, the author used the structure of information gathered to quickly assess the viability of alternative choices without having to redo the same amount of work completed for the 'ideal' candidate.

ACADEMIC PROPOSALS AS A 'SALES PITCH'

The author found that the project proposal one writes for an academic audience needs to be drastically different to the version sent to the target organizations. It needs to be encoded to suit its intended target audience (Creswell, 2007; Paluck & Cialdini, 2014). Effectively, the researcher needs to sell the theoretical idea to an organization. While an academic proposal aims to convince its audience of its merit and contribution to an existing body of literature, including relevant references and citations, the version for the participants needs to promote relevance to the organization under study. It needs to make sense to them, providing "a clear, concise, non-technical explanation of what the project is about" (Collis & Hussey, 2003, p. 34). Academic relevance is very unlikely to be enticing for an organization to grant access. While the author had established that the proposed study was of academic merit, organizations considered data collection on their premises as a source of nuisance for the organization. Academic 'problems' are not the same as those of an organization, and the organization is likely to treat participation as a favor to the researcher (Collis & Hussey, 2003; Hammersley & Atkinson, 2007).

Few publications address the differences between academia and the real, social world in concrete detail, while also providing suggestions of how a researcher can accommodate this existing gap. While newer literature has started to address the emotional toll that attempts to accessing a field site takes, there is little structured advice. Peticca-Harris et al. (2016) include the authors' autoethnographic notes on their own experiences, such as "a massive recruitment campaign via LinkedIn" (p. 393). The aim of understanding the "concept of organization must be studied by observing competent users" (Bittner, 1965, p. 255). When creating her proposal, the author found that she had to transform into a marketer (to 'sell' the study to an organization) and a public relations manager (to facilitate communication between academia and non-academic organization) to organize communications with the proposed host organization (Paluck & Cialdini, 2014). Preparing the first interaction with an organization needs to be done with the target organization in mind, not from the perspective of the proposal writer. Doing so may help to overcome the stark outsider status the academic researcher has to the organization (Peticca-Harris et al., 2016). While Buchanan et al. (1986) state that access is "a game of chance, not of skill," (p. 70) there are steps a researcher can take to increase his/her chances to gain access through open and consensual negotiations (Broadhead & Rist, 1976).

Brady (2009) unpacks how to write a research proposal for different audiences and expected norms for their different competent users. The seemingly simple differentiation between them was eye-opening to the author of this chapter and proved highly useful in navigating obstacles to access. The author of this chapter adapted some of Brady's points to her own experiences.

Table 1. Researcher vs. Organization professional

Academic Researcher	**Organization Professional**
Focus on research gaps in published literature	Focus on organization's problems
Tendency to tell results to others	Expert at asking questions
Detail oriented from start of the project	Supplies details when required
Lexically dense, field related terminology	Reformulated into target language
Difficulty unpacking complex concepts	Simple everyday examples
Focus on research findings	Focus on implementable solution
Research and identity interrelated	Research as tangible product

(adapted from Brady, 2009, p. 170)

Brady (2009) describes the differing orientations a researcher and a professional in the organization have towards a research project and suggests ways in which these can be translated and situated into the organization's social reality. For instance, academic research is linked to exposing gaps in a discipline-specific literature, and such gaps may only be partially relevant to the target organization. The abstract academic problem should be reformulated to at least partially incorporate a problem or challenge the organization may face. Although the researcher's theoretical background knowledge informs the study and its execution, it is important to establish the organization as the expert on the subject. The focus on a fluid and engaged relationship is increasingly highlighted in current research literature (e.g., Cunliffe & Alcadipani, 2016; Riese, 2019). Research is formulated to investigate a particularity within the organization, an observation from a point based on observation and the ubiquitous academic literature with its complex, discipline-specific jargon and terminology. Thus, a research project will frequently be of little consequence to the participating organization. Complex academic concepts need to be expressed in what constitutes relevant everyday examples for the organization, translating why their contribution to the research matters. Findings in research may not necessarily be directly applicable to an organization, at least not within a particular research project. Organizations, however, focus on getting things done to produce solutions which needs to be accounted for in the communication between researcher and potential participant. Brady (2009) notes that research and researcher are deeply connected, maybe even more so for novice researchers; for an organization, research constitutes a product, a means to an end to solve a problem.

WRITING A PROPOSAL

The author's PhD project aimed to explore communication between receptionists and their guests in international hotels. As a qualitative research project, there was no definite research question prior to the data collection, but rather an area of interest. However, hotels are businesses, with the aim to make money, and guests are the only reason why a hotel exists (Ogle, 2009).

The author used her prior training, supervisor's suggestions, and literature to produce her proposal (see also Collis & Hussey, 2003, pp. 33-34). She produced versions in the native languages of the different countries the hotels were located in.

Cover Letter

To provide an overview, the author prepared a cover letter that introducing herself, the subject of her inquiry, a brief summary of the project, and (perhaps most importantly) why she was inviting them to participate. She tried to demonstrate in a sentence that she had done her research on the organization. She was targeting many companies, but she tailored the cover letter for each one, in at least a sentence, and changed the actual company name in the letter. According to recommendations that the author received, the letter should be printed on paper with the affiliate university's letterhead. The author submitted both hard copies and digital versions of her proposal, depending on the location of the hotel, and the information available on accessing gatekeepers.

1. **Proposal content:** Collis and Hussey (2003) suggest that the actual proposal be brief, ideally not more than two pages of A4. This is to ensure that the content is concise and more likely to be read by the recipient. The author of this chapter kept both margins and font size at a standard size to keep easy readability for the receiver. In addition, she used a clear structure and meaningful headings to organize content and to help the reader navigate. She avoided generic headings such as 'introduction,' which might be used in an academic context, and provided the reader with a synopsis of the content of the section instead. She also included information on how she was ensuring to keep participants on all levels safe (the hotel as an organizational body, the organization's employees (receptionists), and the organization's customers (hotel guests), including on how informed consent would be obtained (Paluck & Cialdini, 2014).
2. **Benefits:** Promises regarding benefits for the organization were kept realistic in the proposal. The author offered a brief summary of findings to inform participating organizations of the study results. Stakeholders may need additional assurance that the researcher's presence is supportive to the overall mission of the organization (Paluck & Cialdini, 2014). This also included allowing time for a briefing into and a debriefing from the organization. Such practice means being mindful of the organization's time and resources and providing a tangible end product for them (Buchanan et al., 1986; Peticca-Harris et al., 2016).
3. **Formulation of request:** The proposal clearly stated the resources the researcher required. For this study, the main form of data were videotaped interactions between receptionists and their guests. When researching organizations, and in initial conversations with hotel representatives, it was evident that hotels are highly protective of their staff and their guests. To accommodate this, the proposal was specific in how informed consent would be obtained. Even if a hotel granted access, the staff members were able to decide whether they wanted to partake in the study or not. Consent was not assumed or anticipated through earlier experience. This type of research commonly does not focus on the individuals but on behavior exhibited in a social situation under study; the

specific employees "only happen to be there at the time" (Homan, 2001, p. 39). The duration of the fieldwork was left to the organization and constructed around the individual hotel's needs. The video recorder and microphone were chosen to be small and unobtrusive to minimize impact on interactions at the desk and could be tidied away between recordings. It was also stated that staff would not be involved in the data collection; it was left to the researcher to approach prospective participants and work the recording equipment. Even still, the presence of an outsider has an impact on the day-to-day working environment by simply being there. This needs to be accounted for so that a decision maker can plan for the researcher's presence.

4. **Presentation:** In order to increase the likelihood of the proposal to be read, the author used her prior training in writing business reports and proposals and conducted a search online to find layout templates for the target industry. Online editors, such as Canva, can help with the design process. The author used the Adobe design suite for her proposal. The proposals had a cover page with a title, purpose, and author, including her university affiliation. The proposal was designed to be enticing to the reader. After working in companies over the years, she was aware of the amount of communication decision makers receive. In order to enhance the visibility of the proposal, relevant stock photos were included on the cover. Hard copies of the proposal were bound to make them look professional. In meetings with hotel managerial staff, they often brought the proposal to go over details; in addition, the researchers also brought extra copies of it – e.g., for other participants.

ON GATEKEEPING AND COLD CANVASSING

'Access' is understood here as both the permission granted by an organization to do research, as well as the process of building trust to obtain information once in the field (Cunliffe & Alcadipani, 2016, p. 537). Without established contacts in the target industry, the proposal was sent to organizations deemed suitable for the project. This approach, referred to as 'cold canvassing,' is a common approach in sales to approach prospective clients (Moncrief & Marshall, 2005). Quite often, gatekeepers will make the decision if the document is forwarded to the decision-making person or not. 'Gatekeepers' are understood as individual intermediaries between the researcher and an organization who can grant or deny access to the physical space and the research methods to be used (Clark, 2011; Crowhurst & Kennedy-Macfoy, 2013; Homan, 2001). Both mailing and emailing often means that the proposal is left unanswered. Wherever possible, the author delivered the proposal in person to the organization. Even without access to the actual decision-making recipient, it was a great learning opportunity. Talking to gatekeepers – such as receptionists, administrators, or duty managers – allowed for a sense of how the organization is disposed towards potential cooperation with an outside party. Gatekeeping is a matter of monitoring boundaries (Holmes, 2007), and the gatekeepers at the lower hierarchical level commonly showed great interest in the proposed research and explained in their own words why they felt it is relevant to their industry. In some cases, they displayed the opposite behavior and were defensive over a perceived potential intrusion. This newly gained knowledge was used to tweak the proposal. History and culture are important in understanding how organizations function in their industrial sector that can help in deciding who the gatekeepers are and who has access to the data required for a study (Clark, 2011; Crowhurst & Kennedy-Macfoy, 2013; Cunliffe & Alcadipani, 2016; Terenzini, 1993). Gatekeepers can exert controlling power to access both higher and lower level levels of the organizational hierarchy (Reeves, 2010).

Figure 1. Organizational diagram of gatekeeping and decision-making bodies in hotels

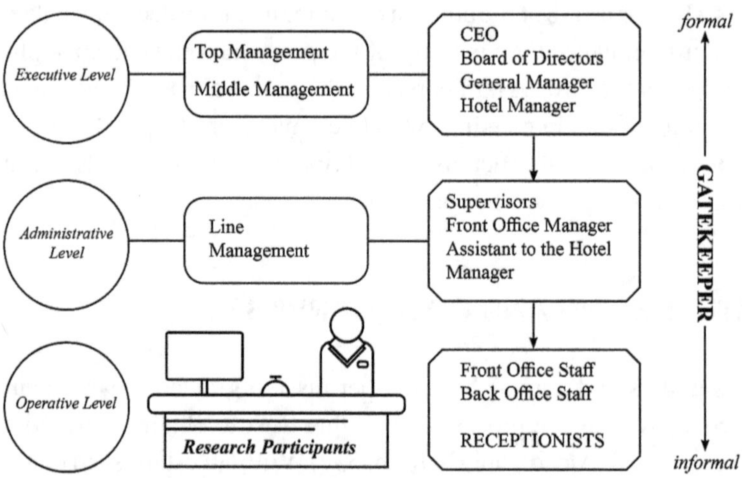

Figure 1 shows that gatekeepers are present at all levels across the hierarchy in an organization. Depending on the particular hotel, not all of them were necessarily present, and access was unique to each one. Potential gatekeepers interpret negotiations within the context of their position in the organization. Researchers typically need permission from influential, formal gatekeepers, but the relationship in the field is also dependent on the informal ones who are not in a position of power (Wanat, 2008). High-quality access depends on ongoing awareness regarding the dynamics of information flow in the organization (Cunliffe & Alcadipani, 2016).

When delivering a proposal in person, an assurance that it will be passed on to the relevant decision-making body, or an indication of when to expect a reply to the query, was often given. However, the nature of cold canvassing is that no reply (i.e., a lack of a response) was more common than a response. The author followed up on proposal submission after one or two weeks. Frequently, no reply was received, even from very promising encounters. Some organizations may have had a practice of blanked refusing all such requests. This may have been due to bureaucratic rigidity, perceived lack of researcher's legitimacy, legal concerns, unwanted intrusion and lack of time, among other reasons (Clark, 2011; Spencer, 1973).

BUMPS IN THE ROAD: DEALING WITH BARRIERS TO ACCESS

In this section, I consider (1) obstacles in obtaining access to an organization and (2) interim suggestions in using creative ways of engaging with organizations that may not want to participate in the study but could provide valuable insights for future negotiations. In my example, I discuss how I used a short-term internship as a way to learn about how the industry works, hierarchical considerations within the organization, the 'lingo' (Calvi, 2001) and expected behavior. Maybe most importantly, I reflect on the humbling experience that this stage was for me. As a novice researcher, fully immersed in the literature, it can be tempting to believe that one already has an in-depth understanding of the site; however, the real-life check that an informal observational activity provides can help a researcher to understand the environment much better. On-site research is iterative, and as such evolving strategies of recruiting methods are common (Mack et al., 2005). A researcher will need to evaluate themselves, deciding if the research proposal can be adapted to a different reality, if the researcher themselves can adapt, and has the resources to do so. It is an active process requiring one to learn about personal abilities and limitations (Peticca-Harris et al., 2016).

While contacting the first organizations, conversations often seemed positive with genuine interest being expressed, as well as a promise to participate in the project. However, in many cases, even when participation seemed to have been agreed upon, the contact person became unreachable and the fieldwork fell through. In addition, this lack of commitment can also impact other factors, such as feasibility of the overall study, time and money constraints. It was easy to feel discouraged and even angry at the lack of communication, and in some cases leading to a feeling of being disrespected. It was important to consider that partaking in a study is not a small undertaking, even more so in a field research setting (Hammersley & Atkinson, 2007). There are considerable costs involved for the participant in the form of emotional effort and time (Homan, 2001). Providing input to a study is an investment, which participants may not realize at the time they agree to partaking in the study. Backing out officially from a study may be linked to a loss of face, so they may choose to stop communicating to save face – on all available channels, such as face-to-face or email[1]. Once the interest of a potential participating organization was secured, it would have been ideal to obtain documentation that covers the agreement. However, the author needed to be mindful of the specific target organization as each can differ from another even within the same sector (Collis & Hussey, 2003; Paluck & Cialdini, 2014). In some instances, requesting more than a verbal agreement put the participating organization's willingness to be part of the research at risk. The more the author learned about working practices in the field and the different countries, the more she understood how to successfully circumnavigate these negotiations. In some instances, it was the organization that requested to enter into a formal contract. Depending on the country, it appeared to be because of their business policies, health and safety considerations, but the decision was sometimes also linked to a requirement by law for them. It was important to understand the terms of the agreement – e.g., to make sure that the researcher owns the data to allow for the data analysis and publications. This concern circles back to the notion of gatekeeping and its potential to influence or control the data (Broadhead & Rist, 1976).

CREATIVE THINKING IN NEGOTIATING ACCESS

It often felt like there was no progress in securing a data collection site, so the author developed an interim solution to learn more about the field and its challenges, without having an organization to commit to participating in the research. The aim here was to create hypothetical field situations to better understand what was preventing the researcher from getting access to the field through entertaining more creative measures (Herskovits, 1950). This form of negotiating access is also reported in some of the literature cited in this chapter (Bryman, 2001; Cunliffe & Alcadipani, 2016; Hammersley & Atkinson, 2007; Silverman, 2013; Van Maanen, 1979, 2011). Unfamiliarity with the research context at a practical level is not a new phenomenon to cause projects to fail (Clarke, 1975). The author had previously been successful in securing data collection sites through temporary jobs and volunteering efforts during other projects. Those options were deemed unsuitable for the target industry and the timeline of the project. Instead, one hotel offered insights into their day-to-day work through a scheme that made sense to them: a short-term internship. This form of 'immersion' into an organization provides an opportunity to be embedded into its inner workings (Cunliffe & Alcadipani, 2016; Reeves, 2010). Even though it was not possible to gather 'official' videotaped data during that time, the author had the opportunity to verify that data collection would be possible in the real world and that the phenomenon under study was indeed occurring at the chosen site. Most importantly, it generated an understanding of how the field site functions – and how removed the site is from academic research and its interests, an observation also made in Reeves (2010) regarding the difference between an academic idea and the real site. This is an element that is often overlooked and neglected in research that is not classed as ethnography, but rather 'happens' to be dealing with people in their natural habitat. However, rules of entering and navigating the field as described in ethnographic methods create awareness and comfort in the data collection environment. Benefits for the main study were seen on two levels: 1) the ability to observe the field without pressure of capturing the data to verify that the research design is valid and that the question can be answered with the planned setup; and 2) the ability to gain insights into the working environment of the field, which will enable much clearer suggestions to future prospective participating organizations, demonstrating a thorough understanding of how they work and how impact on their work could be minimized. This activity of merely being in the field as an observer and researcher demonstrated how it causes a considerable degree of interruption to the normal working pattern of participants. After all, collecting data in the field consists of a strong sentiment of "being there" (Bate, 1997, p. 1163) as a legitimization of processes and hierarchical structures in the field (Sherman, 2005), as opposed to mere "armchair philosophy" (Clarke, 1975, p. 98).

In order for deciding bodies to at least be willing to listen to the proposal, it was key to demonstrate to prospective participating organizations an awareness of challenges the fieldwork may create for the population under study. This practice allowed to learn about the concerns an organization may have regarding a researcher's presence – e.g., how it may affect employees. Hotels with a board of directors further required proof that data collection could be conducted safely. Higher management bodies were presented with similar studies to demonstrate understanding of how fieldwork is ethically conducted in these situations (Paluck & Cialdini, 2014). Being able to refine the purpose of the study and present it to potential participants also helped to generate the necessary rapport and trust to conduct the data collection (Creswell, 2007). Such consideration shows humbleness and reflexivity required to put the well-being of the participating organization and its employees above the benefits of an idealistically constructed study proposal (Homan, 2001).

In addition to spending time outside of the data collection time in a hotel reception environment, the author continued to explore other alternate avenues of interacting with the target organizations. Here the aim was to build up a network of contacts that may be helpful for gaining access to the organization of choice. This included becoming involved with an organization for public relations and communication professionals. Through individuals met through there and the social media platform LinkedIn, connections to hotels were developed. The author was able to identify employees that work at the organizational gatekeeping level relevant to the research project. While this did not lead to access to fieldwork sites directly, the employees were able to share non-committal advice. The same was done by engaging with target relevant associations, events and fairs to build a network. This helped to market the author as an expert who is heavily involved in the non-academic side of the field and can be allowed to add value to the field before asking them to partake in your study. Hammersley and Atkinson (2007) describe gaining access to an organization as a practical matter that requires the researcher to draw upon both intra- and interpersonal connections and strategies acquired in everyday life. Understanding barriers to access and the means to overcome them will provide the researcher with valuable insights about the organization and its sector (Hammersley & Atkinson, 2007; Paluck & Cialdini, 2014). Recognizing patterns that are not working in gaining access is a frustrating experience, often associated with the feeling of failure. Lost resources, such as time, money and personal well-being, are the result (Peticca-Harris et al., 2016). Developing strategies to keep momentum, despite such setbacks through the methods described here, proved to be the key to ultimately conduct a much better study than previously anticipated.

ON THE ROAD: FIRST DATA COLLECTION SITE

In this section, I discuss the idea of letting go of the idea of the 'perfect' participating organization. Even if an original project design may identify a potential perfect form of participation, I suggest that allowing a deviation based on availability can not only provide exploratory data but also prepare the researcher for future data collection in similar sites.

The first organization that finally agreed to participate in the research was far removed from the 'ideal' organization pictured in the original proposal. It was unclear whether the data collection would even be worth the hassle. However, after struggling with securing participating organizations for months, I believe that it is important to give the venue a chance. The author decided to view it as a learning curve and as additional bargaining power with additional organizations, using the experience in collecting the data in one company in addressing concerns another organization may have. She learned to think like the organization (Bate, 1997, p. 1153), for in the words of Fine (1993), "perfection is professionally unobtainable" (p. 285).

In my own experience, I had imagined myself to be collecting data in a large hotel (chain) catering for both business and leisure guests. The first hotel where I collected data was a very small one in the English countryside, at a time of year when they were not busy at all. I drove 70 miles (one way) several times a week over a month period to videotape interactions; sometimes, I recorded one or two conversations in a day. It was a grueling experience, but the data I collected was very much worth it, and I learnt a vast amount about my recording set up, the best way to gain informed consent, how to make sure I kept my intrusion to a minimum and information from the organization, its employees and guests. No matter what kind of data are being collected, interactions in the field are taxing and exhausting (Clarke, 1975; Creswell, 2007), and navigating the social complexities "will make you want to tear your hair out" at times (see preface Kaler & Beres, 2016). Coping with the lived actualities (Appadurai, 1996) of an organization is part of the research experience.

Collecting data involves spending a considerable time with those who provide the data to learn about the rules present in the organization under study. The duration and frequency of such interactions is of a somewhat arbitrary length – a length that is deemed 'sufficient' to gain an understanding of how the social world is constructed. Data collection in this manner is comparable to a gift exchange (Falzon, 2009; Rosen, 1991). Members of any given culture imagine themselves and are imagined by others as group members to develop a sense of social identity, as sense of who we are (Burns & Novelli, 2006; Piller, 2012). As a result, Fine (1993) observes that "Each job includes ways of doing things that would be inappropriate for those outside the guild to know. Illusions are essential for maintaining occupational reputation" (p. 273). Traditions create an organization's value system (Jackson, 1977).

ETHNOGRAPHIC INSIGHTS INTO DATA COLLECTION

As previously indicated, the author believes that there are ethnographic considerations regarding the researcher's role in the field and their insider/outsider status, including entering and leaving and its rituals (Bate, 1997; Fineman, Gabriel, & Sims, 2010; Hammersley & Atkinson, 2007; Rosen, 1991). It concerns issues such as what to wear, how to present oneself, how participants view them, helping out (or not) while in the field, position within the organizational hierarchy and concerns for power, including an insurance that participants have the option to decline participating, even if the company made the decision for them. Van Maanen (1979) speaks of the "continual and careful attention to the details of one's adventures in the field" (p. 548). Decisions on how one presents oneself will influence the type of data the researcher is able to collect on site (Van Maanen, 2011). Wearing similar clothing as receptionists during my data collection was mandated by the participating hotels. While this was not confusing for employees, hotel guests routinely mistook me for a staff member. Since I tried to not engage with guests unless I was recording, they frequently assumed that I was a manager. In extremely busy situations, guests would often refuse my claim of not being an employee and insist on posing their questions to me (see also Hammersley & Atkinson, 2007; Kaler & Beres, 2016 for more examples). The author noticed in the field that she was able to pick up on routine behavior and patterns rather quickly and needed to decide where and how she positioned herself. For example, was she a strict (detached) nonparticipant, a (interactive, emotionally engaged) participant, a marginal member (coolly dispassionate observer), or occupying some middle ground? It then became an ethical choice as to whether I engaged with external participants of an organization to perform menial tasks. In my situation, I decided to 'help out' with tasks that did not involve money (Compare Clarke, 1975; Deckert & Vickers, 2011; Hammersley & Atkinson, 2007; Tedlock, 1991; Van Maanen, 2011, p. 88ff).

Access while in the field will also be informed by the researcher's personality and characteristics and tactful behavior (Buchanan et al., 1986). When presenting herself to her participants, the author considered how she would be viewed by others, such as kind, friendly, honest, precise, observant, unobtrusive, candid, chaste, fair or literary (Fine, 1993). This aimed to help position her in the field and manage expectations participants projected on her (Caretta & Jokinen, 2017). Participants in the present study were struggling to understand what academic research means as they did not attend university. In addition, they were confused that the researcher had not completed an apprenticeship. This generated a notion of 'betweenness' where the researcher is caught between an insider and outsider status (Kalir, 2006; McFarlane-Morris, 2020; Reeves, 2010). The data collection is emergent as to what is deemed acceptable to record in the field (Horwood & Moon, 2003).

The Nature of Organizations as Field Sites

While a field site can constitute of several different types of institutions, this chapter focuses on organizations. The project focused on how interactions are constructed at the international hotel front desk. Arguably, there is no better area for observing a culture in action than in-situ in the business context it occurs in. Cultures tend to reveal themselves in situations where much is at stake (Hooker, 2012, p. 389). Entering a field site reveals how much about a 'place' relies on taken-for-granted mutual understanding and sensemaking about the social life in the environment that is relevant to this particular organization (McCabe & Marson, 2006; Teulier & Rouleau, 2013; Urry, 2005). It is not merely a site, but also an undertaking: "Organizing is a set of social activities through which we try to make our lives more predictable, effective, and stress-free" (Fineman et al., 2010, p. 1). An organization consists of a plethora of interdependent groups that influence the work of organizations, originating both from outside and inside the organization (Jackson, 1977; Mehan, 1991). Organizations have a 'way of doing things,' a locally constructed logic or rationality: knowledge becomes embedded in the way typical courses of actions are undertaken, and it is something that members do routinely to both make and give sense to their identity. Members and insiders can invoke procedures to promote a certain form – or rule – of conduct (Bittner, 1965; Boden, 1994; Fineman et al., 2010; Nielsen, Nielsen, Gravengaard, & Due, 2012; Sarangi & Roberts, 1999; Sharma & Good, 2013; Zimmerman & Boden, 1991).

The way a researcher perceives their site and their study is not static. Pre-understanding (understanding before study) is not the same as understanding (generated during study, constantly updated). It is developed through being socialized into an organizational culture, by learning its rituals and codes (Laws et al., 2006; Nielsen, 2009; Zimmerman & Boden, 1991). Understanding and the point of view of the researcher will inevitably change over the course of the data collection – even if this development, or confession, is not included in the final report (Van Maanen, 2011). Through interaction with the organization in the field, the researcher's understanding of how the company functions will develop over time. This also includes the notion of 'access' and how it is likely to govern behavior in the field, not only in the lead up to the data collection, but throughout. There is a sense of 'protectiveness' that governs behavior within and across organizations that are constructed across various levels of gatekeepers. The type of relationship developed with those gatekeepers will inform the type of data a researcher is likely to be able to collect (Hammersley & Atkinson, 2007). The stay in the field will also raise ethical concerns that are reflective in the field: what type of information will be recorded and what will be left behind in the field. The question becomes not only whose permission is needed, but also who should be asked. This extends to the ownership of the data collected and how this is to be handled (Bengry-Howell & Griffin, 2012; Creswell, 2007; Hammersley & Atkinson, 2007; Mack et al., 2005; Paluck & Cialdini, 2014).

THE JOURNEY CONTINUES: ADDITIONAL DATA COLLECTION SITES

In this section, I consider how I arrived from my original research plan of collecting data at a singular, perfect site at four vastly different, yet cohesive organizations in three countries that in the end greatly improved my initial proposal. I suggest that original barriers or shortcomings in the data collection process may ultimately be of enriching quality for the overarching study. In addition, they may provide an opportunity for growth both personally and professionally. In my case, I was able to create a network that ultimately led to more data collection sites than I was able to work with given the resource constraints of a PhD project. I suggest that data collection in the field is not only beneficial for academic purposes, but also generates expertise and trust in the industry.

While collecting data at the first site, the author continued to search for additional hotels to collect data in. Having the experience of the first location proved invaluable in talking to other hotels and in helping to secure additional participating sites. It demonstrated expertise in knowledge of the industry, the state of research, but also practical aspects of projects and the perceived impact and emotional cost for participants. Going into a second or third organization deepens the initial understanding of the wider industry, even for a small-scale project. The author's study may not be able to generalize to the whole industry, but allowed to uncover working patterns, as well as individual differences between organizations, which were valuable in analyzing the data corpus, but also in differentiating understanding of existing literature and exposing its gaps. It allowed the researcher to better understand how the physical environment constrains field data collection with both visible and invisible boundaries (Fineman et al., 2010).

With each data site the author collected data at, she was able to generate new leads. Having experience of the prior data collection helped to build credibility in the field. I eventually collected data in four different hotels and had secured two additional sites, which I decided to not collect data in because of time and money constraints. I also generated a new network rather easily, where host organizations suggested other companies for me to contact (similar to snowball recruitment strategies). Collecting data from additional sites allows to "follow people, connections, associations and relationships across space" (p. 2) (Falzon, 2009). By the time I had finished my data collection, I had built a far more interesting – and robust – corpus than I had planned for in my original proposal. The decision to allow available 'imperfect' sites to govern the data I collected, forced me to raise to challenges and greatly impacted my ability to react and reflexively correct practices in the field.

Whether the author was an observer or participant researcher differed between sites, and so did the relationships with participants. Setting up equipment and adapting to different corporate rules are notions that transcend an individual research site. They are systematic that build upon one another and are adapted to a new research location. As such, this flexibility shows the implications of what is happening in one location to another and ultimately to locations that are not within the research frame or design (Hannerz, 2003; Marcus, 1995).

RECOMMENDATIONS

- **Be prepared.** Preparation was key in securing organization's participation in the research project. While the academic environment of the project focused on its theoretical grounding, this did not help in conversing with gatekeepers or later with participants. Being able to demonstrate understanding of the industry made a great difference.

- **Be flexible.** A PhD project is often personal, and the researcher was attached to her original plan of data collection. However, once she let go of this notion, she was able to collect much better and richer data than she had initially envisioned. She was able to generate meaningful and original findings. She developed a deep understanding of not only her original topic, but also the existing shortcomings in the literature by not following standard procedure.
- **Be humble.** Academic research can seem like an elitist undertaking, and the theoretical knowledge prior to going into the field can mean that a researcher thinks they know it all. However, there is a reason why fieldwork is conducted and the site was chosen for its informed participants.
- **Be proactive.** When the ideal organization did not want to partake in the project, and it became apparent that others where not particularly interest either, it became paramount to generate creative and alternative ways of engagement with the target industry. Remember: "If plan A doesn't work, the alphabet has 25 more letters" (unknown).

CONCLUSION

Fieldwork should not be disconnected or glossed over in a research project. It should not be treated like a mere gathering exercise, but rather a co-constructed activity in which the field with its actors provide their authentic expertise on the subject matter. As such, it informs the project long after the data has been recorded.

The journey approach enables the researcher to reflect on their data collection processes and aims to point to potential, creative solutions that can help to create a coherent study deeply rooted in its social environment. Fieldwork becomes a part of the project, rather than an isolated element that needs to be done as an integral part to knowledge production (Monahan & Fisher, 2010). Through this approach, even a novice researcher can demonstrate connected insights that are relevant to the study itself, but also the subject of study. Also, the approach may be useful for researchers who feel stuck in an approach – e.g., with regards to a theoretical framework, existing literature and/or simply a preconceived notion of how the study should look like. This means that a researcher can respect the methodological traditions of their field, yet still implement novel and more forward-thinking ideas. Letting go of the idea of solely wanting to collect data from 'the perfect' site can, in my experience, highlight problems in an existing body of literature and provide a new rationale for the importance of the study.

Overall, allowing data collection to become a central part of the research process can make the study extremely rewarding and add an additional layer of accomplishment, revealing surprising results. A creative, multi-layered approach to resources and created opportunities may help increase the feasibility of a study by reacting to and interacting with the social context of the field. Ultimately, an engaged data collection process also teaches how to communicate between academia and industry in addition to a wider range of other potential audiences, which can identify potential practical applications of the study's results (Brady, 2009). This type of journey through data collection can help to develop a thorough understanding of how the phenomena under study is relevant in an academic but also in a professional context. This is by no means a new observation (see Clarke, 1975), but rather a normal process of all fieldwork. Fieldwork is an important contribution to replicability, validity and impact of social research by developing and shaping discourse and practices, and for that reason, it deserves to be treated with such respect (Maner, 2016; Monahan & Fisher, 2010).

REFERENCES

Appadurai, A. (1996). *Modernity at large: Cultural dimensions of globalization*. University of Minnesota Press.

Bate, S. P. (1997). Whatever happened to organizational anthropology? A review of the field of organizational ethnography and anthropological studies. *Human Relations*, *50*(9), 1147–1175. doi:10.1177/001872679705000905

Bell, C. (2020). The Tourism Researcher: Ethical Dilemmas During Fieldwork in Africa, Bali and Myanmar. *Critical Arts*, 1–6. doi:10.1080/02560046.2020.1712445

Bengry-Howell, A., & Griffin, C. (2012). Negotiating access in ethnographic research with 'hard to reach'young people: Establishing common ground or a process of methodological grooming? *International Journal of Social Research Methodology*, *15*(5), 403–416. doi:10.1080/13645579.2011.600115

Berno, T., & Ward, C. (2005). Innocence abroad: A pocket guide to psychological research on tourism. *The American Psychologist*, *60*(6), 593–600. doi:10.1037/0003-066X.60.6.593 PMID:16173892

Bittner, E. (1965). The concept of organization. *Social Research*, 239–255.

Boden, D. (1994). *The business of talk: Organizations in action*. Polity Press.

Brady, J. (2009). From research idea to workplace project: Gaining access to the corporate world. In W. C. V. Bhatia, B. Du-Babcock, & J. Lung (Eds.), *Language for professional communication: Research, practice and training* (pp. 161–174). The Hong Kong Polytechnic University.

Broadhead, R. S., & Rist, R. C. (1976). Gatekeepers and the social control of social research. *Social Problems*, *23*(3), 325–336. doi:10.2307/799778

Bryman, A. (2001). *Social research methods*. Oxford University Press.

Buchanan, D., Boddy, D., & McCalman, J. (1986). *Getting in, getting on, getting out, getting back*. University of Glasgow Department of Management Studies.

Burns, P. M., & Novelli, M. (2006). Tourism and social identities: Introduction. In P. M. Burns & M. Novelli (Eds.), *Tourism and social identities: Global frameworks and local realities* (pp. 1–12). Elsevier.

Calvi, M. V. (2001). El léxico del turismo. In Á. M. Felices Lago (Ed.), *ELE para contextos profesionales*. Cultura e intercultura en la enseñanza del español a extranjeros. Retrieved from http://www.ub.es/filhis/culturele/cont_pro.html

Caretta, M. A., & Jokinen, J. C. (2017). Conflating privilege and vulnerability: A reflexive analysis of emotions and positionality in postgraduate fieldwork. *The Professional Geographer*, *69*(2), 275–283. doi:10.1080/00330124.2016.1252268

Clark, T. (2011). Gaining and maintaining access: Exploring the mechanisms that support and challenge the relationship between gatekeepers and researchers. *Qualitative Social Work: Research and Practice*, *10*(4), 485–502. doi:10.1177/1473325009358228

Clarke, M. (1975). Survival in the field: Implications of personal experience in fieldwork. *Theory and Society*, *2*(1), 95–123. doi:10.1007/BF00212729

Collis, J., & Hussey, R. (2003). *Business research: A practical guide for undergraduate and postgraduate students* (2nd ed.). Palgrave Macmillan.

Creswell, J. W. (2007). *Qualitative inquiry and research design: Choosing among five approaches* (2nd ed.). SAGE.

Cross-Disciplinary Learning. (2012). In N. M. Seel (Ed.), *Encyclopedia of the Sciences of Learning* (pp. 858–858). Springer US.

Crossley, É. (2019). Deep reflexivity in tourism research. *Tourism Geographies*, •••, 1–22. doi:10.1080/14616688.2019.1571098

Crowhurst, I., & Kennedy-Macfoy, M. (2013). Troubling gatekeepers: Methodological considerations for social research. *International Journal of Social Research Methodology, 16*(6), 457–462. doi:10.1080/13645579.2013.823281

Cunliffe, A. L., & Alcadipani, R. (2016). The politics of access in fieldwork: Immersion, backstage dramas, and deception. *Organizational Research Methods, 19*(4), 535–561. doi:10.1177/1094428116639134

Deckert, S. K., & Vickers, C. H. (2011). *An introduction to Sociolinguistics*. Continuum International Publishing Group.

Edmondson, A. C., & McManus, S. E. (2007). Methodological fit in management field research. *Academy of Management Review, 32*(4), 1246–1264. doi:10.5465/amr.2007.26586086

Falzon, M.-A. (2009). Introduction: Multi-sited ethnography: Theory, praxis and locality in contemporary research. In M.-A. Falzon (Ed.), *Multi-sited ethnography: Theory, praxis and locality in contemporary research*. Ashgate Publishing, Ltd.

Feldman, M. S., Bell, J., & Berger, M. T. (2003). *Gaining access: A practical and theoretical guide for qualitative researchers*. Rowman Altamira.

Félix-Brasdefer, J. C. (2015). *The language of service encounters: A pragmatic-discursive approach*. Cambridge University Press. doi:10.1017/CBO9781139565431

Fine, G. A. (1993). Ten lies of ethnography: Moral dilemmas of field research. *Journal of Contemporary Ethnography, 22*(3), 267–293. doi:10.1177/089124193022003001

Fineman, S., Gabriel, Y., & Sims, D. (2010). *Organizing and organizations* (4th ed.). Sage.

Finn, M., Elliott-White, M., & Walton, M. (2000). *Tourism and leisure research methods: Data collection, analysis and interpretation*. Longman.

Gray, D. E. (2014). *Doing research in the real world* (3rd ed.). Sage.

Grönroos, C. (2001). The perceived service quality concept – a mistake. *Managing Service Quality, 11*(3), 150–152. doi:10.1108/09604520110393386

Hall, M. C. (2010). Island destinations: a natural laboratory for tourism: introduction. *Asia Pacific Journal of Tourism Research, 15*(3), 245–249. doi:10.1080/10941665.2010.503613

Hammersley, M., & Atkinson, P. (2007). *Ethnography: Principles in practice* (3rd ed.). Routledge. doi:10.4324/9780203944769

Hannerz, U. (2003). Being there... and there... and there! Reflections on multi-site ethnography. *Ethnography, 4*(2), 201–216. doi:10.1177/14661381030042003

Herskovits, M. J. (1950). The hypothetical situation: A technique of field research. *Southwestern Journal of Anthropology, 6*(1), 32–40. doi:10.1086outjanth.6.1.3628688

Holmes, J. (2007). Monitoring organisational boundaries: Diverse discourse strategies used in gatekeeping. *Journal of Pragmatics, 39*(11), 1993–2016. doi:10.1016/j.pragma.2007.07.009

Holmes, J. (2014). *How do we research workplace talk?* Paper presented at the 17th Warwick International Postgraduate Conference in Applied Linguistics, University of Warwick.

Homan, R. (2001). The principle of assumed consent: The ethics of gatekeeping. *Journal of Philosophy of Education, 35*(3), 329–343. doi:10.1111/1467-9752.00230

Hooker, J. (2012). Cultural differences in business communication. In C. B. Paulston, S. F. Kiesling, & E. S. Rangel (Eds.), *The handbook of intercultural discourse and communication* (pp. 389–407). John Wiley & Sons Ltd. doi:10.1002/9781118247273.ch19

Horwood, J., & Moon, G. (2003). Accessing the research setting: The politics of research and the limits to enquiry. *Area, 35*(1), 106–109. doi:10.1111/1475-4762.00116

Ivanova, M., Buda, D.-M., & Burrai, E. (2020). Creative and disruptive methodologies in tourism studies. *Tourism Geographies*, 1–10. doi:10.1080/14616688.2020.1784992

Jackson, J. M. (1977). The organization and its communication problems. In L. W. Porter & K. H. Roberts (Eds.), *Communication in organizations* (pp. 84–134). Penguin Books.

Johl, S. K., & Renganathan, S. (2010). Strategies for gaining access in doing fieldwork: Reflection of two researchers. *Electronic Journal of Business Research Methods, 8*(1), 42–50.

Kaler, A., & Beres, M. (2016). *Essentials of field relationships*. Routledge. doi:10.4324/9781315429335

Kalir, B. (2006). The field of work and the work of the field: Conceptualising an anthropological research engagement. *Social Anthropology, 14*(2), 235–246.

Kerbrat-Orecchioni, C. (2006). Politeness in small shops in France. *Journal of Politeness Research, 2*(1), 79–103. doi:10.1515/PR.2006.005

Laws, E., Prideaux, B., & Moscardo, G. (2006). Quality and service management perspectives. In B. Prideaux, G. Moscardo, & E. Laws (Eds.), *Managing tourism and hospitality services: Theory and international applications* (pp. 1–14). CABI. doi:10.1079/9781845930127.0001

Lofland, J., Snow, D. A., Anderson, L., & Lofland, L. H. (1971). *Analyzing social settings* (4th ed.). Thomson Wadsworth Pub.

Mack, N., Woodson, C., MacQueen, K. M., Guest, G., & Namey, E. (2005). *Qualitative research methods: A data collector's field guide*. Family Health International.

Maner, J. K. (2016). Into the wild: Field research can increase both replicability and real-world impact. *Journal of Experimental Social Psychology*, *66*, 100–106. doi:10.1016/j.jesp.2015.09.018

Marcus, G. E. (1995). Ethnography in/of the world system: The emergence of multi-sited ethnography. *Annual Review of Anthropology*, *24*(1), 95–117. doi:10.1146/annurev.an.24.100195.000523

McCabe, S., & Marson, D. (2006). Tourist constructions and consumptions of space: Place, modernity and meaning. In P. M. Burns & M. Novelli (Eds.), *Tourism and social identities: Global frameworks and local realities* (pp. 91–108). Elsevier. doi:10.1016/B978-0-08-045074-2.50011-4

McFarlane-Morris, S. (2020). "Home sweet home?" Struggles of intracultural "betweenness" of doctoral fieldwork in my home country of Jamaica. *Area*, *52*(2), 394–400. doi:10.1111/area.12580

Mehan, H. (1991). The school's work of sorting students. In D. Boden & D. H. Zimmerman (Eds.), *Talk and social structure: Studies in ethnomethodology and conversation analysis* (pp. 71–90). Polity Press.

Monahan, T., & Fisher, J. A. (2010). Benefits of 'observer effects': Lessons from the field. *Qualitative Research*, *10*(3), 357–376. doi:10.1177/1468794110362874 PMID:21297880

Moncrief, W. C., & Marshall, G. W. (2005). The evolution of the seven steps of selling. *Industrial Marketing Management*, *34*(1), 13–22. doi:10.1016/j.indmarman.2004.06.001

Nielsen, M. F. (2009). Interpretative management in business meetings: Understanding managers' interactional strategies through conversation analysis. *Journal of Business Communication*, *46*(1), 23–56. doi:10.1177/0021943608325752

Nielsen, M. F., Nielsen, S. B., Gravengaard, G., & Due, B. (2012). Interactional functions of invoking procedure in institutional settings. *Journal of Pragmatics*, *44*(11), 1457–1473. doi:10.1016/j.pragma.2012.06.007

Ogle, A. (2009). Making sense of the hotel guestroom. *Journal of Retail & Leisure Property*, *8*(3), 159–172. doi:10.1057/rlp.2009.7

Okumus, F., Altinay, L., & Roper, A. (2007). Gaining access for research: Reflections from experience. *Annals of Tourism Research*, *34*(1), 7–26. doi:10.1016/j.annals.2006.07.006

Paluck, E. L., & Cialdini, R. B. (2014). Field research methods. Handbook of research methods in social and personality psychology, 81-97.

Peticca-Harris, A., deGama, N., & Elias, S. R. (2016). A dynamic process model for finding informants and gaining access in qualitative research. *Organizational Research Methods*, *19*(3), 376–401. doi:10.1177/1094428116629218

Piller, I. (2012). Intercultural communication: An overview. In C. B. Paulston, S. F. Kiesling, & E. S. Rangel (Eds.), *The handbook of intercultural discourse and communication* (pp. 3–18). John Wiley & Sons Ltd. doi:10.1002/9781118247273.ch1

Reeves, C. L. (2010). A difficult negotiation: Fieldwork relations with gatekeepers. *Qualitative Research*, *10*(3), 315–331. doi:10.1177/1468794109360150

Riese, J. (2019). What is 'access' in the context of qualitative research? *Qualitative Research*, *19*(6), 669–684. doi:10.1177/1468794118787713

Robson, C. (2002). *Real world research: A resource for social scientists and practitioner-researchers* (Vol. 2). Blackwell Oxford.

Rosen, M. (1991). Coming to terms with the field: Understanding and doing organizational ethnography. *Journal of Management Studies*, *28*(1), 1–24. doi:10.1111/j.1467-6486.1991.tb00268.x

Sarangi, S., & Roberts, C. (1999). The dynamics of interactional and institutional orders in work-related settings. In S. Sarangi & C. Roberts (Eds.), *Talk, work and institutional order: Discourse in medical, mediation and management settings* (pp. 1–60). Mouton de Gruyter. doi:10.1515/9783110208375.1.1

Sharma, G., & Good, D. (2013). The work of middle managers: Sensemaking and sensegiving for creating positive social change. *The Journal of Applied Behavioral Science*, *49*(1), 95–122. doi:10.1177/0021886312471375

Sherman, R. (2005). Producing the superior self: Strategic comparison and symbolic boundaries among luxury hotel workers. *Ethnography*, *6*(2), 131–158. doi:10.1177/1466138105057546

Silverman, D. (2013). *A very short, fairly interesting and reasonably cheap book about qualitative research* (2nd ed.). SAGE. doi:10.4135/9781526402264

Spencer, G. (1973). Methodological issues in the study of bureaucratic elites: A case study of West Point. *Social Problems*, *21*(1), 90–103. doi:10.2307/799979

Tedlock, B. (1991). From participant observation to the observation of participation: The emergence of narrative ethnography. *Journal of Anthropological Research*, *47*(1), 69–94. doi:10.1086/jar.47.1.3630581

Terenzini, P. T. (1993). On the nature of institutional research and the knowledge and skills it requires. *Research in Higher Education*, *34*(1), 1–10. doi:10.1007/BF00991859

Teulier, R., & Rouleau, L. (2013). Middle managers' sensemaking and interorganizational chage initiation: Translation spaces and editing practices. *Journal of Change Management*, *13*(3), 308–337. doi:10.1080/14697017.2013.822674

Touchstone, E. E., Koslow, S., Shamdasani, P. N., & D'Alessandro, S. (2017). The linguistic servicescape: Speaking their language may not be enough. *Journal of Business Research*, *72*, 147–157. doi:10.1016/j.jbusres.2016.10.008

Urry, J. (2005). The 'consuming' of place. In A. Jaworski & A. Pritchard (Eds.), *Discourse, communication and tourism* (pp. 19–27). Channel View Publications. doi:10.21832/9781845410216-004

Van Maanen, J. (1979). The fact of fiction in organizational ethnography. *Administrative Science Quarterly*, *24*(4), 539–550. doi:10.2307/2392360

Van Maanen, J. (2011). *Tales of the field: On writing ethnography* (2nd ed.). The University of Chicago Press. doi:10.7208/chicago/9780226849638.001.0001

Veal, A. J. (1997). Research methods for leisure and tourism: A practical guide (2nd ed.). London: Ilam.

Wacquant, L. (2004). Following Pierre Bourdieu into the field. *Ethnography*, *5*(4), 387–414. doi:10.1177/1466138104052259

Wanat, C. L. (2008). Getting past the gatekeepers: Differences between access and cooperation in public school research. *Field Methods*, *20*(2), 191–208. doi:10.1177/1525822X07313811

Zimmerman, D. H., & Boden, D. (1991). Structure-in-action: An introduction. In D. Boden & D. H. Zimmerman (Eds.), *Talk and social structure: Studies in ethnomethdology and conversation analysis* (pp. 3–21). Polity Press.

KEY TERMS AND DEFINITIONS

Access: Being allowed to conduct research in an organization, dependent on ongoing cooperation.

Cold Canvassing: Terminology borrowed from sales; approaching a prospective organization without having prior ties.

Gatekeeping: Individuals at all levels of hierarchy that regulate access by outsiders to the organization.

Organization: Here: a for-profit business as a site for data collection.

Sales Pitch: Here: translating an academic proposal into the language and conventions of the target industry.

Site: Place where data collection takes place; encompasses the notion of the organization and the participating individuals.

ENDNOTE

1 Thank you to Kristina Stoewe for the reminder that this can (and will happen) in all conceivable situations, and is unfortunately not limited to cold canvassing, but also happens with prospective participants with whom you have already seemingly established rapport.

Section 4

Gender and a Cultural Impact on Research

Chapter 7
On the Way to Researching Universities:
The Case of Higher Education in Turkey

Nazlı Ayşe Ayyıldız Ünnü
Ege University, Turkey

ABSTRACT

This chapter aims to shed light on the significant experiences and challenges that have occurred during a university-funded project, exploring the dynamics of academic employment in the light of gender roles. The data for this mixed methods study comes from a research on Turkish universities, including 505 questionnaires and 46 in-depth interviews with academics from 39 different universities, located in 20 different cities of Turkey. The stratified sample of quantitative research represents the distribution of gender, department, managerial experience, position, and development level of the city, where the universities are located. In-depth interviews are used along with the survey to better understand the nature of academic employment and the implications of quantitative data. This chapter addresses the challenges, such as accessing the universities, ethical and emotional considerations, the effect of Turkish culture, censorship, and the ways that they affect the research process itself and the researchers.

INTRODUCTION

It is uncommon for social sciences scholars to write on their personal research experiences, nevertheless that might not be the case with gender studies (see Pini, 2002; Wilkinson & Kitzinger, 2013; Chege, 2019). For instance, Wilkinson &

DOI: 10.4018/978-1-7998-5826-3.ch007

Kitzinger (2013) discusses about the problem of objectivity in gender studies in the light of insider researcher concept. Additionally, Chege (2019) shares her research experiences and challenges as a pregnant woman among male beach workers. On the other hand, some studies are examining the issues that researchers face when they choose to undertake research projects that involve collecting data from universities (see Stavretis & Boucher 2001; Nobahar, Nobahar, & Hamidi 2014). However, studies on research processes have quite recently started in Turkey (see Oğuz, 2012; Şentürk, 2016). When the studies in the Turkish context are examined thoroughly, researchers can notice that the fieldwork of studies in Turkish universities has not been discussed yet.

Therefore, in this chapter, the author, as a member of the research team, presents her reflections on a university–funded project, exploring the dynamics of academic employment in the light of gender roles. Conducting a research on the gendered nature of academia and dynamics of academic employment, sheds a new light on some research issues, such as consent, censorship, rapport building, ethical considerations, and cultural barriers. Additionally, other challenges and commitments, the research team has to deal with during the data collection process, are discussed.

The chapter consists of five sections, and each presents a critical aspect of the researcher's journey while conducting this project. The first section covers the general perspective, objectives, and background of the research. It also presents some details regarding the research team and discusses the originality of the study. The second section aims to familiarize the readers with the Turkish higher education system and provides a summary of the system. The third section explores the methodology used, gives an overview of the research phases, and presents the initial challenges faced by the research team in each phase. The fourth section further presents the challenges faced by the research team throughout the different stages of the fieldwork, such as accessing the participants, dealing with ethical and security considerations. The fourth section reflects on the impact of Turkish culture on conducting research studies and presents the findings of the study. Finally, concluding marks are given in the last section. The intent is to help graduate students, and novice researchers smoothly conduct their future research studies in the Turkish context.

BACKGROUND

On the Way to Researching Universities: What Triggered Us?

We are two female researchers and coworkers from Turkey at various stages of their career journey. Like other female colleagues, many of our conversations goes around our gender roles, such as our familial duties, relations between parents and husbands, and also academic challenges that we face on a daily basis. This by itself served as a driving force to carry this research. For instance, Standing (1998) claimed that research does not take place in isolation, since researchers cannot separate themselves from their values, opinions, and past experiences. It can be said that "all research is ideological" (Letherby, 2003, p. 5). So, it went far beyond what is known as academic curiosity for us. Thus, the main motivation of this research was an attempt to understand and explain our own experiences as professional females in a Turkish academic context and employment in the light of gender roles.

Firstly, we tried to understand the paradox of women's academic employment in Turkey. Despite the relatively low levels of female participation in the total labor force and notable discrimination against women in many parts of the social life, women constitute a remarkable part of academic labor in Turkey. Besides, there is a higher proportion of female academics than in some other European Union countries.

Turkey has a low rank in international gender equality indexes. For instance, Turkey ranks 59th among 189 countries in the Gender Inequality Index (United Nations Development Programme, 2019). It is the 130st country, according to Global Gender Gap Index 2020, among 153 countries (World Economic Forum, 2019). It has the lowest female employment rate (31.4%) among the OECD countries, whereas the average of OECD countries' female employment rate is 44.6% (OECD, 2019).

However, the ratio of female academics is 44.4% in Turkey (YÖK, 2017), whereas the ratio is relatively low in EU-28 countries (41%) based on She Figures 2015 Report (European Commission, 2016). Similarly, the results of a research highlight an apparent disparity in the ratio of male to female academics in nearly all of the countries assessed. Data, provided voluntarily by institutions ranked in the top 400 of THE's World University Rankings, was collated in 2013. The widest gender gap was found in Japan, where women make up just 12.7% of the academics at the country's top-rated universities. This problem persists even in Scandinavian countries, such as Sweden (36.7%), Norway (31.7%), and Denmark (31%). However, Turkey, in which 47.5% of staff at the top five universities were females, came close to achieving an equal gender split (Grove, 2013).

On the other hand, the ratio of women at senior/managerial positions is significantly low in Turkey. For instance, only 9 out of 174 universities have female rectors (Kadın Adayları Destekleme Derneği, 2015). Besides, the percentage of female professors is 30.2%, whereas the female research assistants and lecturers are 51.72% and 57.09%, respectively (Devlet Personel Dairesi Başkanlığı, 2015). Additionally, female academics are mostly clustered in academic fields of nursing, pre-school teaching, and special education, which are constructed as a woman's job by society. So, we decided on researching the gendered nature of Turkish academia thoroughly.

The field studies on female academics and gender roles in Turkish context include a single university (see Dikmen & Maden, 2012, Ergöl, Koç, Eroğlu, & Taşkın, 2012; Poyraz, 2013), comparison of two universities (see Özkanlı & White, 2008; Kahraman, Kahraman, Ozansoy, Akıllı, Kekillioğlu, & Özcan, 2014) or a single department (see Özgüç, 1998). On the other hand, the theoretical studies (see Acar, 1993; Bildirici, Anafarta, & Aykaç 2003; Özbilgin & Healy, 2004) explain the visibility of women in academia in the light of Kemalist modernity. The Kemalist modernity project used the public visibility of women as a strategic mechanism to show Turkey's new secular character. Women were perceived as the carriers and symbols of secularism and modernity. Thus, "women's rights, liberation and visibility" became critical dimensions of Kemalist discourse (Gole, 2004; Özçetin, 2009). However, explaining the visibility of women in academia in the light of Kemalist modernity is not enough to have a detailed insight thus, we aimed to find answers to the following questions by utilizing a representative sample of 505 academics and 46 in-depth interviews.

- Why are women employed in academia at a higher rate than other occupational sectors?
- What are the effects of gender roles at universities?
- Are the reasons behind the choice of an academic job is different for men and women scholars?
- How can we overcome the problem of under-representation of women on boards and heads of institutions?
- What can be done to promote equal access for women to decision making bodies?
- How gender segregation of academic careers can be eliminated?

A BRIEF NOTE ON TURKISH HIGHER EDUCATION SYSTEM

One of the influences on women's academic employment involves the institutional level. So, an overview of the Turkish higher education system is given to familiarize the readers with the context. The Council of Higher Education (CoHE, YÖK in Turkish) is the main body responsible for higher education in Turkey. It decides and coordinates main administrative and financial issues such as appointment and dismissal of academic staff, student intake, supervision of university budgets, disciplinary issues etc. Interuniversity Council and the Ministry of National Education, with limited impact, are the other coordinating mechanisms in the Turkish Higher Education System (Yağcı, 2010, pp. 588-589).

There are both public and private universities in the higher education system. Faculty members include professors, associate professors, assistant professors, and the other ranks, which are nominated as ancillary academic staff, such as research assistants, lecturers, and instructors (Gürüz, 2012, pp. 298-299).

The remuneration of academic staff that has civil servant status in public universities, is regulated by legislation, whereas in private universities, pay scales are set by the university administration. So, in public universities, there is no pay difference between men and women who hold the same position and title. Furthermore, in public universities, assistant professorships are temporary positions, whereas professors and associate professors have tenure. Private institutions usually function by calling part-time teaching staff members in public universities to work with them. They may also hire full-time teaching staff with a different social insurance mechanism, but with the same career development structure. In general, private universities do not offer permanent positions.

Moreover, considering the current status of the Turkish higher education system, there is an increase in number of universities and students as well as the need for academics. However, the number of academics remains insufficient due to inadequate opportunities for scholarships. Since becoming a Bologna Process country in 2001, the number of universities has increased from 77 to 206 in 2018, and the number of students in the first (including the associate degree) and second cycles has risen from 1,573,281 in 2000/2001 to 7,465,221 in 2017/2018 (Ayyıldız Ünnü & Şentürk, 2020: 73). However, the total number of academics, including both faculty members and ancillary staff, is only 158.098 (Akyol & Tanrısevdi, 2018, pp. 132). This leads to an increase in the academics' teaching and supervision workloads and gives a teaching-oriented character to Turkish academia (Akyol & Tanrısevdi, 2018, pp. 135-136).

Meanwhile, the shift from teaching-based universities to research-based ones all around the world has ended up with a project of CoHE and the Ministry of Development, called "Mission Differentiation and Specialization Aimed at Regional Development" in 2015 (CoHE, 2018). The major aim of the project is to increase the regional contribution of universities and promote them to specialize in specific fields by transforming them into research-based universities. Thus, competitive allocation by the government has lead to increasing concentration of resources in so-called "research-intensive universities." (Ayyıldız Ünnü & Şentürk, 2020: 72). Additionally, public universities have started to add on an extra criterion for promotions with the aim to increase the quality level of their academic staff and enhance their efforts to be in the list of research-intensive universities. In parallel, public universities have recently announced norm cadre practices, which determine the quantity and quality of the optimal staff that a university and its units need to carry out its previously determined tasks. This puts great pressure on both university management and academics, yet these practices would foster competition among academics to get permanent university positions (Ayyıldız Ünnü & Şentürk, 2020: 72).

Additionally, the Regulation on Academic Incentive Allowance was put into effect since 2015 to increase the academic productivity of researchers (Yücel & Demir, 2018, p. 39). Within the scope of the regulation prepared in accordance with the Higher Education Law, a maximum total of 100 points can be obtained with a maximum of 30 points from each of the following activities: project, research, publication, design, exhibition, patent, citation, conference participation, and award. Since 2016, financial support has been provided to the academic staff at different levels and the rates were indicated in the regulation. An academic can get a maximum of 30 points from each activity and a maximum of 100 points in total. Evaluating committees are established in every higher education institution to evaluate the contributions of academics, regardless the academics are not always satisfied with the commission's practices (Yücel & Demir, 2018, p. 39).

In the light of these changes and developments, Turkish academics find themselves in a tense academic competition, while dealing with complex and multidimensional tasks. They face a critical dilemma of choosing between "dealing with consistently increasing teaching, supervision, administrative workloads" or "focusing on research activities," and ending up with an increased workload and time constraints.

METHODOLOGY

An overview of the research phases and the initial challenges, faced by the research team are presented in this section. Besides, the solutions generated by the researcher(s) are emphasized as well for novice scholars. Table 1 gives a brief information on the research design.

Table 1. An overview of the research

	Qualitative	Quantitative	Details
Methodology	In-depth interviews (n=46)	X CAPI (N=505)	Qualitative data was analyzed via MAXQDA 12 program Quantitative data was analyzed via SPSS 20
Location	20 cities, 39 universities – Turkey		
Sampling Method	Stratified sampling for Quantitative, Maximum variation sampling for Qualitative part of the research		
Data Collection Method and Period	Surveys and in-depth interviews / 19.02.2016- 07.06.2016		

Sampling, Not an Easy Task: Thanks to Stratified Sampling

Firstly, we had to decide on the sampling method. Thus, documentary data on academics' gender distribution to the universities, departments, ranks, administrative positions, and cities were collected from the annual reports of Student Selection and Placement Centre (Ölçme, Seçme ve Yerleştirme Merkezi in Turkish) and Association for the Support and Training of Women Candidates (Kadın Adayları Destekleme Derneği in Turkish). The first challenge had come on the scene in this preliminary phase. Some of the critical data were missing, so the websites of the universities were used to source missing data. In some cases, researchers directly contacted universities to get relevant information. Throughout this phase, we searched for the current status of public and private universities. The universities, which were founded after 2013, had not started education and training, and/or their staff policies had not been clear at that time-point. Thus, we decided to focus on the universities, which were founded before 2014 to serve the context of our research. The number of female academics was around 45.500 (Ölçme, Seçme ve Yerleştirme Merkezi, 2013a) so, we assumed that the sample size should be about 500. However, the distribution of female scholars among universities, departments, and positions were not equal. Besides, there was a limited number of female academics at some universities that are located in small cities (Ölçme, Seçme ve Yerleştirme Merkezi 2013b), whereas as the number was very high in some cities. This was another challenge and might have ended up with the *risk of overrepresentation and/or underrepresentation* of female dominant departments, universities, and cities. As a result, we segmented the faculty members based on the numerical data provided by SSPC to overcome this problem. Mutually exclusive sub-groups of academics were generated based on the female academics' ratio of universities, departments, positions, development level of the cities, and university types. Finally, 20 cities, 39 universities were determined based on stratified sampling, and at least one university was targeted from each region (See Figure 1 and Table 2, respectively).

Surveying and Interviewing in the Shadow of Time and Mobility Limitations: Take Help, but Do Not Lose Control

After the sampling phase, a survey was conducted with 505 academics (266 female, 239 male) in 39 different universities, located in 20 different cities of Turkey. Academics were asked about their demographic and familial information, academic and managerial experiences, workload, motivational factors, opinions regarding vertical and horizontal segregation in academia and the effect of gender roles. Scales were generated based upon previous measures, review of the relevant literature, and preliminary interviews.

Figure 1. Cities and the universities targeted based on stratified sampling

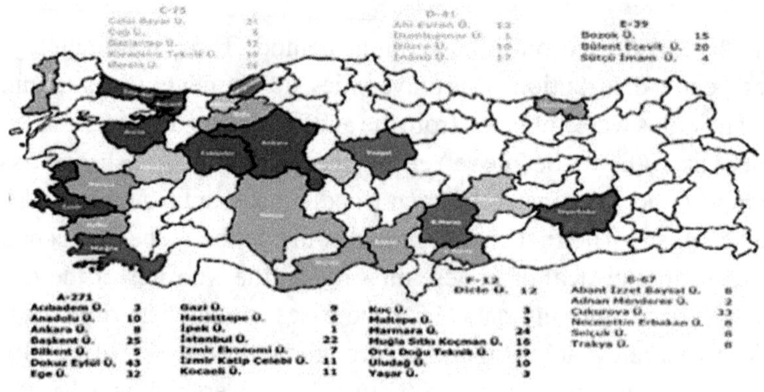

Table 2. The groups, universities, disciplines, positions, number of male/female academics targeted based on stratified sampling

Group	University	N F:Female M:Male	Natural Sciences	Social Sciences and Art	Medical & Health Sciences	Administrative Sciences & Law	Engineering	Educational Sciences
A	ODTU, Istanbul, Ege, Mugla, Bilkent, Koc, Ozyegin, Baskent, Fatih, Nisantasi, Izmir Ekonomi, Anadolu (N=12)	130 F, 170 M	Res Asst: 15 Asst: 7 Assoc: 4 Prof: 4	Res Asst: 35 Asst: 15 Assoc: 10 Prof: 10	Res Asst: 50 Asst: 20 Assoc: 10 Prof: 20	Res Asst: 25 Asst: 15 Assoc: 5 Prof: 5	Res Asst: 10 Asst: 5 Assoc: 2 Prof: 3	Res Asst: 10 Asst: 10 Assoc: 5 Prof: 5
B	Cukurova, Selcuk, Abant, Trakya (N=4)	30 F, 40 M	Res Asst: 2 Asst: 1 Assoc: 1 Prof: 1	Res Asst: 5 Asst: 5 Assoc: 3 Prof: 2	Res Asst: 10 Asst: 4 Assoc: 2 Prof: 4	Res Asst: 5 Asst: 5 Assoc: 3 Prof: 2	Res Asst: 2 Asst: 1 Assoc: 1 Prof: 1	Res Asst: 3 Asst: 4 Assoc: 2 Prof: 1
C	Karadeniz Technical Uni, Mersin, Gaziantep, Toros (N=4)	19 F, 26 M	Res Asst: 2 Asst: 2 Assoc: 1 Prof: 1	Res Asst: 3 Asst: 3 Assoc: 2 Prof: 1	Res Asst: 5 Asst: 2 Assoc: 1 Prof: 2	Res Asst: 3 Asst: 2 Assoc: 2 Prof: 1	Res Asst: 2 Asst: 1 Assoc: 1 Prof: 1	Res Asst: 2 Asst: 3 Assoc: 1 Prof: 1
D	Cumhuriyet, Hitit, RTEi Duzce (N=4)	17 F, 23 M	Res Asst: 2 Asst: 1 Assoc: 1 Prof: 1	Res Asst: 3 Asst: 2 Assoc: 2 Prof: 1	Res Asst: 5 Asst: 2 Assoc: 1 Prof: 2	Res Asst: 2 Asst: 2 Assoc: 2 Prof: 1	Res Asst: 1 Asst: 1 Assoc: 1 Prof: 1	Res Asst: 2 Asst: 2 Assoc: 1 Prof: 1
E	Ataturk, Ordu, Aksaray (N=3)	13 F, 17 M	Res Asst: 1 Asst: 1 Prof: 11	Res Asst: 2 Asst: 2 Assoc: 1 Prof: 1	Res Asst: 3 Asst: 2 Assoc: 1 Prof: 2	Res Asst: 2 Asst: 2 Assoc: 2 Prof: 1	Res Asst: 1 Asst: 1	Res Asst: 1 Asst: 2 Assoc: 1 Prof: 1
F	Dicle, Sirnak (N=2)	7 F, 8 M	Res Asst: 1 Asst: 1	Res Asst: 1 Asst: 1 Assoc: 1	Res Asst: 2 Asst: 1 Assoc: 1 Prof: 1	Res Asst: 1 Asst: 1	Res Asst: 1	Res Asst: 1 Asst: 1

In the following phase, 46 (24 female, 22 male academics) semi-structured in-depth interviews were used along with the survey to have a better understanding of the given situation of the gendered nature of academic employment in Turkey and the implications of quantitative data collected for this research. A diverse group of academics was targeted in terms of professional and managerial experiences, gender, age, marital status, academic ranks, disciplines, and public/private university experiences to reach various perspectives, consistent with the maximum variation sampling method. I suggest the use of a wide range of informants as this is one way of triangulating via data sources. Here individual viewpoints and experiences can be verified against others and, ultimately, a rich picture of the attitudes, needs or behavior of those under scrutiny may be constructed based on the contributions of a range of people.

I strongly recommend conducting pilot interviews as well. Thus, 4 pilot interviews were conducted to develop the final interview structure and to revise the question list. Data collection continued until theme saturation was reached. As the growing publication pressures, administrative work, long teaching and supervision hours, and the bureaucratic nature of universities have ended up with heavy workloads for Turkish academics, and it was challenging to schedule a time for the interview. Thus, to complete the sample and to fill gaps in the participant characteristics, snowball sampling was also used. All interviews were conducted based on the consent of the respondents. Interviews lasted between 30 and 65 minutes and were audio-recorded with the permission of participants and fully transcribed by the authors. MAXQDA 12 was used for qualitative content analysis. Directed content analysis approach was adopted as the codes were derived from both theory and relevant research findings (Hsieh & Shannon, 2005).

Due to the scarce time, limits of physical mobility, and the impossibility of visiting so many academics from different cities, the help of a research company was taken during the realization of the second and third research phases. The responsibility of this company was to conduct both the surveys and interviews with the academics that the researchers could not visit.

Although the research company members were experts in surveying and interviewing, they were given information and trained to control and ensure the consistency of survey and interview approaches. Besides, an interview protocol, which guided the implementation of the interviews to ensure consistency, was used and shared with the research company members to increase the reliability of the findings.

Self-reflexivity at Every Stage: Take Notes and Observe Your Reactions

As self-reflexive researchers, we tried to examine our impact on the research design. In doing so, we tried to find out which types of knowledge are readily available, and are likely to be shielded or hidden by academics. Thus, the in-depth interviews and ongoing verbal/non-verbal communication processes were significant for us.

In this regard, the majority of the in-depth interviews were undertaken by us. The reflective notes are taken during and shortly after each interview to document our observations related to the research and research questions. These notes enabled us to observe and be aware of our thoughts, feelings, and reactions during data collection and analysis. These notes are shared with the other research members to track critical concepts, themes, key quotes, and similarities across the data. The discussions based on reflective notes helped us to see the patterns, appearing to emerge, in the data collected.

OTHER CHALLENGES AND RECOMMENDATIONS FOR NOVICE RESEARCHERS

In this section, the other challenges, such as accessing the universities, ethical considerations, cultural barriers, building rapport, and censorship, are discussed. Besides, the solutions generated by the research team and recommendations for novice researchers are presented. As culture plays a significant role in interpreting the research findings and shaping the behavioral patterns, attitudes, values, and personalities of people and organizations, the effect of Turkish culture is discussed in a more detailed manner. So, it is aimed to emphasize the importance of culture and to guide novice researchers, who are planning to conduct research in the Turkish context as well.

Accessing the Universities and Participants: Thanks to the "Acquaintances"

Researchers and the external research company members had major obstacles in entering the university campuses. One has to pass the security checkpoint to enter the Turkish universities. If the person is not a student or employee of the university, a university staff should call the security checkpoint or escort the visitor. Neither our university ID cards nor ethical approval documents were sufficient to gain access to private universities and some of the public universities. The greatest obstacles were faced in entering the private universities located in Istanbul, specifically. There were security problems in Istanbul due to recent attacks in different parts of the city at that point in time. As well as that the private universities, operating as businesses, face higher competition in the higher education market when compared to public universities. Thus, these universities might have been more suspicious when compared to public ones and tried to keep their gates closed due to security, sustainability, and legitimization concerns. Besides, it was really very hard to reach participants from Medical and Health Sciences as they were out of the office generally. The academics from social sciences were more willing to participate in this research, as they know how challenging fieldwork can be and as they were more familiar with the research questions

We mainly followed two strategies to overcome these problems. The first one was to send an e-mail, explaining our aim to do the survey to all possible faculty members according to the strata and wait for the answer. When we got a response, we could have gained access through the respondent. Although the number of responses was very low, we did not have much to do. The second strategy was to find an "acquaintance" in the determined universities. In this case, we kindly asked our acquaintance to help us in entering the university, and when we entered, we started to search for the academics according to our sample and hold the surveys. We realized that acquaintances were very important to reach our targets in the academic setting. When our research was introduced as "a research of a friend," unknown academics were more willing to participate. As Dubetsky (1976) suggested, the behavior of people to strangers or to the people whom they were close was not supposed to be the same. The request of an acquaintance feels more secure than a stranger, and people feel more responsibility in responding to it. Thus, we had the chance to build rapport with the help of acquaintances.

In this context, I suggest that scholars and novice researchers should broaden their networks and increase the number of acquaintances by attending conferences, workshops, seminars, project groups that offer participants the opportunity to be more visible in academic contexts, and create long-term relationships.

Ethical and Security Considerations: "Tailored" Documents for the Authorities

We had applied for ethical approval document, and it took three weeks to prepare all the necessary documents. We got the document, declaring that our research meets the ethical standards, and then we tried to enter the universities with the scanned copies of it. Despite of our expectations about the sufficiency of ethical approval in gaining access to universities, we were asked for different type of documents just for being able to enter to the university campuses and the buildings at each level of the academic administrative hierarchy (from the security people at the entrance to the dean of the faculties). At some universities, security staff wanted to directly communicate with the other members of the research team. The university authorities, whom were contacted in advance to request their help to reach their staff, had asked for the details of the research design. They not only asked about the research questions, but also the explicit and implicit expected results. As we were surprised by such demands, these were more than enough to show the self-security concerns and worries of the academics and university management.

As emphasized above, we found an acquaintance in the determined universities, and we also prepared detailed documents to answer the questions of university authorities. Each document was restructured to meet the questions of the university contacts. Thus, I recommend novice researchers to be aware of the general environmental conditions (e.g., political, economic, socio-cultural, legal) that may enable or disable the research process. Therefore, they should be flexible and adapt their research style to the requirements of the organizations, authorities, and participants.

Censorship: Assure Confidentiality

Academics, whom we did not contact beforehand, were quite skeptical about answering our questions. Actually, we were surprised at first, as we thought that universities are arenas, in which people dare to think, challenge, and question freely. However, we realized that the national culture in general and the in-group collectivistic values in specific had a great influence on the daily behavioral and thinking patterns of people, regardless of their profession and/or work setting. To overcome this skepticism and censorship concerns, we did not ask their names. Although we did not request their names or wrote them on the surveys, academics kindly asked us not to use their names to protect their job status. Moreover, we received calls and e-mails from the participant academics, interviewed by the research company, to verify the research team members. We again assured them that the information they gave us is confidential, and their identity will not to be revealed. So, the participants could contribute ideas and talk about their experiences without fear of losing credibility in the eyes of administrators and/or the organization. I used my verbal and non-verbal communication skills that have progressed due to many years of teaching to encourage participants to share information. Besides, it should be made clear to participants that they have the right to withdraw from the study at any point, and they should not even be required to disclose an explanation to the investigator.

Time Constraints Due to Heavy Workloads: Show That You Are One of Them

A large group of academics stated that they do not have time to participate in this survey. Many of them also asked about not only the types and categories of questions, but also the number of questions. Thus, some critical findings of the research regarding heavy workloads of Turkish academics are given in this sub-section. The goal is to familiarize the readers with the time pressures emphasized by all the participants. For instance, lecturing appeared as the most time-consuming academic activity, with a mean of 3.99, as the academics were asked to score a set of academic activities on a scale 1 through 5 (score 1 indicates never takes time, 5 indicates takes most of my time). In Turkish higher education system, academics are not allowed to lecture before completing their Ph.D. Following their Ph.D., the minimum hours for lecturing have to be 10 hours, and they are paid for each extra hour above 10 hours. In many of the cases, the academics are not able to decide their lecturing loads since the departments have to cover graduate and undergraduate programs with a limited number of academics (Ayyıldız Ünnü & Şentürk, 2020, p. 75).

Additionally, 30% of the respondents among the faculty members hold an administrative position, and 32% of them said they used to have an administrative position in universities. For example, compulsory teaching hours, graduate students' supervision, and responsibility of publication remain for the faculty members who hold administrative positions such as the chair or vice-chair of a department, institute, research centers, or faculties. Academics, with such managerial positions, mentioned the difficulty of conducting research, making publications, and teaching simultaneously with their administrative jobs. 74.5% of the participants said that they work on their research and publication in their homes as they do not have time since teaching and administrative roles and managerial duties take all their time during working hours at university. This situation leads to a very limited time for social and family life and other activities. When academics were asked about their activities after their work, 44% said that they do not have time for other things or people (such as household tasks, socio-cultural activities, activities with family members and/or friends, etc.) due to their academic tasks (Ayyıldız Ünnü & Şentürk, 2020, p. 76).

Professors and associate professors have tenure; meanwhile, assistant professorships are employed on a contractual basis in Turkish academic employment system. The contract is renewed based on publication performance, and private universities set more demanding requirements for contract renewal.

In this context, growing publication pressures arising from demanding criteria for contract renewal, administrative work, teaching and supervision hours, the unpredictability of academic positions, and the bureaucratic nature of universities were reported as the factors, leading to heavy workloads in Turkish academia in this research. Surveys and in-depth interviews clearly suggested that academics, regardless of their department, university and academic post, have to limit their social and family life for the aim of fulfilling the publishing requirements as they have apparently no time for to do so in work hours since their daily agendas are filled with teaching, supervision and administrative tasks (Ayyıldız Ünnü & Şentürk, 2020, p. 80). They also have to limit their voluntary activities, such as being a research participant.

While conducting the surveys and interviews, we realized that some of the academics were getting really stressed and impatient due to the aforementioned time constraints. So, to overcome their impatience, we tried to create emotional bonds and give examples of our time constraints as well to encourage them to keep on answering the questions. We also emphasized how challenging fieldwork could be and showed our gratitude to all participants, who made it possible. Showing empathy, being transparent, kind, grateful, and allowing autonomy to the participants are recommended for novice researchers.

Cultural Barriers: Know the Cultural Context

As mentioned earlier, acquaintances are very important for potential participants to feel more secure and respond to our surveys. This result is quite reasonable, considering the effect of our national culture. In various studies, it is found out that the most dominant characteristic of Turkish culture is collectivism (Hofstede, 1991, 2001; Kabasakal & Bodur, 2007; Aycan et al. 2000; Paşa et al. 2001). In collectivistic cultures, the identity of members is shaped primarily by their group. In comparison, with individualistic cultures, personal choices, and achievements shape the identity of members (Hofstede, 1980; House et al., 1999; Sargut, 2001). In collectivistic societies, the relationships have a moral base (Sayar, 2003), and collectivists place a premium on maintaining relationships, place more emphasis on obligation and loyalty (Sullivan, Mitchell, & Uhl-Bien, 2003). The collectivistic nature of Turkish culture has also been attributed to the religion of the population. In Turkey, 99% of the population is Muslim, and approximately 77.5% of which is Hanafi Sunni (International Religious Freedom Report, 2018). The social and ethical obligations of a Muslim are based on the belief that the Islamic community is a brotherhood, and this notion of collectivity is stressed in the Sunni doctrine (Pellegrini & Scandura, 2006, p.265). Thus, social exchanges in general, and leader-member exchanges in specific become vital for Turkish society (Wasti, 2003; Pellegrini & Scandura, 2006). Self is developed as a relational one in a collectivist context so, shame originates from the internalization of cultural moral standards. Thus, shame mainly operates as a social-control mechanism (Bedford & Hwang, 2003, p. 135).

Thus, these aspects of Turkish society helped us to access a huge number of participants as they would feel shame when they did not respond positively to the demands of their acquaintances to maintain the relationships.

Besides, Turkey gets low scores on assertiveness, that focuses on competition, success, confrontational social relationships, and limited emotional involvement with others (Hofstede, 1980; House et al., 1999). This might also explain why participants (who were not willing to participate in the research in the first place) responded positively when the acquaintances were involved in the process. We also realized that participants were even afraid of sharing their names and were always asking the same question of whether their personal information was going to be kept secret or not. The effect of Turkish culture also plays a great role in interpreting that fear and distrust atmosphere as uncertainty avoidance level is very high in Turkish context (Hofstede, 1980; House et al., 1999). Uncertainty avoidance level refers to the extent, which individuals feel threatened by and try to avoid uncertain and ambiguous situations (Hofstede, 1980). As there is an alignment between national and organizational cultures, organization charts resemble a pyramid-like structure, and there is almost no formal horizontal communication among employees in Turkey (Sargut, 2001). In these systems, uncertainty is reduced through high-power distance, and each member accepts authority without question. This uncertainty avoidance reflects the society or organization, as the authorities' directives are accepted without question. In situations where uncertainty cannot be avoided, individuals then turn to God or the military to reduce the negative impact of uncertainty on society (Sargut, 2001). Similarly, in our study, participants and security checkpoint staff turned to authorities to act in accordance with the set directives and rules to avoid personal risk.

Another significant dimension of national culture is long-term orientation, which is added as a fifth dimension to compare national cultures (Hofstede & Bond, 1988). Its opposite pole, short-term orientation, which is highly apparent in Turkish culture, refers to fostering of values related to the past and present, specifically respect for tradition and the fulfillment of social obligations (Hofstede, 2001). We realized that even scholars behaved in a short-oriented way and were not willing to participate in a study, designed by another group of scholars, aiming to suggest future-oriented solutions to problems and challenges in Turkish academia. Kabasakal & Bodur (2007) report that Turkey is also below the world average on a similar dimension called as future orientation. Future orientation is the extent to which the society encourages and rewards future-oriented behaviors, such as delaying gratification, planning, and investing in the future (Pellegrini & Scandura 2006).

While the in-depth interviews were conducted and decoded by the research team, it was realized that that the titles and the way we talk to, especially professors (such as "Professor X", Dear Esteemed Professor") were very important for the participants. This was not also surprising for us as Turkish society, still maintains its high-power distance level (Hofstede, 1980, 2001, Aycan et al., 2000; Sargut, 2001; Bodur & Kabasakal, 2002) and the structures are designed to emphasize the power differences to overcome high uncertainty. That is why in most of the universities, either private or public, the structures are pyramid-like ones.

Due to high power difference levels, organizations, and society at large maintain inequality among members by stratification of individuals and groups with respect to power, prestige, status, wealth, and authority. Titles are used when addressing others, who are not intimate friends (Fikret Pasa et al., 2001, Kabasakal & Bodur, 2002). In organizations generally, and in most of the universities, specifically dining places, coffee rooms, parking spaces, and restrooms are separated according to the status of the employees. The way people address each other in society also reflects status differences. Lower-status people are addressed directly by their first names, whereas for higher-status people, different prefixes are added to their first names (e.g., Ali, a staff worker responsible from security in the faculty, will refer to Ayşe, an administrative staff, as Ms. Ayşe, whereas Ayşe will call him simply Ali). As well as that in most of the cases, academic staff may call administrative staff members simply with their names.

Turkish culture is also characterized by relatively higher levels of paternalism (Aycan et al. 2000; Pasa et al., 2001): This might be due to the Turkish family structure, where members are expected to comply with the decisions and directions of the father without question. Norms, which are created in the family, are extended to other institutions in society and promote acceptance of inequalities in power distribution (Pellegrini & Scandura, 2006, p. 267). Paternalism is also congruent with the values of collectivistic cultures. In the Turkish context, being concerned with the personal problems of employees is an important aspect of effective leadership, and the leader's involvement in employees' personal life is desired (Paşa et al., 2001). In paternalistic cultures, people in authority assume the role of parents and consider it an obligation to provide protection to others under their care. Subordinates, in turn, reciprocate such care and protection of the paternal authority by showing loyalty, deference, and compliance (Pellegrini & Scandura, 2006). In line with this cultural orientation, we realized that especially the academics, who have or had managerial roles, were mentioning about the welfare of their followers throughout the in-depth interviews. They were both implicitly and explicitly emphasizing that they accepted these managerial roles (Dean, Vice-Dean, Chair, Vice-Chair, Head of Institute.) for their scholars' and institutions' wellbeing, not for themselves.

RECOMMENDATIONS

Suggestions for scholars and novice researchers, who are willing to conduct research on Turkish universities and the higher education system specifically, and the Turkish context generally are offered in previous sections. However, this section speaks directly to novice researchers and graduate students and, tells them about do's and dont's of conducting gender studies and collecting data from a Turkish context.

- **Identify what triggers you:** Identify clearly what triggers you? Ask yourself these critical questions. "What value will your work have in 5 years, 10 years, 100 years?", "What serves as a driving force to carry this research." It will keep you motivated throughout the research process.
- **Identify your beliefs and define your borders:** Once you have focused your questions, sort out your beliefs about knowledge and reality, and decide on the level or levels of your gender analysis. Then you can design your research

process. Avoid superficial analysis and collecting irrelevant or exhaustive data as it endangers the research quality.

- **Develop a reflexive approach to create meaning:** Bear in mind that the field is a messy cultural context. Be ready for intensive problem solving and data collection. Convert challenges to positive field experiences through a reflexive approach. Remember that the interpretation of collected data requires reflexivity and creativity. So, take reflective notes during and after each interview and survey.
- **Know the context, but be flexible and ready for the surprises:** Be aware of the general environmental conditions (e.g., political, economic, socio-cultural, legal) that may enable or disable the research process. Be flexible and adapt your research style to the requirements of the organizations, authorities, and participants. Be open to the possibilities that can arise, remember that interesting ideas and new insights might emerge. Engage in different strategies, approaches, paradigms to learn the research process and work with the data better.
- **Establish privacy, confidentiality and anonymity:** Be aware of the academic environments to which you belong and present your research results in such a way as to prevent any future complications. Take research ethics seriously. Keep in mind that research ethics is one of the main priorities of the researcher throughout the research process.
- **Examine your impact on the research:** Seek out and report personal bias and interpretations, which differ from those of the participants studied. Refrain from making claims, which are not supported by the research methods, the data collected and the epistemology adopted.
- **Create effective communication and emotional bonds:** Reduce anxiety of the participants by means of a calm intonation, mutual understanding and eye contact. Show empathy, be transparent, kind, grateful, and allow autonomy to the participants.
- **Get support:** Get support from the acquaintances, supervisor(s), friends, and colleagues. Get the help of mentors, that share their knowledge to help you develop your research skills, and recommend extra readings, discussions, and walk-throughs of the process.

CONCLUSION

There is a growing interest in researching universities and academic staff (Ernø-Kjølhede, Husted, Mønsted & Wenneberg, 2000; Hansson & Mønsted, 2006). The primary factor that contributes to the increasing amount of research being undertaken on academic staff is the growing tendency for researchers to include their own experiences as part of their data. Besides, more and more studies examine the challenges that researchers face, who choose to undertake research projects that involve collecting data from universities (Stavretis & Boucher 2001; Nobahar, Nobahar, & Hamidi 2014). However, there is a limited number of scholars, who write on their own research processes and the challenges they face, in Turkey. So, this chapter addresses the significant experiences and challenges that had occurred during a mixed methods research while exploring the dynamics of academic employment in the light of gender roles in Turkey.

The mixed methods research method, which is identified as a useful approach in similar studies (see Johnson, Onwuegbuzie, & Turner 2007), is adopted for our study. This method combines quantitative and qualitative research, and these combinations enable confirmation or corroboration of each other through triangulation. Besides, combinations are used to enable or to develop analysis in order to provide richer data and initiate new modes of thinking by attending to paradoxes that emerge from the two data sources (Rossman & Wilson, 1985). This property is well-aligned with our interest since there was little research on the nature of academic employment in Turkey. Narrative accounts of the academics provided valuable and in-depth information about the nature of Turkish academia and universities, whereas more generalizable results were offered by numerical data.

Due to the effects of national culture in general and the nature of Turkish academia in specific, it became apparent that acquaintances are critical to reaching the participants. Accessing the participants via acquaintances could have destroyed our sample; however, we had used a stratified sampling method to overcome this probable challenge. We also realized that more formal and tailored documents should have been prepared to access the universities easily due to censorship and security concerns.

While conducting surveys face to face, the criticisms regarding questions, the design of the survey, even the length of it had arisen. In some cases, where the researcher is a student, a new researcher, or a "professional researcher" (just like in our research design), the participants may think they know more about research than the researcher. They may critique the research design, interrogate the researcher around validity, and give all kinds of advice. This may be done with the aim of helping or educating the researcher, but it may also be a way of exercising power over the researcher. A researcher may face similar problems not only in qualitative research but also in quantitative one, just as we had experienced. So, we could have only been as transparent, kind, and grateful as we can be, and we did so to encourage them to complete the surveys. Besides, I didn't take it personally and tried to see it as a help rather than destructive criticism.

Besides, as Turkish academics have limited time due to multidimensional workloads, the surveys could have been designed shorter. As well as heavy workloads, many of the participants did not want to have surveys or interviews due to fear and censorship. This situation was attributed to the job security problem arising from short-term contracts, in private universities.

Additionally, knowing about the national cultural context generally, and the higher education system and universities specifically helped us to foresee, how we can adapt to the demands of participants and authorities.

Consequently, scholars and novice researchers, who are willing to conduct research on Turkish universities and the higher education system specifically, and the Turkish context generally should first of all design their research as flexible as possible to be able to adapt new conditions, sudden changes in institutions, rules, and expectations. They should be aware of the context and cultural/political barriers and know how to deal with emotional, security, and censorship concerns to encourage participants to contribute. They should also know about the dynamics of academic employment and the nature of institutions (such as the differences between private and public universities). As it became apparent in our research process, networks and acquaintances are really powerful mechanisms in helping to conduct research. So, novice researchers should collaborate knowledgeable researchers from various academic settings, who can provide an opportunity to build up networks. Finally, they should examine their impact on the research design and phases and take reflective notes, especially while conducting interviews.

REFERENCES

Acar, F. (1993). Women and university education in Turkey. *Higher Education in Europe*, *18*(4), 65–77. doi:10.1080/0379772930180406

Akyol, B., & Tanrısevdi, F. (2018). Academicians in Turkey: An Evaluation of Current Status of Academic Staff in Higher Education. *Higher Education Studies*, *8*(4), 129–138. doi:10.5539/hes.v8n4p129

Aycan, Z., Kanungo, R. N., Mendonca, M., Yu, K., Deller, J., Stahl, G., & Khursid, A. (2000). Impact of Culture on Human Resource Management Practices: A Ten Country Comparison. *Applied Psychology*, *49*(1), 192–220. doi:10.1111/1464-0597.00010

Ayyıldız Ünnü, N. A., & Şentürk, B. (2020). All-in-One Academics: Mental Workload in Turkish Academic Employment. In Evaluating Mental Workload for Improved Workplace Performance (pp. 69-87). IGI Global.

Bedford, O., & Hwang, K. K. (2003). Guilt and shame in Chinese culture: A cross-cultural framework from the perspective of morality and identity. *Journal for the Theory of Social Behaviour, 33*(2), 127–144. doi:10.1111/1468-5914.00210

Bildirici, M., Anafarta, N., & Aykaç, E. (2003). Türkiye üniversitelerinde kadın emek gücü. *İktisat, İşletme ve Finans, 18*(206), 92-108.

Bodur, M., & Kabasakal, H. (2002). Türkiye-Arap Kümesinde Kurumsal Kültür. *Journal of Management Research, 2*(1), 2–22.

Chege, N. (2019). 'If you give me time I can love you': A Pregnant Researcher among Male Beach Workers on Kenya's Liminal South Coast Beaches. *Anthropology Matters, 19*(1), 1–34. doi:10.22582/am.v19i1.494

CoHE. (2018, June). *Mission Differentiation and Specialization Aimed at Regional Development Program.* Council of Higher Education. https://bolgeselkalkinma.yok.gov.tr/hakkinda/proje-hakkinda

Cordingley, P., & Needham, K. (2010). Chapter 15: School leaders using inquiry and research. In Campbell & S. Groundwater-Smith (Ed.), Connecting Inquiry and Professional Learning in Education: International perspectives and Practical solutions (pp. 192-199). Routledge.

Devlet Personel Daire Başkanlığı. (2015). *Üniversitelere ilişkin sayısal veriler.* Retrieved May, 5, 2017, from http://www.dpb.gov.tr/trtr/istatistik/universitelere-ilskin-sayisal-veriler

Dikmen N., & Maden, D. (2012). Kadın akademisyenlerin görünmeyen emeği üzerine bir araştırma: Ordu Üniversitesi örneği. *İstanbul Ticaret Üniversitesi Sosyal Bilimler Dergisi, 11*(21), 257-288.

Dubetsky, A. (1976). Kinship, primordial ties, and factory organization in Turkey: An Anthroplogical View. *International Journal of Middle East Studies, 7*(3), 433–451. doi:10.1017/S0020743800025745

Ergöl, Ş., Koç, G., Eroğlu, K., & Taşkın, L. (2012). Türkiye'de kadın araştırma görevlilerinin ev ve iş yaşamlarında karşılaştıkları güçlükler. *Yükseköğretim Dergisi, 2*(1), 43–49.

Ernø-Kjølhede, E., Husted, K., Mønsted, M., & Wenneberg, S. B. (2001). Managing university research in the triple helix. *Science & Public Policy, 28*(1), 49–55. doi:10.3152/147154301781781679

European Commission. (2016). *She Figures 2015 Report.* Retrieved October, 5, 2014, from https://ec.europa.eu/research/swafs/pdf/pub_gender_equality/she_figures_2015-final.pdf

Fikret Pasa, S., Kabasakal, H., & Bodur, M. (2001). Society, organisations, and leadership in Turkey. *Applied Psychology, 50*(4), 559–589. doi:10.1111/1464-0597.00073

Gole, N. (2004). Visible women: Actress in the public sphere. *New Perspectives Quarterly, 21*(2), 12–13. doi:10.1111/j.1540-5842.2004.00653.x

Grove, J. (2013). *Global Gender Index, 2013.* Retrieved May, 10, 2012, from https://www.timeshighereducation.com/cn/features/global-gender-index-2013/2003517.article

Gürüz, K. (2012). Remuneration of Academic Staff in Turkish Universities. In P. G. Altbach, L. Reisberg, M. Yudkevich, G. Androushchak, & I. F. Pacheco (Eds.), *Paying the Professoriate: A Global Comparison of Compensation and Contracts* (pp. 297–307). Routledge.

Hansson, F., & Mønsted, M. (2006). *Leadership Dilemmas and Dynamics in research projects – entrepreneurial leadership?* MPP Working Paper, No. 6.

Hofstede, G. (1980). *Culture's Consequences: International Differences in Work-Related Values.* Sage Publications.

Hofstede, G. (2001). *Culture's Consequences: Comparing Values, Behaviors, Institutions, and Organizations Across Nations* (2nd ed.). Sage Publications.

Hofstede, G., & Bond, M. H. (1988). The Confucius Connection: From Cultural Roots to Economic Growth. *Organizational Dynamics, 16*(4), 4–21. doi:10.1016/0090-2616(88)90009-5

House, R. J., Hanges, P. J., Ruiz-Quintanilla, S. A., Dorfman, P. W., Javidan, M., Dickson, M. W., Gupta, V., & … . (1999). Cultural Influences on Leadership and Organizations: Project GLOBE. In W. H. Mobley, M. J. Gessner, & V. Arnold (Eds.), *Advances in Global Leadership* (pp. 171–233). JAI Press.

Hsieh, H.-F., & Shannon, S. E. (2005). Three Approaches to Qualitative Content Analysis. *Qualitative Health Research, 15*(9), 1277–1288. doi:10.1177/1049732305276687 PMID:16204405

International Religious Freedom Report. (2018). *Executive Summary Turkey.* Retrieved 03, November, 2019, from https://www.state.gov/wp-content/uploads/2019/05/TURKEY-2018-INTERNATIONAL-RELIGIOUS-FREEDOM-REPORT.pdf

Johnson, R. B., Onwuegbuzie, A. J., & Turner, L. A. (2007). Toward a definition of mixed methods research. *Journal of Mixed Methods Research, 1*(2), 112–133. doi:10.1177/1558689806298224

Kabasakal, H., & Bodur, M. (2002). Arabic cluster: A bridge between East and West. *Journal of World Business, 37*(1), 40–54. doi:10.1016/S1090-9516(01)00073-6

Kabasakal, H., & Bodur, M. (2007). Leadership and Culture in Turkey: A Multi-Faceted Phenomenon. In J. Chhokar, F. Brodbeck, & R. House (Eds.), *Managerial Cultures of the World: A GLOBE Re- port of In-depth Studies of the Cultures of 25 Countries* (pp. 835–874). Lawrence Erlbaum Associates.

Kadın Adayları Destekleme Derneği. (2015). *Kadın İstatistikleri.* Retrieved January, 10, 2014 from http://www.ka-der.org.tr/tr- TR/Page/Show/400/kader-istatistikleri.html

Kahraman, L., Kahraman, A. B., Ozansoy, N., Akıllı, H., Kekillioğlu, A., & Özcan, A. (2014). Nevşehir Hacı Bektaş Veli Üniversitesi toplumsal cinsiyet algısı araştırması. *Turkish Studies, 9*(2), 811–831. doi:10.7827/TurkishStudies.6244

Letherby, G. (2003). *Feminist Research in Theory and Practice.* Open University Press.

Nobahar, N., Nobahar, M., & Hamidi, M. (2014). Investigation of Research Barriers and Challenges in University from the Perspective of Faculty Members of Kermanshah City. *IAU International Journal of Social Sciences, 4*(4), 15–21.

OECD. (2019). *OECD Labour Force Statistics 2019.* Advance online publication. doi:10.1787/g2g9fb3e-en

Oğuz, H. Ş. (2012). Alanda Bir Kadın. *Fe Dergi, 4*(1), 64–78. doi:10.1501/Fe0001_0000000064

ÖSYM. (2013a). *Öğretim elemanları özet tablosu.* Retrieved December, 3, 2012, from http://www.osym.gov.tr/Eklenti/5190,2ogretimelemanlarisayozettablosupdf.pdf?0

ÖSYM. (2013b). *Yüksek Öğretim istatistikleri, 2012-2013 öğretim yılı lisans eğitimi veren yükseköğretim programlarında görevli öğretim elemanlarının öğretim alanlarına göre sayıları.* Retrieved December, 3, 2012, from http://www.osym.gov.tr/dosya/1-69402/h/13ogretimalanlisansogrencisay.pdf

Özbilgin, M. F., & Healy, G. (2004). The Gendered Nature of Career Development of University Professors: The Case of Turkey. *Journal of Vocational Behavior, 64*(2), 358–371. doi:10.1016/j.jvb.2002.09.001

Özçetin, H. (2009). 'Breaking the silence': The religious Muslim women's movement in Turkey. *Journal of International Women's Studies*, *11*(1), 106–119.

Özgüç, N. (1998). Türkiye Üniversitelerinde Kadın Coğrafyacılar. In N. Arat (Ed.), *Aydınlanmanın Kadınları* (pp. 177–201). Cumhuriyet Kitap Kulübü.

Özkanlı, Ö., & White, K. (2008). Leadership and strategic choices: Female professors in Australia and Turkey. *Journal of Higher Education Policy and Management*, *30*(1), 53–63. doi:10.1080/13600800701745051

Pasa, S. F., Kabasakal, H., & Bodur, M. (2001). Society, Organizations, and Leadership in Turkey. *Applied Psychology*, *50*(4), 559–589. doi:10.1111/1464-0597.00073

Pellegrini, E., & Scandura, T. (2006). Leader-Member Exchange, Paternalism and Delegation in the Turkish Business Culture: An Empirical Investigation. *Journal of International Business Studies*, *37*(2), 264–279. doi:10.1057/palgrave.jibs.8400185

Pini, B. (2002). Focus Groups, Feminist Research and farm women: Opportunities for empowerment in rural social research. *Journal of Rural Studies*, *18*(3), 339–351. doi:10.1016/S0743-0167(02)00007-4

Poyraz, B. (2013). Akademi kadınların cenneti mi? Ankara Üniversitesi örneği. *Ankara Üniversitesi Sosyal Bilimler Enstitüsü Dergisi*, *4*(2), 1–18. doi:10.1501beder_0000000057

Sargut, S. (2001). *Kültürlerarası Farklılaşma ve Yönetim*. Verso Yayıncılık.

Sayar, K. (2003). *Kültür ve Ruh Sağlığı*. Metis Yayınevi.

Şentürk, B. (2016). Çokuz Ama Yokuz: Türkiye'deki Akademisyen Kadınlar Üzerine Bir Analiz. *ViraVerita E-Dergi*, (2), 1-22.

Standing, K. (1998). Writing the voices of the less powerful: research on lone mothers in Feminist dilemmas in qualitative research: Public knowledge and private lives. In J. Ribbens & R. Edwards (Eds.), (pp. 186–202). Sage Publications.

Stavretis, L., & Boucher, C. (2001). *Researching the researchers: Comments on the experience of doing research in universities*. RMIT Business, School of Management. Working Paper Series.

Sullivan, D., Mitchell, M., & Uhl-Bien, M. (2003). The New Conduct of Business: How LMX can Help Capitalize on Cultural Diversity. In G. Graen (Ed.), *Dealing with Diversity* (pp. 183–218). Information Age Publishing.

United Nations Development Programme. (2019). *Human Development Report, Gender Inequality Index*. http://hdr.undp.org/en/composite/GII

Ünnü, N. A. A., & Şentürk, B. (2020). All-in-One Academics: Mental Workload in Turkish Academic Employment. In *Evaluating Mental Workload for Improved Workplace Performance* (pp. 69–87). IGI Global. doi:10.4018/978-1-7998-1052-0.ch004

Wasti, A. S. (2003). The influence of cultural values on antecedents of organisational commitment: An individual-level analysis. *Applied Psychology*, *52*(4), 533–554. doi:10.1111/1464-0597.00150

Wilkinson, S., & Kitzinger, C. (2013). Representing our own experience: Issues in "insider" research. *Psychology of Women Quarterly*, *37*(2), 251–255. doi:10.1177/0361684313483111

World Economic Forum. (2019). *Global Gender Gap Report 2020.* http://www3.weforum.org/docs/WEF_GGGR_2020.pdf

Yağcı, Y. (2010). A different view of the Bologna process: The case of Turkey. *European Journal of Education*, *45*(4), 588–600. doi:10.1111/j.1465-3435.2010.01456.x

Yücel, A. G., & Demir, S. B. (2018). Academic Incentive Allowance: Scientific Productivity, Threats, Expectations. *International Online Journal of Educational Sciences*, *10*(1), 39–55.

KEY TERMS AND DEFINITIONS

Council of Higher Education: The main body, which is responsible for higher education in Turkey. It decides and coordinates the administrative and financial issues (e.g., student intake, appointment and dismissal of academic staff, supervision of university budgets, disciplinary issues, etc.).

MAXQDA: A program designed to facilitate and support qualitative, quantitative and mixed methods research projects. It enables to import, organize, analyze, and visualize various forms of data such as interviews, surveys, tables, pictures, etc.

Ministry of Development: An organization, which plans and guides Turkey's developmental processes with a macro approach and focuses on the coordination of policies and strategy development.

Mission Differentiation and Specialization Aimed at Regional Development: A project of CoHE and the Ministry of Development to increase the regional contribution of universities and promote them to specialize in specific fields by transforming into research-based universities in Turkey.

Mixed Methods Research: A form of research, which combines quantitative and qualitative research as such combinations enable confirmation or corroboration of each other through triangulation.

Norm Cadre: The quantity and quality of the optimal staff that a university and its units need in order to carry out its tasks.

Stratified Sampling Method: The sampling method, which is conducted by dividing the whole population into sub-populations, and then applying random sampling to each sub-population to select the representative samples.

Triangulation: The use of multiple methods, theories, data and/or researchers in the research process.

Section 5

Counseling and Psychology Education Research

Chapter 8
When Participants' Trauma Becomes Mine:
How I Used "Countertransference" to Enhance the Quality of My Study

Sigalit Gal
McGill University, Canada

ABSTRACT

This chapter is a reflection on the author's work in the context of trauma-focused qualitative research entitled "Risk and protective factors for the mental health consequences of childhood political trauma (Argentina 1976-1983) among adult Jewish Argentinian immigrants to Israel." By examining the author's emotional reactions during the process of the data collection and analysis of her doctoral study, the author will explore the challenges that she faced, as well as the solutions she employed (both the effective and ineffective). More specifically, using the lens of the psychoanalytical term "countertransference", she will discuss the manifestations of her positionality as a qualitative researcher and its impact on her engagement with her study. The author will elaborate on different strategies that she used for her study, and propose qualitative researchers to use "countertransference" as a way to understand and address the complexity of a researcher's positionality in narrative research.

DOI: 10.4018/978-1-7998-5826-3.ch008

INTRODUCTION

This exploration addresses a gap in the literature about the intersection of qualitative research and psychoanalysis by looking at my experience of conducting narrative research, and my professional history as a clinical counsellor.

Despite the similarities between the conducting of narrative research and clinical counselling, there is very limited knowledge on the intersection of the two fields of practice. There is some literature on the subject that focuses on the therapeutic effect of the biographical narrative (Vajda, 2007; Rosenthal, 2003), showing that for participants, it is a positive and healing experience, especially in the context of trauma (Bar-On, 2000; Harvey, et al. 2000), a reflection of a psychotherapist on their interactions with participants as a narrative researcher (Josselson, 1996), and the benefits of storytelling in psychoanalysis therapy (Ferro, 2006). After presenting the topic of my study and explaining the ways the narrative inquiry overlaps with talk-based therapy, definitions of psychoanalysis and the term countertransference will be provided. Then, by going through the different phases of my research (e.g., study design, data collection, data analysis) I will explore some of the personal and professional dilemmas, which have been part of this fascinating journey, as I examine the manifestation of countertransference in the context of narrative research. In this reflection, which is partially based on journaling for my doctoral study, and my reflections on my experience during my postdoctoral study, I will specifically discuss issues relative to positionality and reflexivity (Bradbury-Jones, 2007; Stronach et al., 2007; Pillow, 2003) and their contribution to the quality of the qualitative study. Finally, Strategies of managing and using my countertransference will be specified, and ways of using "countertransference" as a way to understand the complexity of a researcher's positionality in narrative research will be suggested.

REASONS FOR CHOOSING THE NARRATIVE METHOD FOR MY DOCTORAL DISSERTATION

Before I discuss the complexity of my positionality, it is important to understand the reasons for choosing the narrative method approach, and the ways the narrative inquiry overlaps with talk-based therapy, and specifically with psychoanalysis therapy.

The aim of my doctoral study was to explore the manifestation of mental health consequences of traumatic childhood experiences in the case of adult Jewish immigrants to Israel, who experienced the military dictatorship in Argentina (1976-1983) as children (Gal, 2019). As the purpose of my retrospective research was to expand our understanding of my exploratory research question, it was clear that I would need to use a qualitative methodology that captured the unique and subjective experiences of traumatized adult immigrants to Israel in historical context. My choice to use the narrative approach for my doctoral research was not coincidental, as it seems to be most suitable to a trauma-focused study, which included immigrants as my participant population. Indeed, the literature shows that studies in the fields of mental health increasingly use narrative methodologies (Davidson, 2003; Kirkpatrick, 2008). These methodologies are also employed in studies about immigrants (Polkinghorne, 1991; Knaifel & Mirsky, 2015; Morrow & Smith, 2000), as this method is effective in illustrating the subjective experiences of participants, and it is the closest research approach to talk-based therapy.

Following formal approval from the Research Ethics Board at McGill University to conduct this research, I met each of my 15 participants in Israel, for an average of 90 minutes, and similarly to clinical sessions, I assured that the interview location offered the participant confidentiality and privacy. When using the narrative method (Denzin & Lincoln, 2005) for my doctoral study, I used a two-phase narrative interview (Rosenthal, 1993) to collect my data. The first phase was an open life-story interview (main narrative) with minimal involvement by me as an interviewer. This part began with me asking the interviewee: "Please tell me your life story." This strategy of minimizing the interference of the interviewer is similar to a clinical setting, in which other than one or two opening sentences which are provided by the therapists at the beginning of the session (e.g., "So, why you are here today?"; "How do you feel today?") while leaving "the stage" to the clients by allowing them to raise topics they chose to raise, in the order, and in the ways they chose to discuss them.

Another similarity to clinical settings, is the underlying meaning to what the client chooses to mention (or not to mention), as well as is the *way* it is mentioned. The aim of this phase in my study was to identify what themes participants spontaneously raised in relation to their childhood, and in relation to their behaviors and emotions after they immigrated as adults to Israel, and the ways in which these themes were raised (e.g., particular use of words; sentences, silences; tone of voice).

The second phase in my narrative interview was a more structured interview, including some clarification questions about the first section when necessary, and a list of possible semi-structured and structured questions on topics that as the researcher I felt were not addressed during the first part of the interview. These types of questions were also similar to a clinical session, in cases when the therapist needs to clarify particular topics which remain unclear to them. Finally, demographic information was collected.

My thematic textual analysis was centered on the behavioral and emotional manifestations of these immigrants' past experiences in their current lives as adult immigrants in Israel. Like the way in which the therapists may combine the reported information with observations on clients' behavior in order to obtain an insight, I used triangulation of verbal and non-verbal data collected during the interviews in order to draw conclusions. The findings of my study revealed that participants' present-day experiences included both negative and positive long-term emotional and behavioral reactions, which are located between two extreme outcomes (Gal, 2019). At the negative end of the scale was Complex Post-Traumatic Stress Disorder (Van der Kolk, 2005), while at the positive end was Post-Traumatic Growth (Berger & Weiss, 2009). These seemingly contradictory present-day reactions are correlated with previous studies that confirm the coexistence of negative and positive long-term outcomes in this context, such as longitudinal and retrospective studies on child Holocaust survivors (Lev-Wiesel & Amir, 2006).

COUNTERTRANSFERENCE IN THERAPY AND NARRATIVE RESEARCH

There is a wealth of knowledge on psychoanalysis therapy and extensive literature on narrative research; however, the intersection of the two fields of practice is an area that has to date received little attention.

Apart from multiple similarities between the narrative interview and clinical sessions mentioned above, in both cases, a professional (the therapist or researcher) who is responsible for creating a safe environment, which will allow the narrator (the client or participant) to share intimate and personal information with an authoritative figure they trust. Therefore, it should not be a surprise that within this encounter, there may also be similar inter-personal dynamics, which may include countertransference.

The Definitions of Countertransference in Therapeutic Relationships

The term countertransference, is taken from the field of psychoanalysis. Psychoanalysis is a theory and a therapeutic technique developed by Freud and his followers (e.g. Fromm, 2013; Birnbach, 1961) for the purpose of assisting individuals' mental functioning and wellbeing (Milton, Polmear and Fabricius, 2011). It is based on the exploration of unconscious conflicts in the shape of suppressed memories thoughts, feelings and desires that may have a negative impact on individual's emotional and behavioral function (Freud and Bonaparte, 1954; Milton, Polmear and Fabricius, 2011). Apart from the therapeutic techniques of identifying the unconscious conflicts (e.g.: hypnosis, dream analysis and free association) and interpreting them (Milton, Polmear and Fabricius, 2011; Freud and Bonaparte, 1954), one of the foundations of psychoanalysis is the analysis of the therapeutic relationships and the dynamic between the therapist and the client. This dynamic includes the term "transference", which describes the unconscious ways in which the client transfers emotions for one's parents or other significant relationships onto the therapist (Freud and Bonaparte, 1954; Jung, 1969). These emotions can be negative (e.g.: anger or hostility towards the therapist) or positive (e.g.: expression of affection, love and friendliness towards the therapist). Working through transference is an important tool of making progress in the therapy (Freud and Bonaparte, 1954; Kapelovitz, 1976; Jung, 1969).

Countertransference is defined by Bouchard et al. (1995) as "the entirety of the [therapist's] emotional reactions to the patient within the treatment situation" (p. 719). In other words, the therapist's emotional and behavioral response to the client. In a Freudian context, countertransference is seen as an obstacle to therapists' objectivity, and as an emotional reaction that should be eliminated; however, recent research sees it as inevitable. As put by Dalenberg (2000):

Almost all reactions of the therapist contain both objective and subjective features, both reactions that are dependent on the patient and reactions independent of him or her, both realistic reactions and fantastic-magical-conflictual beliefs, wishes, and emotions. (p. 8)

Furthermore, if therapists can identify their own experience of countertransference, psychoanalysis sees it as a sign of successful therapy (Rudd & Joiner, 1997; Sandler, 1976). Literature shows that therapists' awareness of their countertransference is vital not only to regulate their emotions in their therapeutic role, but also as an important tool for them to explore what clients are trying to convey, and their unconscious feelings and aspirations, such as when a client irritates the therapist, tries to seduce the therapist, or manifests other unconscious provocations (Etchegoyen, 2005). As such, countertransference is important data for the therapist that can contribute to successful treatment while using therapists' reflection on their own emotional reactions (negative or positive) toward the clients, and subsequently mirroring these feelings back to them in a constructive way (Dalenberg, 2004). Therefore, the modern approach does not emphasize how therapists should eliminate countertransference, but how countertransference can be beneficial for clients. This is specifically significant when treating traumatized clients (Dalenberg, 2004).

Additional components of countertransference also take into consideration that in some cases, therapists' emotional reactions may not always be the result of clients' unconscious provocation, as they could be also related to therapists' experiences, such as their own emotional history, or their personal preference of one client over the other. Therefore, in these cases, it is advised that therapists seek counselling themselves, in order to "reset" their professional ability (Freud, 1957). In cases where they cannot manage countertransference, therapists are advised to refer the client to another therapist.

MY EXPERIENCE OF COUNTERTRANSFERENCE DURING MY WORK ON MY DOCTORAL STUDY

Understanding countertransference in my narrative research helped me as a researcher to be aware of my subjectivity and positionality. In order to understand the immense emotional impact that meetings with the participants has had on me as a researcher, it is important to understand the emotional dynamics of these interviews.

Containing Intense Emotions for Thirty Hours of Interviews

Given that the topic of my doctoral study was the long-term impact of childhood trauma, the content of the interviews was focused on two periods in the participants' lives: childhood, during which they were exposed to the political trauma during the military dictatorship in Argentina (1976-1983), and immigration to Israel (Gal, 2019). Participants' narratives were not told chronologically, but rather as a back and forth movement between the past and their present lives.

For most participants, interviews for this study marked the first time they had revealed many details which they had kept to themselves for decades. Therefore, for most participants, this was an intense emotional experience during which some of them cried and at times became emotionally detached when describing traumatic events. Other participants also exhibited somatic reactions during interviews, such as complaints of feeling cold or tired or experiencing sudden headaches. In some cases, participants' strong emotional reactions continued after the interviews were finished.

Although listening to traumatic stories was not new to me as a clinical counsellor, it was not something, which I had to address in every session, or every day. Therefore, listening to 15 different individuals who shared traumatic experiences, for more than 30 hours, was not an easy thing for me to do emotionally, nor the research tasks, which I had to conduct after the data collection has been completed.

As if I Was "Them", as if I Was "There"

Despite the long time that had passed since they experienced these traumatic events, most of participants' descriptions were conveyed in great detail, and with immense emotional intensity. The manifestations of this intensity were explicit (e.g., sobbing) and less explicit (e.g., silence, avoidance). As a result, during the interviews I was often able to imagine the horrendous events described. I could also feel the fear, the anxiety and the helplessness that the participants conveyed strongly and vividly during the interviews, despite the fact that they had experienced these events more than four decades ago. I often found myself "there" with them during "that time". Whether it was a nine-year-old boy observing a public execution from a crack in his bedroom window, or a seven-year-old child who had been randomly taken by the Junta [The Argentinian security forces who systematically persecuted the left-wing underground organizations and their alleged supporters during the dictatorship] on his way to school and put in prison for the purpose of interrogation, without his parents knowing, for several days with no food or water. I could also imagine the horror of a six-year-old child whose father "vanished" one day, after being taken away by the Junta, as a result of a mistake of identification.

While listening to participants' stories of trauma and survival, I often asked myself, what I would have done in the same extreme situations of political violence and persecution that participants experienced as children, and if I would have reacted in the same way. For example, in my reflective notebook I wrote: "What it feels like for a little girl to escape with your parents without being able to say goodbye to her friends at school and How can a child keep such a big secret?"

Another example was my reaction after interviewing one of the participants who was severely traumatized during the dictatorship, and to this day does not leave his home without "a survival kit". At the end of his interview, he showed me the balcony at his apartment, while explaining that "this is the only place where l feel that I am alive, rather than just surviving, because here I can look at the horizon". Shortly after the interview ended, I was making my way home, gazing from the big train windows as the city's lights passing by, wondering what makes a specific place safe enough for one to feel "alive" and how one knows where and how to find it. I wrote in my notebook: "[He]carries survival kit...where is the place where I feel most like me, where do I feel most alive, where do I feel safe?"

The Bittersweet Taste of My Nostalgia

As noted earlier, countertransference can also be the result of the therapist's own "emotional baggage". As my research population was immigrants, having spent some time in Canada while completing my studies enabled me to take on the role of an "insider", which included the benefit of investigating the familiar (Padgett, 2008 Hellawell, 2006; Dwyer, & Buckle, 2009). At the same time my identification with participants' experiences as cultural minority suggested some bias. Listening to participants' immigration experiences almost automatically made me think about my own experience as a minority in Canada, and raised questions which I wrote in my reflective notebook:

How did my personal experience as an Israeli native in Canada compare to my participants' experience in Israel? Was my experience easier or harder, similar or different compared to theirs? Can I even compare our experiences considering the different circumstances?

Nonetheless, despite participants' unique immigration circumstances, I found myself empathizing with participants' descriptions of their feelings of being or labelled as "other", and with their experience of being a part of a cultural minority. Other topics that I have found myself relating to, were the sense of being a part of two cultures at once and the sense of being simultaneously "both here and there": feeling like a "Canadian" when visiting Israel and feeling like an "Israeli" when visiting Canada.

In addition to the above, while monitoring my emotional reactions during the interviews I noticed that discussion of Israel had a strong impact on me, given my strong attachment to my homeland, and my identity as an Israeli-Jewish person. Although during the interviews I could at times theoretically agree with some of my participants' criticism about Israel at present time, at times I felt uncomfortable. These strong emotions were also raised during the process of interview transcription and the data analysis. When trying to analyse my reactions, I realized, that I had been nostalgic about my homeland.

When the Political Violence in Argentina and in Israel Becomes One

During the data collection and analysis, I also realized that the emotional intensity of participants' narratives evoked in me strong emotions in relation to my own personal history of situations during which I felt an extensive fear, anxiety, and helplessness, for example, in the context of my childhood experience in Israel of the Yom Kipur War (1973), and in relation to my daily exposure as a journalist to death and bereavement during the 1st intifada in Israel (1987-1993). In a way the impact of the political violence during the military dictatorship in Argentina, as described by the participants, had "merged" with my memory as a child and as an adult and triggered it.

Furthermore, I found that the strong impact that participants' narratives has had on me extended beyond the time during which I was directly engaged with the data. Some nights I dreamed about what participants described to me, and during the day, I found myself thinking about their stories even when engaging with other tasks unrelated to my study. During data collection and analysis, I also found it very difficult to be engaged with leisure activities, such as meeting with my friends and discussing topics, which were unrelated to my research work.

While I was content that I was so deeply engaged and passionate about my study, I was also concerned that my emotional involvement in it may blind me from seeing and identifying elements which deserved attention during my analysis and that this may have an impact on my conclusion and eventually on the dissemination of the results.

Who Am I to Judge Anyone?

An example of such a concern was the different levels of empathy which I felt toward each of the participants. As explained earlier, in clinical settings, therapists' countertransference can be impacted by their level of comfort with specific clients. As some clients may evoke (unintentionally) in the therapist specific negative or positive emotions, each one of us may connect better with some individuals than with others. My extensive emotional investment in the data collection and analysis also included feelings of either anger or an admiration toward particular participants, whether it was due to the way they described their reactions, the nature of their reactions, and even the tone of their voice. Also, during the interviews, I felt that I was able to show more empathy with participants who were closer to my age than with participants who were much younger. Furthermore, as a woman, I felt more attentive during interviews that were conducted with female participants.

This reaction made me feel "professionally guilty", as I expected myself to be more objective. I was therefore concerned that these personal preferences of one participant over the other, or one narrative over the other, may impact my interpretation and my judgment during the analysis and on my decision to include in my dissertation specific sections from one interview while leaving out or dismissing another. I expressed my concern in my reflective notebook: "Who am I to judge anyone? Every experience of each of the participants has had an impact on someone lives". It was only later that I realized that given the limited length of my dissertation and my articles, it would have been impossible to provide an equal representation to all the fifteen participants, and for me to equally empathize with each of the participants. I was also glad to see that later, during the process of the process of my work, these subjective empathy barriers during some of the interviews had become less and less significant and impact (whether consciously or unconsciously) during my data collection, and in the process of analysing the data and retrieving conclusions, as my analysis included other criteria than my emotional reactions to the participants and their narratives (e.g.: relevance of the narrative to the thematic textual analysis etc.).

Strategies of Managing and Using Countertransference

In order to use my countertransference for the purpose of enhancing the quality of the study, I had to employ various strategies that would help me to be aware of, master, modify, and regulate the impact of the interviews on me during my work on my study.

Strategy 1: managing my reactions by using talk-therapy skills

Similar to the way therapists should make sure to focus on their clients at all times, while being aware of their countertransference, I was able to retain awareness of my emotional reactions and behavior during interviews, while making sure that the focus in these interviews stayed on the narrators. As explained by Bar-On (1996), in the context of narrative research and ethics, interviewers should distance themselves emotionally from the narrators and their narration, yet at the same time to also be empathetic and critical. Although I do not think that distancing myself emotionally from the narrators and their narrations would be useful or even realistic, a creative way for me to address my subjectivity and positionality was to alternate simultaneously between my awareness of my emotional reactions and my ethical responsibility as a researcher in the interview room. During the interviews, I was aware of each participant's personal journey, as well as my own journey. I kept this self-analysis to myself mainly by writing notes in my notebook after interviews, while maintaining the focus of the interviews on the interviewees. Although I could learn from the interviews about myself, it was clear to me that these interviews were not therapy sessions for me, and that addressing my own memories and reflections should be done in a different setting. I therefore used my skills as a clinical counsellor and put an effort in to control verbal and non-verbal reactions that may have indicated my emotions, although I am sure that some participants could feel that I was touched.

While managing reflections on my own emotional history, my attention was also directed toward the objective of my research, and my ethical responsibilities in it. Therefore, I used my verbal and non-verbal counselling skills to encourage and induce participants to share information about the long-term impact of their childhood experiences. However, given the sensitivity of the topic and its relation to trauma, and given that the purpose of the research interview was not to provide therapy for the participants, the aim was not to encourage the participants to "dig deep" into their traumatic painful memories, but rather to give them the freedom to choose what narratives to share, if they chose to share at all, while I tried to maintain focus on the structure and the empowering nature of a research interview.

Although these interviews were not aimed to be therapy sessions, I often found myself using my theoretical and practical experience in providing psychosocial support, by providing a holding environment (Winnicott, 1960), in order to address participants' reactions during these interviews. As specified earlier, some participants relived childhood trauma during interviews while exhibiting strong emotional and behavioral reactions.

Strategy 2: turning a little notebook into my confidante

My reflective notebook, or its more official name: "the researcher's reflective journal", was a good way not only to document my thoughts and insights, but also a tool to observe and being aware of my reactions during the data collection and analysis.

A short time before meeting my research participants, I prepared a notebook, which initially was intended to remind me of the names of my participants, their addresses and important questions I had for them. It was a plain 10x15 cm notebook with blank white pages which has been distributed to students at the Humanities library at Tel Aviv University, carrying on its front blue page the caricatures of celebrities such as Pavarotti, Woody Allen, Napoleon, and Salvador Dali. Yet, during and after conducting the interview, this notebook has become useful during my data collections, data analysis, and even during the writing of my dissertation. It has been filled with my observations, thoughts, and feelings during the long hours of my travelling by trains and busses to different remote places in Israel. This little notebook and the other three, which followed it, have witnessed tears and triumphs, both mine and participants', and insights that I wrote, at times in the middle of the night. What started as a notebook that included only technical reminders on blank white pages, has turned in time into an important way for me to express my feelings and dilemmas throughout my work. For example, while being aware of my strong emotional identification with some of participants' narratives, I also realized that my individual experiences, might have not necessarily been similar to my participants', as there are many other variables at work. This awareness saved me from the risk of interpreting and analyzing the data as an Argentinian version of my life story, instead of the participants' stories.

Strategy 3: using the self-talk technique to regulate my emotions

My second-hand experience of participants' traumas was not manifested only during the interviews but have also emerged long after interviews had been completed. I have found myself re-experiencing the emotional impact of the description of the painful sights participants' witnessed, the sounds of their crying, as well as their silence. This experience emerged each time I engaged in data analysis: when listening to the audio recording of the interviews in order to transcribe them, when conducting the textual analysis, and even when writing my dissertation and articles for journals long after the data had been collected and analysed. My memory of participants' voices, sighs, tears, and silence have been as vivid and clear to me as their descriptions of their traumatic experiences during the interviews, which were conducted months ago. In a similar way to the "vitality" of the traumatic memory of the participants after more than four decades, the impact of conducting these interviews on me was as strong and intense as when heard it for the first time during the interviews.

In order to regulate my emotions and not act upon my reactions, I prepared a "to do list" in my reflective notebook. This is what I wrote:

This is not about you. It's about them. Emotions. What makes me feel good? walking in the park, going to the beach, watching comedies on TV, listening to relaxing Jazz and classical music, not to think now about dieting, eat what I wish to eat. Getting the tea that I like. Meeting with old friends. Helping someone. Talking with Julia [my doctoral field supervisor]. Breath. It is not about you. It is about them. Stick to your work plan. Focus-Focus-Focus. One step at a time. You'll get there.

Strategy 4: taking a break, sharing my experience

During my work on the data transcription and analysis I often found myself pausing, by turning my eyes away from the computer screen, or by getting up to make myself a cup of tea, as I felt that I needed to take some moments to myself in order to process what I had just read or transcribed. I was aware that I sighed frequently when feeling sadness or compassion. At other times, I felt anger toward the cruelty of humanity, and the injustice in this world. In contrast, when the participants were describing something funny, I laughed, and when their description illustrated their inner strength, I smiled. Furthermore, I noticed that my strong reactions to participants' descriptions recurred each time I interacted with the recordings or transcripts, even after going through the same text multiple times. Taking these "tea breaks", helped me to do data analysis work while observing my own reactions.

At times, sharing with my field supervisor in Israel some of the sections in the interviews, which affected me the most, and receiving empathy, provided meaningful relief. It was helpful to have my supervisor's ongoing support, as she was able to be my sounding board during this difficult process of data collection and analysis.

Strategy 5: listening to the soundtrack of hope and resilience

Another coping strategy was found in the locations that I chose to do the transcription, the analysis and the writing of my dissertation. I often preferred to do my work at the library, where there were always people around me. Another favorite location was at an apartment balcony where I could see trees, and hear children playing in the playground. This was a soundtrack of life that carried hope, a background noise which helped me to balance the heavy weight of participants' descriptions of death and horror.

At times, themes of life, hope and resilience were also found in participants' descriptions, despite their traumatic experiences. For example, the way their childhood traumatic experience was used by some of them to obtain "survival skills" as new immigrants in Israel, a country which is characterised by ongoing political instability (Gal, 2020). It was also a heartening, elevating experience for me to hear these individuals who, despite suffering and facing human cruelty, still believe in hope and goodness in our world (Gal & Hanley, 2019; Gal, 2020a). Their incredible resilience was inspirational.

TAKING COUNTERTRANSFERENCE A STEP FURTHER

During my research work on my postdoctoral study data collection, I could see how the different strategies that I used during my doctoral study have helped me to achieve a greater level of reflexivity in relation to my countertransference as a researcher, such as using a reflective journal, decreasing emotional reactions, confirming the resemblance of narrative interview to talk-therapy, and learning from my previous experience.

The Crucial Role of My Reflective Journal During the COVID-19 Pandemic

For my postdoctoral study, I had to conduct 45 narrative interviews with children of traumatized Argentinian immigrants in Israel, in order to explore the manifestations of intergenerational trauma transmission. Some of these interviews were conducted during the COVID-19 pandemic, which changed my methods of data collection. Given the obligation to be in quarantine during that time, some interviews, which were supposed to take place at participants' homes, had to be conducted over the phone. This had implications not only on the dynamic of the interview, but also on the way that I conducted it. As a video call was impossible due to bad internet connection, I was not able to conduct observations on participants' emotional and behavioral reactions during the interviews. Therefore, I have had no visual data to confirm or dismiss the verbal data. As a result, I had to rely solely on my ability to be attentive to the participants over the phone and how they chose to articulate themselves. At the same time, I had to be tremendously aware of my own verbal reactions at any given time, and to create a safe interview environment using only my voice, as I had no way of mediating to the participants my reactions in a visual way.

During my postdoctoral study, the little old 10x15 cm notebook that accompanied my work on my doctoral study, turned into a digital journal, as I have found it easier to access the journal through my cell phone. Writing in my digital notebook was a very useful tool during the data collection of my postdoctoral study, and particularly during the COVID-19 pandemic.

During that stressful time, I used the opportunity to express my thoughts and my feelings in my journal almost every day. I have realized that writing helped me not only to observe and be aware of my own reactions to the pandemic, but also to make an important distinction between my own anxiety and the anxieties expressed by some participants. A virtual space in which I can express my emotions in relation to the research during those "Corona days" also enhanced my ability to contain participants' experiences of helplessness, lack of control, fear and uncertainty, while putting my own feelings about that daily surreal reality aside.

My reflective journal also made me aware of the emotional intensity that the pandemic could have in one's personal life, therefore enabled me to show empathy and understanding to the variety of the emotional reactions which were exhibited by the participants during the interviews. It also allowed me to identify the complexity of participants' experiences during that stressful time, considering the context of their experience of "old" trauma.

Decreasing the Intensity of My Emotional Reactions

In contrast to my experience during my work on my doctoral study, during my postdoctoral research I was able to manage my reactions more effectively. More specifically, I felt that I had obtained better skills to observe and control the expression of my emotional reactions during the interviews while exhibiting empathy to the participants, and attending my ethical responsibilities. As an example, during one of the interviews I experienced a strong emotional reaction when one of the participants shared her experience of growing up without a father. I felt during the interview that I could strongly relate to her experience, even though the circumstances of the father's absence were quite different. Despite my strong emotional identification, I utilized my awareness of my own experience as a fatherless daughter, in order to ask her questions that can explore the unique nuances of her experience (e.g.: to what extent, she has been carrying her father's absence in her life). Following my emotional experience in that interview, I allowed myself to find the right opportunity to process my reflection on my personal experience as a fatherless daughter without transferring my perceptive on the participant's during the interview.

Furthermore, in contrast to my experience of a secondary trauma during my work on my doctoral study, I find that during data collection for my postdoctoral study my emotional reactions were not as intense. At times, I found myself thinking about what had been said during the interviews, usually on the same day as the interview. In addition, unlike my experience during my doctoral research, the content of my thoughts did not involve "re-experiencing" of the emotional impact of participants' descriptions of their painful experiences. Also, this time during the data collection, I did not avoid leisure activities that provided me with pleasure and joy.

Confirming the Resemblance of Narrative Interview to Talk-therapy

Finally, similarly to my doctoral study, the interviews conducted for the purpose of my postdoctoral study had therapeutic impact on the participants. Some participants described the interview as a tool that had helped them to realize things they were not aware of, mainly in relation to their relationships with their parents, and the ways that the parental trauma had affected their lives. Some said that following the interview they wished to initiate for the first time a talk with their parents about their past trauma. Others elaborated on the "healing experience" that they had during the interview. These reactions confirmed once again the resemblance of narrative interview to the setting of talk-therapy, and therefore the similarities to the power balance and the special dynamic between the researcher and participant, and clinical counsellor and client. As I look at my recent experience during my work on my postdoctoral research, I realize I have managed to improve my awareness of countertransference and my strategies to address it.

RECOMMENDATIONS

The description above of my experience as a narrative researcher doing trauma-focused study, suggests important implications for other qualitative researchers during data collection and analysis. It does not only illustrate the complexity of one's positionality, but also demonstrates the various mechanisms of managing and utilising a researcher's subjectivity to enhance the quality of the study, using the following strategies:

- **Awareness of one's reactions during the data collection:** Being "here" and "there" at the same time means to retain awareness of your emotional reactions and behavior during interviews, while making sure that the focus stays on the narrators, and while keeping in mind your ethical responsibilities.
- **Keep in mind: the interview is not your therapy:** Despite the many similarities to talk-therapy sessions, and although some topics raised during the interview may correspond with your own personal life experiences, the narrative interview is not a therapy session for the researcher, nor for the research participants. It is a methodological tool that helps to retrieve participant's subjective experiences and answers your research question.
- **Create a reflective notebook**: Use a "researcher's reflective journal." You do not need to be worried about grammar, sentence structure or form, just write what comes into your mind, and express freely your thoughts and emotions. This is an excellent way not only to document your insights, but it is also a tool to be aware of your reactions during data collection and analysis. It provides a safe outlet for your emotions and thoughts that may come up during the research work, and it can help you to observe your reactions instead of acting upon them during your work.
- **Use self-talk and self-soothing:** Self-talk is our inner voice, an internal dialogue, which can help one to master and regulate one's strong emotional reactions. You can achieve this by writing in your reflective notebook, or by "speaking" to yourself in first person. For an example:

It is so sad what the participant just shared with me. The world is sometimes so cruel and unfair! It reminds me of the loss that I have experienced in my family. But now I am conducting the interview, so I should be focused on the interviewee. I will talk with my therapist next week about my strong reaction today and the sad memories it evoked in me during the interview. Everything is alright. I am focused.

- **Take a break:** In any research work, and specifically during your work on a trauma–focused study, you may feel overwhelmed. Take a break. It could be a short coffee or tea break, taking a walk at the park or at the beach, or, when needed, taking the whole day off, for activities unrelated to your research. Your work will wait patiently for you until you will be back.
- **Find an empathetic "sounding board":** At times, sharing your experiences with someone who is patient, wise and empathetic can be a real asset to your wellbeing during your research. This could be your research supervisor, a good friend, a therapist, or your spouse or partner. Expressing your emotions

and thoughts to another person not only provides an emotional relief, but may also help you to obtain new perspectives and insights on your interpretations during data analysis and analysis.

- **Look at the bright side of life, find the balance:** Research work can at times be an isolating experience, especially if you are the principal researcher, and not part of a research team. In addition, the content of your trauma-focused research can be at times "dark" or "heavy". Find a working environment that can help with this sense of isolation, and improve your emotional wellbeing. Some prefer to work at home, others in public. Find the environment that works for you, and which provides not only a place where you can concentrate on your work, but also a space that you find as pleasant, comfortable and comforting.
- **Practice, practice practice:** The more reflect and observe during your research work, the ability to be reflexive will improve. Practicing countertransference in research is not an easy thing to do, but the more you do it, the easier it gets.
- **Self-care is your best friend, it is not a waste of time:** Last but not least, during your research work, especially when it is trauma-focused study, it is also important to eat well, sleep well, exercise, socialize and do leisure activities. Self-care is your friend: when you are rested, calm and have sense of both physical and emotional wellbeing, the quality of your judgment, observations and the activities you perform as a researcher will improve!

DISCUSSION AND CONCLUSION

This reflection contributes to the evolving discussion about the role researcher's subjectivity has in qualitative research practice, and the growing awareness of the potential negative and positive impact that researcher's emotions, personal histories, opinions and beliefs may have on the research process and outcomes. Given the many similarities between narrative research approach and talk-therapy, in terms of the interview structure and the interpersonal dynamic in the interview room, it seemed only natural to use the concept of "countertransference" as a lens through which one can explore the manifestation of researcher's positionality and reflexivity in the context of narrative research, and in this case a trauma-focused study.

Using Researcher's Subjectivity as Knowledge Production

In contrast to Goldstein (1994) and Mason (2002), who argue that researcher subjectivity prevents objectivity or detachment from the data, in a similar way to countertransference experienced in therapeutic relationships, subjectivity in qualitative research cannot be avoided. As put by Peshkin (1988): "one's subjectivity is like a garment that cannot be removed. It is insistently present in both the research and non-research aspects of our life" (p.17).

In the same manner that therapists are engaged with self-reflection when identifying their countertransference, our concern as qualitative researchers should not be how to eliminate our subjectivity, but rather how to handle it in a way that can contribute constructively to the quality of our research. In the context of qualitative research, countertransference cannot be treated as an obstacle or "contaminating" factor, but rather as a very important source of information about unconscious processes (Holmes, 2014). Therefore, countertransference can be a source of insight and understanding about one's research question, and used as an integral part of research data. In a similar way that countertransference is one of the ways to achieve successful therapy by providing therapists knowledge about the state of their patients, developing awareness of researchers' subjectivity in narrative research practice is a vital component that helps achieve trustworthiness in qualitative research (D'Cruz et al., 2007; Gerstl-Pepin & Patrizion, 2009; Ahmed Dunya et al., 2011). Our experiences of countertransference in the context of qualitative research can too, enrich our understanding of the experiences of our research participants, and therefore help to capture their genuine experiences, as expressed through interviews.

During my work on my doctoral study followed by my postdoctoral study, I experienced overflow from participants' traumatic experiences into my own emotional history. My goal was to gain some control over my reactions in a way that would allow me to critically observe the meaning of my reactions in the context of the experiences of my research population. For example, during my work on my doctoral study, I could see how my strong emotional reactions during data collection and analysis clearly indicated that my participants' childhood traumatic experiences are very much alive even after more than four decades. My ability to observe my reactions and to separate them and the data was eventually used as evidence for the long-term impact of participants' trauma. In a similar way to therapists who use awareness of their own feelings in relation to their clients as significant data, this chapter illustrated how my subjective way of knowing, which resonated with my own personal emotional history, was a significant part in the process of knowledge production. This way of knowing and the use of the self, its fluidity during the research process and my awareness of it, enhanced the quality of my study.

The "Workout" of Researcher's Self-monitoring

This chapter also illustrated the strategies taken to be aware of countertransference, and using it as a valuable source of knowledge. Acknowledging, recognizing, recording and "managing" our countertransference in the context of conducting qualitative research, can help us to treat our emotional reactions and positionality from the stance of "observation" and as a source of information. Peshkin (1988) argues that "researchers should systematically seek out their subjectivity, not retrospectively when the data have been collected and the analysis is complete, but while their research is actively in progress" (p.17). This "formal, systematic monitoring of self" (Peshkin, 1988, p.20), is described by the same author as:

[...] a necessary exercise, a workout, a tuning up of my subjectivity to get it into shape. It is a rehearsal for keeping the lines of my subjectivity open--and straight. And it is a warning to myself so that I may avoid the trap of perceiving just that which my own untamed sentiments have sought out and served up as data [...] By monitoring myself, I can create an illuminating, empowering personal statement that attunes me to where self and subject are intertwined, l do not thereby exorcise my subjectivity. I do, rather, enable myself to manage it. (p.20)

In accordance with the above statement, throughout the process of my work on my doctoral and postdoctoral studies, I engaged with an on-going reassessment of my positionality and countertransference that may have impacted my attitude, responses and interpretations throughout my studies. This on-going reassessment has helped me to repeatedly reposition myself throughout my work on my research projects (Creswell, 2007).

I used several strategies to enhance my self-reflection and awareness of my subjectivity. By using my skills as a clinical counsellor, I expressed my reactions to participants and their narratives only after the completion of the interviews in my journal and later during my meetings and email correspondence with my field supervisor. I took notes during every step of data collection and analysis, and then used them to retrace these steps. In addition, by keeping a journal that documented my thoughts and feelings, I was able to reflect on reactions that may have included bias, resistance responses, or other emotional content.

During my doctoral research work I tried to be transparent about my bias and weaknesses, while explaining and specifying my position as a researcher and the way that I addressed these biases (Gal, 2019). My biases include my age, gender, ethnicity, level of education, emotional history, social location, personal experiences, opinions based on professional experience, political and ideological biases, theoretical academic knowledge, cultural subjectivity, and my emotional responses to my research participants (Hamzeh and Oliver, 2010; Padgett, 2008; Bradbury-Jones, 2007; Kosygina, 2005).

In addition to the above statement, I used research guidelines and rationale as indicators during different stages of my doctoral research and to verify whether I was being influenced or positioned by my own personal bias and preferences.

Furthermore, during analysis, I maintained a skeptical point of view until I was able to make definitive conclusions. When drawing conclusions, and during the process of analysis, I did not dismiss data that did not "fit" with my desired results. I also did not try to "objectively" prove my personal opinions, but also included different viewpoints (Dudley, 2005). Finally, the doctoral study results were reported after using "thick description", allowing other readers to review my interpretation (Riessman, 1993; Berg, 2001).

As stated at the beginning of this chapter, although there are similarities between the conducting of narrative research and talk therapy, there is limited knowledge on the intersection of the two fields of practice. The purpose of this chapter was to focus specifically on the concept of "countertransference" and its intersection with researcher's subjectivity in the context of narrative research. In an era where the boundaries between one discipline to the other are becoming increasingly vague, specifically in the context of the contribution of psychoanalysis to other domains (Gobrin, 2020), it is only natural that more studies and literature should explore the intersection between qualitative methodology and the world of psychoanalysis.

REFERENCES

Bar-On, D. (1996). Ethical Issues in Biographical Interviews and Analysis. In *Ethics and Process in the Narrative Study of Lives* (Vol. 4). Sage. doi:10.4135/9781483345451.n2

Bar-On, D. (Ed.). (2000). *Bridging the gap. Storytelling as a way to work through political and collective hostilities*. Körber-Stiftung.

Berg, B. L. (2001). *Qualitative research methods for social sciences*. Allyn & Bacon.

Berger, R., & Weiss, T. (2009). The Posttraumatic Growth Model: An Expansion to the Family System. *Traumatology*, *15*(1), 63–74. doi:10.1177/1534765608323499

Birnbach, M. (1961). Neo-Freudian Social Philosophy. Stanford University Press.

Bouchard, M. A., Normandin, L., & E-Hélène Séguin, M. (1995). Countertransference as instrument and obstacle: A comprehensive and descriptive framework. *The Psychoanalytic Quarterly*, *64*(4), 717–745. doi:10.1080/21674086.1995.11927473 PMID:8584597

Bradbury-Jones, C. (2007). Enhancing rigor in qualitative health research: Exploring subjectivity through Peshkin's I's. *Journal of Advanced Nursing*, *59*(3), 290–298. doi:10.1111/j.1365-2648.2007.04306.x PMID:17590210

Creswell, J. W. (2007). *Qualitative inquiry & research design: choosing among five approaches*. Sage.

D'Cruz, H., Gillingham, P., & Melendez, S. (2007). Reflexivity, its meaning and relevance for social work: A critical review of the literature. *British Journal of Social Work, 37*(1), 73–90. doi:10.1093/bjsw/bcl001

Dalenberg, C. (2000). *Countertransference and the treatment of trauma*. American Psychological Association. doi:10.1037/10380-000

Dalenberg, C. J. (2004). Maintaining the safe and effective therapeutic relationship in the context of distrust and anger: Countertransference and complex trauma. *Psychotherapy (Chicago, Ill.), 41*(4), 438–447. doi:10.1037/0033-3204.41.4.438

Davidson, L. (2003). *Livening outside mental illness: Qualitative studies of recovery in schizophrenia*. New York University Press.

Denzin, N. K., & Lincoln, Y. S. (2005) Methods of collecting and analyzing empirical materials. In The Sage Qualitative Research, (3rd edition (pp.641-642). Thousand Oaks, CA: Sage.

Dudley, J. R. (2005). *Research methods for social work: becoming consumers and producers of research*. Pearson.

Dunya, A., Lewando, H. G., & Blackburn, C. (2011). Issues of gender, reflexivity and positionality in the field of disability: Researching visual impairment in an Arab society. *Qualitative Social Work: Research and Practice, 10*(4), 467–484. doi:10.1177/1473325010370188

Dwyer, S. C., & Buckle, J. L. (2009). The space between: On being an insider-outsider in qualitative research. *International Journal of Qualitative Methods, 8*(1), 54–63. doi:10.1177/160940690900800105

Etchegoyen, R. H. (2018). *The fundamentals of psychoanalytic technique*. Routledge. doi:10.4324/9780429481864

Ferro, A. (2006). *Psychoanalysis as therapy and storytelling*. Routledge.

Freud, S. (1957). The future prospects of psycho-analytic therapy. In J. Strachey (Ed. & Trans.), The standard edition of the complete psychological works of Sigmund Freud, (Vol. 11, pp. 141-151). London: Hogarth Press. (original work was published in 2010)

Freud, S., & Bonaparte, P. M. (1954). *The origins of psychoanalysis* (Vol. 216). Imago.

Fromm, E. (2013). *The revision of psychoanalysis*. Open Road Media.

Gal, S. (2019) *Risk and protective factors for the mental health consequences of childhood political trauma (Argentina 1976-1983) among adult Jewish Argentinian immigrants to Israel* (Unpublished doctoral dissertation). McGill University, Montreal, Canada.

Gal, S. (2020). "Israel is not Switzerland": The impact of the political violence in Israel on historically traumatized Argentinian immigrants. *Journal of International Migration and Integration*.

Gal, S. (2020a). Childhood experiences of Anti-Semitism in Argentina (1976-1983): Stories of trauma, resilience and long term outcome. In H. S. Moffic, J. Peteet, A. Hankir, & M. V. Seeman (Eds.), *Psychiatry and Anti-Semitism - Recognition, Prevention, and Interventions*. Springer.

Gal, S., & Hanley, J. (2019). The long-term impact of political activism on mental wellbeing in the context of adult Jewish immigrants who experienced childhood political trauma. *Journal of Immigrant and Minority Health*.

Gerstl-Pepin, C., & Patrizion, K. (2009). Learning from Dumbledore Pensive: Metaphor as an aid in teaching reflexivity in qualitative research. *Qualitative Research*, *9*(3), 299–308. doi:10.1177/1468794109105029

Gobrin, A. (2020) *The cultural sofa: Psychoanalysis in the Spaces of Literature, Philosophy and Society*. Narkisus Publication.

Goldstein, H. (1994). Ethnography. Critical inquiry and social work practice. In E. Sherman & W. Ried (Eds.), *Qualitative research in social work* (pp. 42–51). Columbia University Press.

Hamzeh, M. Z., & Oliver, K. (2010). Gaining research access into the lives of Muslim girls: Researchers negotiating muslimness, modesty, inshallah, and haram. *International Journal of Qualitative Studies in Education: QSE*, *23*(2), 165–180. doi:10.1080/09518390903120369

Harvey, M. R., Mischler, E. G., Koenen, K., & Harney, P. A. (2000). In the aftermath of sexual abuse: Making and remaking meaning in narratives of trauma and recovery. *Narrative Inquiry*, *10*(2), 291–311. doi:10.1075/ni.10.2.02har

Hellawell, D. (2006). Inside–out: Analysis of the insider–outsider concept as a heuristic device to develop reflexivity in students doing qualitative research. *Teaching in Higher Education*, *11*(4), 483–494. doi:10.1080/13562510600874292

Holmes, J. (2014). Countertransference in qualitative research: A critical appraisal. *Qualitative Research*, *14*(2), 166–183. doi:10.1177/1468794112468473

Josselson, R. (1996). On writing other people's lives: Self-analytic reflections of a narrative researcher. In R. Josselson (Ed.), The narrative study of lives, Vol. 4. Ethics and process in the narrative study of lives (p. 60–71). Sage Publications, Inc.

Jung, C. G. (1969). *The psychology of the transference* (Vol. 20). Bollingen Foundation.

Kapelovitz, L. H. (1976). *To love and to work: A demonstration and discussion of psychotherapy*. Grune & Stratton.

Kirkpatrick, H. (2008). A narrative framework for understanding experiences of people with severe mental illnesses. *Archives of Psychiatric Nursing*, *22*(2), 61–68. doi:10.1016/j.apnu.2007.12.002 PMID:18346562

Knaifel, E., & Mirsky, J. (2015). Interplay of identities: A narrative study of self-perceptions among immigrants with severe mental illness from the former Soviet Union. *Transcultural Psychiatry*, *52*(1), 74–95. doi:10.1177/1363461514552585 PMID:25315487

Kosygina, L. V. (2005). Doing gender in research: Reflection on experience in field. *Qualitative Report*, *10*(1), 87–95.

Lev-Wiesel, R., & Amir, M. (2006). Growing out of ashes: Posttraumatic growth among Holocaust child survivors—is it possible. In L. G. Calhoun & R. G. Tedeschi (Eds.), *Handbook of posttraumatic growth: Research and practice* (pp. 248–263). Psychology press.

Mason, J. (2002). *Qualitative researching* (2nd ed.). Sage.

Milton, J., Polmear, C., & Fabricius, J. (2011). A short introduction to psychoanalysis. *Sage (Atlanta, Ga.)*.

Morrow, S. L., & Smith, M. L. (2000). Qualitative research for counseling psychology. Handbook of Counseling Psychology, 3, 199-230.

Padgett, D. K. (2008). *Qualitative Methods in Social Work Research*. Sage.

Peshkin, A. (1988). In search of subjectivity—one's own. *Educational Researcher*, *17*(7), 17–21.

Pillow, W. S. (2003). Confession, catharsis, or cure? Rethinking the uses of reflexivity as methodological power in qualitative research. *International Journal of Qualitative Studies in Education: QSE*, *16*(2), 175–196. doi:10.1080/0951839032000060635

Polkinghorne, D. E. (1991). Narrative and self-concept. *Journal of Narrative and Life History*, *1*(2-3), 135–153. doi:10.1075/jnlh.1.2-3.04nar

Riessman, C. K. (1993). *Narrative analysis*. Sage.

Rosenthal, G. (1993). Reconstruction of life stories: Principles of selection in generating stories for narrative biographical interviews. *The Narrative Study of Lives, 1*(1), 59-91.

Rosenthal, G. (2003). The healing effects of storytelling: On the conditions of curative storytelling in the context of research and counseling. *Qualitative Inquiry, 9*(6), 915–933. doi:10.1177/1077800403254888

Rudd, M. D., & Joiner, T. (1997). 'Countertransference and the therapeutic relationship: A cognitive perspective. *Journal of Cognitive Psychotherapy, 11*(4), 231–250. doi:10.1891/0889-8391.11.4.231

Sandler, J. J. (1976). Countertransference and role-responsiveness. *The International Review of Psycho-Analysis, 3*, 43–47.

Stronach, I., Garratt, D., Pearce, C., & Piper, H. (2007). Reflexivity, the picturing of selves, the forging of method. *Qualitative Inquiry, 13*(2), 179–203. doi:10.1177/1077800406295476

Vajda, J. (2007). Two survivor cases: Therapeutic effect as side product of the biographical narrative interview. *Journal of Social Work Practice, 21*(1), 89–102. doi:10.1080/02650530601173664

Van der Kolk, B. A. (2005). Developmental trauma disorder: Toward a rational diagnosis for children with complex trauma histories. *Psychiatric Annals, 35*(5), 401–408. doi:10.3928/00485713-20050501-06

Winnicott, D. (1960). The theory of the parent-child relationship. *The International Journal of Psycho-Analysis, 41*, 585–595. PMID:13785877

Chapter 9
Reflections of Challenges and Resolutions in Fieldwork Research on Quality of Life of Cancer Patients:
Counseling and Psychology

Kavita Gupta
The Maharaja Sayajirao University of Baroda, Vadodara, India

ABSTRACT

Challenges and resolutions in fieldwork have become a dominant focus of interest for academicians and researchers conducting researches in social sciences. The objective of the chapter is to describe and reflect the researcher's experience in conducting a research study on anxiety, depression, and quality of life of the cancer patients that could be beneficial to novice researchers in the field of cancer studies. The author encountered these experiences in 2016 as part of her doctoral thesis that she had conducted in a tertiary care hospital based in Delhi. The chapter discusses the context of study with respect to the challenges faced in selecting a field site and acquiring access, recruiting, and building trust and rapport with the research participants, concern of confidentiality and emotional and psychological wellbeing of the researcher. Further, the chapter focuses on the resolutions and recommendations to deal with compassion fatigue, burnout, and secondary traumatic stress to avoid and manage great amount of emotional demands on the researchers.

DOI: 10.4018/978-1-7998-5826-3.ch009

INTRODUCTION

Fieldwork is best described as a primary data collection carried out to accomplish the objectives of study stated by a researcher. This practical information is conducted in an innate social set-up with the resources at hand and within a specific time frame. The efficacy of fieldwork builds upon capabilities of the researcher in resolving the challenges encountered. A researcher's dedicated approach is the motivating factor for dealing with any impasse while moving ahead with the research process. Fieldwork forms an integral part of any research project. We shall discuss here how the hardships faced were dealt with in the research study that was conducted on the anxiety, depression and quality of life of cancer patients. In this chapter the target is the confrontation of the challenges. The progression steps executed in a social science research process should be placed in a flawless sequence; as such study has specific significance certainly.

In the data collection process, unexpected hurdles may crop up that might hinder the completion of the research study. Therefore, the researcher should always be mentally prepared to handle such obstacles adeptly and make best use of the appropriate solutions whenever necessary. The challenges experienced in a fieldwork are neither mutually exclusive and nor are totally distinct. However, they need to be deciphered as per the potential requirements analogous with the aim and objectives of research study.

The fieldwork challenges are never forthright. One may require interrogating with the right question in order to reach an appropriate answer having the solution. In the available literature there is a lack of substantial discussions on these issues that the novice researchers and graduate students need to know. Therefore, this chapter is specifically about the reflective descriptions of a doctoral researcher's fieldwork experience. This is with respect to the challenges faced during her research on the cancer patients at a tertiary care hospital based in New Delhi, India. This study was pertaining to the patients' quality of life during their therapy sessions. The researcher exemplifies 'reflective description' with the practical strategies adopted and appropriate actions employed to deal with similar challenges.

The researcher selected the hospital as a field study site in order to delve into the relationship amongst the variables through mixed method (quantitative and qualitative method). For data collection, the researcher designed a set of questionnaires as the valid and reliable tools for her research study. After gaining approval from the ethics committee, the researcher acquired access to the site for collection of data through administration of questionnaires to the potential participants. The process of data collection from the research participants required building of trust, rapport and maintaining the confidentiality. This chapter discusses about the fieldwork experience during the research study on quality of life of cancer patients who were undergoing chemotherapy and radiotherapy sessions. The cancer patients are considered sensitive and vulnerable population, the researcher, therefore, had to handle concerns about their uneasiness, nervousness, worries and anticipated fear and their disrupted emotions while undergoing chemotherapy or radiotherapy sessions.

BACKGROUND

To begin with, writing a research problem statement and pointing the objectives of the research study is only the start of a research journey. The rest of it focuses on data collection from the fieldwork, data analysis and interpretation, and conclusion with recommendations. The selection of the research tool for a data collection is a tedious process. After validating the questionnaire, problems may arise in administering the same. This could be due to lengthy questions or touching upon the sensitive aspects and the consequential response scales. Further, short-listing the patients for such studies is another big challenge. For instance, you might shorten the list of your participants for many different reasons such as low literacy rate among some participants, or/and the lack of collaboration with the researcher and the familiarity with the research. Also, the patients may feel bothered about their well being and the information provided could be inaccurate or the contact information could be missing too. Other issues that might arise in such studies are due to the patients' miscommunication with their health care providers, which would result in the lack of understanding the diagnosis, treatment and prognosis. To protect such vulnerable population, the Institutional Review Board (IRB) is more likely to put specific requirements upon those researchers who intend to include participants from vulnerable groups/special population. Practical approaches to resolving all these issues are discussed in the ethical guidelines of this chapter.

The quality of life of cancer patients represents in majority the suffering, helplessness, and loss. This is in addition to the demystifying beliefs about the disease and its treatment. The treatment strategies in the form of chemotherapy and radiotherapy sessions often result in changes in the patient's body image and social activities. Buetto, Scatralhe, Zago, and Fontão (2015) discussed about the perception of cancer patients concerning their physical and psychological health, level of functionality, sociability, and relationships with environmental characteristics. All of the above mentioned factors and more might result in the loss of control over life and a limitation of the ability to work.

This chapter, therefore, intends to unravel various difficulties associated with fieldwork during a research process that involves working with vulnerable population. It is important to note that as per the official lawful procedure in India, any research study that involves humans and vulnerable population have to go through approval from the Institutional Review Board Ethics committee first.

NARRATIVE

The researcher is going to share her doctoral research experience where she learned a lot, and gained lots of expertise in that field of study. From the very beginning, the process of working on a doctoral study was not easy at all. For instance, the drafting of the research proposal, the involvement in the data collection and analysis process, and writing the thesis as a whole were a tedious task. The theoretical perspective was very different from the practical experience due to the sudden ups and downs faced by her. According to the researcher, a thorough and intense review of literature must be done before defining a problem of statement. This cannot be achieved without devoting ample time to this process, so that the guidelines for a concrete introduction with a definite aim and objectives can be framed. This will help guide the researchers while working on their thesis.

The researcher conducted her study to assess the correlation of anxiety and depression with the domains of quality of life of cancer patients during inpatient chemotherapy sessions and outpatient radiotherapy sessions. Although the researcher did not have previous practical background, she saw her mother struggling with cancer and undergoing the therapies. The researcher understood the applied value of this significant topic as she had seen her own mother suffering for many years. Therefore, she was firm to go ahead with the research, regardless of the long time the pre-requisites took. She drafted the problem statement with pronounced reasoning and determination. She had been a caregiver already and, henceforth, had a vision to do a research in this very sensitive area. By approaching such a topic, she intended to add a valuable input to the society and to the field through her practical observation and the theoretical knowledge she gained.

The researcher continued to examine and engage in the reflexive nature of conducting her research through creating transparency and maintaining research integrity throughout the study. In the following section, the researcher reflects upon four main challenges that she encountered in carrying out research with the cancer patients in the field; the strategies used to address and resolve these challenges have also been discussed.

REFLECTIONS ON THE FIELDWORK CHALLENGES

When making her entry to the fieldwork, the researcher adhered to the standard operating procedures and investigation protocol. Despite this, she faced certain major challenges during the fieldwork research process. The researcher's field for investigation and data collection from the potential research participants was IPD (Inpatient Department) for cancer patients receiving chemotherapy, and OPD

(Outpatient Department) for cancer patients receiving radiotherapy at a tertiary care hospital based in New Delhi city in India. A tertiary care hospital is unfortunately a real world where there is a reversal of normal life due to the ongoing treatment. The symptoms in the patients undergoing chemotherapy and radiotherapy always keep on fluctuating day-by-day as treatment progresses ahead.

The fieldwork challenges experienced when undertaking the research with cancer patients included issues related to selection and gaining access to the suitable site, short-listing patients for the study and then gaining mutual trust, concern of confidentiality emotional and psychological well being of the researcher herself. The supporting literature with certain ground rules was available to help in developing the strategies for fixing up the troubles mid way as the work was gaining momentum.

Selection and Gaining Access to the Suitable Field Site

This section discusses the challenges in selecting and gaining access to a field site, which includes considerable obstacles such as accessing the research field and getting the IRB approval. Strategies used to overcome these included initial contacts to identify a resource person, followed by approaching the Department of Academic Affairs & Continuing Education (AARCE).

Per accessing the research field, it was tedious for the researcher to contact AARCE department of the hospital. After the initial contact, resource persons at various levels were identified and approached. The researcher met the senior consultant and the head of the respective departments prior to submission of the research protocol and IRB applications. The meeting with the head of the respective departments allowed the researcher to engage with the junior staff and key members of the functional team (e.g., technician, clinical research staff, nurse and desktop executives). She handed them a copy of her research questionnaire and discussed the objectives of the research project along with the risks versus benefit ratio in a positive, clear, and respectful manner.

Prior getting the IRB approval, the researcher was guided to present her research proposal before the Institutional Review Board Ethics Committee panel, for necessary approval, as the sample population included vulnerable population who is cancer patients. All essential documents were submitted to the department office prior to presentation. It took about two months to get the IRB approval. After that, the date for research proposal presentation was approved. Subsequent to the presentation, there was meticulous discussion and question-answer round. Additionally, further editing was recommended. After the necessary legalities and the final approval from the research ethics committee, the research proposal for data collection was documented with the required terms and conditions.

Thus, after spending around six months with the Institutional Review Board, permission was finally granted to gain access to the hospital and to become familiar with cancer patients. The researcher aimed to build a rapport with the potential participants and to help them develop an understanding of the nature of the research. Thummapol, Park, Jackson and Barton (2019) showed in their previous work that choosing an appropriate study site and gaining access to participants is a combination of careful planning; academic knowledge; practical research experience; researcher knowledge, skills, and commitment; and luck.

To sum up, initially the researcher did not face any challenges in locating a suitable place to conduct her research study. However, after selecting a suitable field site, a few hurdles came up in gaining entry and acquiring access to the field site. Thummapol et al. (2019) discussed in their research that negotiating access and gaining entry to the field can be challenging, and in some situations it can be quite problematic. Additionally, novice researchers need to know the importance of becoming familiar with the IRB procedure before entering into field site.

Recruitment and Building Trust and Rapport With the Potential Participants

This section highlights the challenges in the recruitment of the research participants, as well as in trust and rapport building. The research design and sampling techniques were prepared prior to the field study for data collection. The researcher utilized her intellect and expertise for the recruitment of potential participants for her study. The layout comprised of communication with the concerned officials and executives, and of interactions with participants for building trust and showing empathy. She was flexible with the day time schedule for filling up questionnaires. She also ensured briefing adequate information prior to the participants' consent. As per the guidelines, no reimbursements were offered to the participants; hence, the researcher had to substitute this limitation by listening to them sympathetically and expressing her emotions for their morale support. The researcher did this not just for data collection work, but also for the sake of humanity. She listened to their grievances about the quality of life before diagnosis, after diagnosis, before treatment and after treatment and its impact on some other underlying conditions.

Firstly, it was very challenging for the researcher to recruit and invite participants for the study. As it is not a simple process, but a complicated one and it demands careful review since the respondent most of the time felt fatigued. However, the researcher managed this with her patience. She motivated herself to tackle this challenge by being gentle to the patients and greeting them every day. Thummapol et al. (2019) had discussed in their study that recruiting and retaining participants can be challenging and time-consuming depending on the skill of the researcher in communicating, negotiating, and building a sense of trust and mutual respect with each participant. As Duncombe and Jessop (2012) pointed out that close rapport enables participants to disclose their experiences and speak more freely and thus enhances the gathering of reliable and valid data. This could be better understood with the help of Denial, Anger, Bargaining, Depression, Acceptance model often abbreviated as DABDA, which explains why it was difficult to recruit the potential participants for the researcher's study.

The fieldwork is quite challenging for the researcher since she had to collect the data through questionnaire. The patients undergoing chemotherapy and radiotherapy were not always ready for such debriefings, obviously out of frustration and sufferings. This can be better explained through the model DABDA which was first proposed by Elisabeth Kubler-Ross (1969), a Swiss psychologist who explained how patients become unique as the focus of their care shifts from becoming well to preparing for death. As mentioned by the authors, C. Corr, D. Corr, and Doka (2019) mentioning Kubler-Ross theory, patient suffering from any terminal illness progresses through 5 stages: Denial, Anger, Bargaining, Depression, and Acceptance. To further elaborate and reflect, the DABDA model can be better understood in the following way:

1. **Denial:** This is been defined as a stage often accompanied by shock or speechlessness where the patient denies to have been diagnosed with the underlying disease (Kubler-Ross, 1969). The patient believes that there is a blunder in the diagnosis done, or inaccuracy in the test results performed. They may doubt the credibility of their report wherein assuming that there was carelessness in methods, observation and diagnosis followed by analysis (C. Corr, D. Corr, and Doka, 2019).
2. **Anger:** This stage represents the patient's frustration and irritation further followed by anger and aggression which might get displaced to friends, family, and healthcare providers (Kubler-Ross, 1969). Here, in this stage, the patient often questions himself/herself and to others as well wondering and asking themselves that they have never done anything wrong to anyone throughout their life. These statements often lead to emotional burst out of the patients (C. Corr, D. Corr, and Doka, 2019).
3. **Bargaining:** This stage represents the bargaining process (Kubler-Ross, 1969). Patients show faith that there is higher power in their religion to cure the disease or prolong their life if they reform their lifestyle. This stage is much more comparable to a barter exchange system where patient does this in terms of religious and spiritual manner. It might happen sometime that an atheist might transform into theist all of a sudden and starts praying in order to bargain for mercy and life extension for self or fulfilment and accomplishment of responsibilities or both (C. Corr, D. Corr, and Doka, 2019).

4. **Depression:** This stage often characterized by sadness, fatigue, self-harms, self-pity, refusal for medications and treatment or social avoidance (Kubler-Ross, 1969). The patient is in such a state of despair that s/he declares nothing matter anymore as her/his death is approaching and it is inevitable (C. Corr, D. Corr, and Doka, 2019).
5. **Acceptance**: This stage is often accompanied by acceptance in the patient of the underlying condition thereby resulting in calm and stable emotions (Kubler-Ross, 1969). They try to start living in the present moment irrespective of the time they are being left with. Spending time with loved ones can create pleasant memories for survivors to reflect on (C. Corr, D. Corr, and Doka, 2019).

The above-narrated phases of grief model explained how the diagnosis and prognosis of cancer could have a devastating effect on them. Life through these different stages keeps on fluctuating with no definite characteristics and symptoms. It is also important to note that these stages cannot be considered as sequential; rather any of the stages can appear first followed by the other one. Kubler-Ross (1969) was criticized for this theory and for her stages of grief, death and dying. However, according to the researcher's fieldwork experience, the patients under treatment after diagnosis did not show characteristics of all the stages. The inpatients who are receiving the chemotherapy are more affected as they are in constant and daily communication with the other patients in the hospital. The patients share their grief with each other and this impacts one another. On the contrary, the outpatients who are receiving radiotherapy suffer in terms of the wait time in the hospital and in their turn to receive the treatment. So, regardless, the treatment patients receive, life is difficult after diagnosis, prognosis and treatment, which need resilience and coping up a. (C. Corr, D. Corr, and Doka, 2019).

The researcher had been observing all of this, which had a significant impact on her emotional and psychological well being. This practical experience was very different according to her, since it was not previously mentioned in any literature review how it feels to meet the cancer patients with varying degrees of sufferings. After the fieldwork experience, the researcher felt that she should have done the field survey prior to her drafting of synopsis in order to gain an in-depth evaluation of the quality of life of patients undergoing the therapy sessions. The researcher here faced a major intellectual and emotional loss.

The researcher informed all the participants prior to filling up of the questionnaires that they could freely refuse to answer any questions or withdraw from the study at any time without giving any reason. However, there was no evidence of any participant's discomfort or withdrawal from the study. Potential participants were identified with the help of desktop executives and were contacted. They were requested, then, to provide further information about the study and ascertain their willingness to participate. They were not given any token gift/reimbursements, since this provision is a controversial issue and could have raised ethical problems.

Among the different methods used for enlisting participants, the help of desktop counselors and assistance of consultants worked efficiently due to mutual trust. This reflects the importance of integrating the resource skills and public relations in a research project. The novice researchers in this field need to remain optimistic and show resilience to not give up and discontinue with the fieldwork when faced by any adverse situation or the pitiful sufferings shared by the participants.

Concern of Confidentiality

This section presents the challenges involving the concern of confidentiality of the data collected by the potential research participants. The questioning mostly took place in the afternoon and evening hours. All were held in the counseling room provided to the researcher by the hospital management team. At times, undesirable bystanders who were not even the patients' caretakers or family members, amassed due to curiosity of what was going on. These people further increased the inevitable technical distractions (e.g., noise from conversation around, telephones and etc.). The researcher was aware that such presence of strangers around could be a source of discomfort for the participants and would divert their minds while answering the questionnaire. This could affect the very credibility of the process. Still, asking people to leave the place would be rude and inappropriate. It was crucial to determine an 'ethical line' as discussed in the work of Thummapol et al. (2019). Even though, the participants did not get agitate at this, the researcher found it frustrating. She was concerned about the potential breaches of privacy and confidentiality, especially as she was a psychology counselor. As a result of these experiences, the researcher was left with questions about whether confidentiality and privacy can also be assured alongside maintaining sensitivity towards cancer patients.

Emotional and Psychological Well-being of Researcher

This section is about the issues related to researcher's emotional and psychological well being when working with people suffering from traumatic pain all around her. The risk of compassion fatigue and secondary traumatic stress being imposed on her could not be overlooked. The emotional involvement of the field researcher has received very little attention. Ethically, the researchers have to consider the respondents only as a source of data. But human factor corroborates while interacting with the respondents. The researchers tend to get attached to the field respondents (Gangopadhyay, 2019). In this context, symptoms of secondary traumatic stress could be seen.

Secondary traumatic stress is caused by repeated indirect exposure to aversive trauma details. This may occur in such professional settings where the counselors, healthcare professionals and therapists are consistently working with the clients suffering acutely. Similar was the situation in the present scenario due to close association with suffering patients; the researcher could get into an emotional turmoil. In addition to emotional and mental toll upon the researcher, occasional symptoms of secondary traumatic stress and compassion fatigue interfered with the daily activities. This cast an adverse impact on the completion of the fieldwork process within preliminary phase of the determined timeframe.

As discussed by Brown (2020), the researchers at University of Kentucky Center on Trauma and Children were awarded a grant from the Substance Abuse and Mental Health Services Administration (SAMHSA) to develop and test interventions and methods to decrease traumatic stress symptoms and improve functioning in professionals who have high doses of indirect trauma exposure.

Conducting research involving cancer patients can create a number of threats to the physical, psychological and emotional safety of the researchers, particularly novice and/or graduate student researchers. The need to secure relief backing for researchers is mentioned in the work of Thummapol et al. (2019), yet there is paucity in discussing challenges and issues related to researchers who could be in vulnerable positions. The paperwork at every stage required intricate detailing of the steps to ensure that the research would not harm the safety, dignity or privacy of any research participant. While the researcher agreed with the need for a thorough consideration to protect the participants as a primary focus, she felt that a section was missing in the process which should be accounting for any potential risks, discomfort or harm to the student researcher. She wondered why a researcher's safety and vulnerability is taken for granted and not mentioned anywhere in the ethical guidelines.

Despite having adequate supervisors' guidance and support, the researcher's fieldwork was an emotionally challenging experience. Although she became familiar about the anxiety and depression levels that affect the quality of life of cancer patients, listening to the sorrow of people affected by cancer burdened her own emotional stress. The researcher's vulnerability became clear and it was increasing. This was noticeable as she often struggled with emotional responses while listening to some of the patients' concerns and catharsizes.

Secondary traumatic stress is accompanied by symptoms like difficulty in sleeping, anxiousness, fear, physical sicknesses, irritability, burnout, hopelessness, and compassion fatigue. This could be defined as an emotional turmoil of an individual after listening to someone else's shared traumatic experience (Sprang, Clark and Whitt-Woosley, 2007). The symptoms of secondary traumatic stress often lead the researchers to undergo compassion fatigue, thereby ultimately affecting their cognition and emotions at times. This issue needed attention and thus some significant steps and actions were followed to address and resolve the same by emphasizing the importance of having an internal and external focus upon enhancing researcher's compassion, and patience (Stockwell, 2017).

The researcher found that regular meetings with her supervisors were effectively helping her to cope-up with such issues and provided her with some sort of emotional support. Sherry (2013) proposed that 'quality supervision' is essential as it brings researchers' emotions into spotlight drawing professional support. The audio/video conferences were done on a regular basis as when required through social networking sites that provided her with clear guidance. There were regular interactions and periodic meetings with consultants of respective departments to discuss alternative plans, any support needed and to debrief. At times, the researcher was reluctant to acknowledge her feelings about the fact of being under emotional stress. She pretended that the process did not affect her at all; however, later on she realized that suppressing her emotions had affected her health and well being. She had headaches and was occasionally sleepless at some nights, which were symptom of compassion fatigue and secondary traumatic stress.

Despite the emotional challenges experienced, there were some very positive aspects of undertaking research with cancer patients. The researcher could identify the actualities gradually. She substantiated that every individual suffering from cancer copes-up differently with the anxiety, depression and deterioration in the quality of life. Some patients do have the resilience power to adapt to the circumstances. The researcher learned that death is inevitable and certain, and the fear and pain of the cancer patients under therapy is inescapable, yet they can be trained to regulate their emotions simultaneously to make situations better through mind-management skills as discussed by Baer (2003). Unfortunately, the social, financial, and environmental factors added fuel in the fire, thereby aggravating and worsening their quality of life.

The support received throughout the data collection process helped the researcher to delve deep in the fieldwork. She learned about the inadequacy regarding the physical, psychological, social and environmental support in the lives of her participants. Many of them were living alone with minimal financial aid and dealing with multiple drawbacks. Despite the emotional challenges and limitations, the researcher's attitude, willpower and motivation acted as the force that kept her going ahead. Profoundly, she was filled with feelings of gratitude and blessings to those cancer patients who actively participated in her research study during the inpatient chemotherapy and outpatient radiotherapy sessions.

Similar studies were conducted with respect to the nurses working with oncology patients. These studies provide recommendations and means for reducing symptoms of secondary traumatic stress, compassion fatigue and burnout. Throughout the literature review, it was observed that there is a dearth of studies regarding the emotional and the psychological well being of student researchers of oncology.

Consequently, considering this limitation, the researcher has advocated and highlighted the challenges and resolutions for nursing officials who are working with cancer patients. These nurses are dealing with the patients on regular basis are acting as counselors as well. For instance, Duarte and Pinto-Gouveia (2016) conducted an experimental study on a sample consists of nurses with an objective to reduce compassion fatigue. As an intervention, the researchers carried out fifteen minutes to two hours of mindfulness and meditation exercises for six weeks. Results showed that nurses in the experimental group scored less in compassion fatigue and burnout, thereby progressing towards high compassion satisfaction. In a similar study conducted by Hevezi (2016), it was observed that there was a marked reduction in oncology nurses' compassion fatigue and burnout scores and a remarkable increase in the levels of compassion satisfaction and self-care by short-term meditations practice. A compact disk with the required instructions and demo was provided to them for that purpose. Thus, it was observed that various coping methods were quite effective in increasing the compassion satisfaction level.

From the above studies, it could be stated prospectively that "mind management skills" as discussed by Baer (2003), could help the novice researchers in maintaining their psychological and emotional well being during their fieldwork related to oncology patients. A strategy the researcher followed herself to enhance her psychological well being. This practice reduces the perceived stress and emotional tension acquired from the researchers and practitioners daily contact with patients in a delicate state of health, thus enhancing clarity of thoughts, intelligence level, greater compassion, gratitude, and inner peace within self.

RESOLUTIONS

The emotional turmoil of a researcher is not mentioned anywhere in the research ethics. However, it is important that researchers should be trained in the 'mindfulness based cognitive therapies' (Kabat-Zinn, 1990) beforehand, so that they do not get distressed themselves as tough circumstantial conditions may encroach during any sensitive research process. Therefore, self-care is an important and focal issue to be taken care of to resolve issues posited by 'secondary traumatic stress'.

The authors of Bloomquist, Wood, Trainor and Kim (2015), Eastwood and Ecklund (2008), and Killian (2008) discussed in their research that physical care can be achieved by eating healthy balanced diet instead of consuming caffeinated beverages, junk food and fast food meals followed by sufficient sleeping hours, exercising, taking time for vacations, engaging in hobbies and seeking medical care as needed are all equally important. On the other hand, emotional self-care includes but it is not limited to spending time with entertaining people, laughing, nurturing healthy/trusting relationships, reading a favorite book and seeking a therapy when required (Bloomquist et al., 2015; Eastwood & Ecklund, 2008). Furthermore, psychological self-care is recommended in practicing mindfulness as discussed by Christopher and Maris (2010), and conscious breathing and putting attention on the present moment (Kabat-Zinn, 1990), thereby feeling calm and more grounded. Journal writing, personal goal setting, and taking time for reflection are also beneficial. Additionally, spiritual self-care can be useful when spending time in a spiritual community or amidst nature, prayers, inspirational reading, meditation and yoga that might reinforce positivity and optimism in the fieldwork. It has also been revealed in the works of Cox and Steiner (2013), Harrison and Westwood (2009), and Killian (2008) that meditation, spiritual connect and prayers play significant part in self-care and are valuable in managing workplace trauma and stress, further reducing despair and isolation. It is also recommended to be proactive and build a strong foundation support with family, friends, supervisors, guide, colleagues and other professionals of the department.

The authors of Bloomquist et al. (2015), Eastwood and Ecklund (2008), Killian (2008) MacNeil (2010), Miller (2015), discussed in their research that deepening self-awareness, meaning, pleasure, enjoyment, altruism, relational connections, physical exercise, nutrition, spending time with family and friends, self-expression, maintaining a spiritual connection and mindfulness are all factors that play a vital role in the management of internal and external focus to orient oneself towards self-care, thereby distressing themselves and becoming motivated.

All of the above-mentioned factors and recommendations enhanced the researcher's feelings of calmness, developed her confidence in her abilities, increased her sense of tolerance of ambiguity and enhanced her ability to relinquish the control and the judgmental approach. This ultimately made her more aware and less reactive, and helped her gain compassion satisfaction and acceptance for herself.

RECOMMENDATIONS

From the vast field experience of the researcher, it is understood that there is a strong need for a support structure of counselors and therapists who can aid in increasing motivation and boosting self-confidence of the researchers who are carrying out research that involves cancer patients undergoing therapy sessions. The writing of journals and discussions with the peer groups and supervisors are certainly beneficial to some extent. However, exercising mindfulness, emotion management and emotion regulation further help in distress tolerance while dealing with the cancer patients.

The focus on the emotional and the psychological well being of the researchers should be well addressed and taken into account. Efforts must be geared towards training new researchers through internship programmes prior to the start of a similar research project. Such essential efforts will undoubtedly boost the robustness of social science researches.

The present chapter highlighted the fieldwork challenges in collecting sample data from sensitive population (cancer patients), matching with the theme of the book to overcome challenges in fieldwork, thus aiding novice researchers and graduate students in the field of counseling and psychology in particular. In addition to what has been covered throughout the chapter, here are certain recommendations from the experience of the researcher herself that might help novice researchers to some extent. It is recommended that researchers take breaks in between, to de-stress, talk with colleagues and other officials regarding the research study, and find a balance between the research study process phase and one's personal life.

Furthermore, it is important for researchers to be provided with opportunities for on-going training to learn appropriate skills of dealing with secondary traumatic stress, burnout and compassion fatigue during the fieldwork exposure in the research study. Killian (2008) had proposed and recommended some strategies to recognize the warning signs of compassion fatigue and burnout, as this recognition can enable to take action towards prevention and/or recovery.

Above all this, one thing should always be noted. It is the fact that all researchers try to make a blueprint of their research along with the timelines; however, in the practical scenario, the goal is not always achievable. This is mainly when it is the researcher's first time experience in the field. It will be noteworthy if the researchers make use of a Gantt chart and be flexible in conducting quality research.

In the end, the researcher learned from the fieldwork experience in dealing with the cancer patients that understanding life can counter the undergoing of any suffering. Healing of self is of prime importance, so that other participants also see the ray of hope that they will be healed. Therefore, it is important for the novice researchers in this field to learn to relax, cultivate social support (familial and collegial), and most importantly focus on self-care, in order to refuel and restore energy when confronted with secondary traumatic stress, burnout and compassion fatigue. Thus, by adopting the recommendations for the challenges faced during fieldwork, the prevention and dissolution of cognitive biases, if any, can be done efficiently. This is of prime importance when dealing with a sensitive case such as cancer patients for the research work project.

CONCLUSION

According to the researcher's experience, the success in fieldwork process is inevitable to guarantee the research success. Following the ethics is essential and integral, particularly with the collection of data from sensitive population group. This reflective account of fieldwork experiences provide a glimpse into those fundamental issues of any (doctoral) research project, which are often overlooked or taken for granted. Such was the need to bring up the challenges encountered in carrying out a research with cancer patients who were undergoing chemotherapy and radiotherapy sessions at a hospital in New Delhi, India. This chapter addressed personal emotional aspects (good ones and otherwise as well) that surfaced during the research process, which are rarely written about, or are less spoken about in the literature. The reflective suggestions and recommendations can be the resource for novice researchers and graduate students particularly in the field dealing with or involving cancer patients as the potential participants. The study of concepts involving fieldwork research is not direct and simple as it involves the adept mindset to ask the right questions in order to arrive to appropriate answers. Therefore, gaining specific academic and research techniques that are practiced skillfully by graduate students pursuing research-oriented courses or modules are mandatory (e.g., research proposal, dissertation, thesis, and etc.)

The researcher discussed the challenges faced in order to reach to the stage of data collection through meticulous IRB approval, followed by recruiting the potential participants. The researcher has expressed gratitude to the officials, IRB committee members, desk executives, academic professionals, guides, consultants and research participants who helped her to accomplish her effortful task of data collection. Last but not the least, the researcher felt emotional turmoil due to her exposure to the live scenario of life of cancer patients undergoing chemotherapy and radiotherapy sessions, which caused her compassion fatigue. Nonetheless, as the time passed by, she got recovered and started to work in the academics with other students in a normal way. She can now cultivate moral values and compassion among others by narrating her close interaction incidents, and thus move others towards taking care of the elderly and those suffering from a terminal illness.

As stated by Yu and Gatua (2014), there is an argument that one can never anticipate all the possible challenges and issues prior to undertaking fieldwork. Prior to the actual fieldwork, the researcher was continuously reviewing literature extensively. However, after her practical entry into the field for data collection, many things got unfolded. She learned that facing the challenges is cropping up with simultaneous resolutions as she got engrossed day by day into the process. The novice or student researchers must be guided to prepare themselves even better to deal with both, the anticipated and the emerging challenges as well as the work progresses. Writing a reflective account of research experience not only serves as an aid to the other researchers working in similar circumstances, but also helps the researcher to alleviate and manage their own secondary traumatic stress symptoms. In conclusion, it is noteworthy that emotional and psychological well being plays a vital role in the journey towards the quest for knowledge acquisition, pursued in the social sciences disciplines. In this branch of study, those researchers who are in the pursuit of knowledge exploration often undergo a compassion fatigue and secondary traumatic stress in order to complete a doctoral research project. The success of the fieldwork highly depends on the researcher's ability in gaining access to the suitable site, building rapport and trust, maintaining confidentiality and privacy, as well, and protecting the indispensable emotional and psychological well being of the researcher.

The fieldwork experience taught the researcher that those who are most empathic are also the most vulnerable to the secondary traumatic stress with compassion fatigue and burnout. The researcher's sensitive perception made her feel the gloomy little world of the patients suffering from cancer. She sensed powerlessness while observing their monotonous stay in IPD or waiting for long hours at OPD in the hospital. The chapter portrays how the researcher's experiences in the medical sector opened her eyes towards different aspects of the hospital care, the patients' sufferings and illnesses, which she had not known earlier at all. Thus, every experience that the researchers improvise and gain during fieldwork exposure could later on becomes a path and lesson for other researchers in that specific field of study, just as in the case of the present researcher.

It can thus be concluded that the book covers a series of practical experiences, all-encompassing the expected as well as abrupt and tough challenges encountered during the completion of different research studies. It presents sound resolutions and remarkable recommendations with an objective to add significant values to the human understanding.

REFERENCES

Baer, R. A. (2003). Mindfulness training as a clinical intervention: A conceptual and empirical review. *Clinical Psychology: Science and Practice*, *10*(2), 125–143. doi:10.1093/clipsy.bpg015

Bloomquist, K. R., Wood, L., Friedmeyer-Trainor, K., & Kim, H. W. (2015). Self-care and Professional Quality of Life: Predictive Factors among MSW Practitioners. *Advances in Social Work*, *16*(2), 292–311. doi:10.18060/18760

Brown, H. (2020). *Researchers Awarded Grant to Study Secondary Traumatic Stress in Health Workers*. Retrieved August 5, 2020, from http://uknow.uky.edu/research/researchers-awarded-grant-study-secondary-traumatic-stress-health-workers

Christopher, J. C., & Maris, J. A. (2010). Integrating mindfulness as self-care into counseling and psychotherapy training. *Counselling & Psychotherapy Research, 10*(2), 114–125. doi:10.1080/14733141003750285

Corr, C. A., Corr, D. M., & Doka, K. J. (2019). *Death & dying, life & living* (8th ed.). Cengage.

Cox, K., & Steiner, S. (2013). Preserving commitment to social work service through the prevention of vicarious trauma. *Journal of Social Work Values and Ethics, 10*(1), 52–60.

Duarte, J., & Pinto-Gouveia, J. (2016). Effectiveness of a mindfulness-based intervention on oncology nurses' Burnout and compassion fatigue symptoms: A non-randomized study. *International Journal of Nursing Studies, 64*(1), 98–107. doi:10.1016/j.ijnurstu.2016.10.002 PMID:27744228

Duncombe, J., & Jessop, J. (2012). Doing rapport and the ethics of faking friendship. In Ethics in qualitative research (pp. 108–121). London: Sage. doi:10.4135/9781473913912.n8

Eastwood, C. D., & Ecklund, K. (2008). Compassion fatigue risk and self-care practices among residential treatment center childcare workers. *Residential Treatment for Children & Youth, 25*(2), 103–122. doi:10.1080/08865710802309972

Gangopadhyay, J. (2019). The ethical challenges of field research: A personal narrative. *Indian Journal of Medical Ethics, 5*(1), 34–38. doi:10.20529/IJME.2019.081 PMID:32103818

Harrison, R. L., & Westwood, M. J. (2009). Preventing vicarious traumatization of mental health therapists: Identifying protective practices. *Psychotherapy (Chicago, Ill.), 46*(2), 203–219. doi:10.1037/a0016081 PMID:22122619

Hevezi, J. A. (2016). Evaluation of a meditation intervention to reduce the effects of stressors associated with compassion fatigue among nurses. *Journal of Holistic Nursing, 34*(4), 343–350. doi:10.1177/0898010115615981 PMID:26598000

Kabat-Zinn, J. (1990). *Full Catastrophe Living: Using the Wisdom of Your Body and Mind to Face Stress, Pain, and Illness*. Bantam Doubleday Dell.

Killian, K. D. (2008). Helping till it hurts? A multimethod study of compassion fatigue, burnout, and self-care in clinicians working with trauma survivors. *Traumatology, 14*(2), 32–44. doi:10.1177/1534765608319083

Kubler-Ross, E. (1969). *On death and dying*. Macmillan Publishing.

Miller, J. (2015). Seven Self-Care Strategies. *Reflections: Narratives of Professional Helping*, *21*(1), 52–58.

Newell, J., & MacNeil, G. (2010). Professional burnout, vicarious trauma, secondary traumatic stress, and compassion fatigue: A review of theoretical terms, risk factor, and preventative methods for clinicians and researchers. *Best Practices in Mental Health*, *6*(2), 57–68.

Sherry, E. (2013). The vulnerable researcher: Facing the challenges of sensitive research. *Qualitative Research Journal*, *13*(3), 278–288. doi:10.1108/QRJ-10-2012-0007

Sprang, G., Clark, J. J., & Whitt-Woosley, A. (2007). Compassion fatigue, compassion satisfaction, and burnout: Factors impacting a professional's quality of life. *Journal of Loss and Trauma*, *12*(3), 259–280. doi:10.1080/15325020701238093

Stockwell, A. (2017). *Effectiveness of Self-Care in Reducing Symptoms of Secondary Traumatic Stress*. St. Catherine University. Retrieved from https://sophia.stkate.edu/msw_papers/797

Thummapol, O., Park, T., Jackson, M., & Barton, S. (2019). Methodological Challenges Faced in Doing Research With Vulnerable Women: Reflections From Fieldwork Experiences. *International Journal of Qualitative Methods*, *18*(1), 1–11.

Yu, L., & Gatua, M. W. (2014). Methodological considerations for qualitative research with immigrant populations: Lessons from two studies. *Qualitative Report*, *19*(30), 1–16.

KEY TERMS AND DEFINITIONS

Burnout: It is characterized by the symptoms such as, emotional exhaustion, pessimism, and feelings of ineffectiveness at work.

Cancer: Cancer is defined as a generic term for a large group of chronic diseases that is known to be pathologically characterized by abnormal rapid growth of cells thereby affecting any part of the human body of both genders.

Chemotherapy: Chemotherapy is defined as a concentrated and repeated treatment drug regimen for the management of cancer and its related symptoms. It is found to be associated with adverse reactions such as, hair loss, nausea, vomiting, and diarrhea, thereby leading to extended periods of treatment and repeated admissions to the hospital eventually affecting the QOL of cancer patients.

Compassion Fatigue: It can be defined as a stressful situation that leads to emotional turmoil by exposure and re-experience of difficult events.

Confidentiality: It is defined as one of the key aspect of ethical consideration of keeping the participant's information, such as name, socio-demographic details, identity, medical history and records as a secret and not disclosing it to anyone except for research and publication purposes.

Emotional Well Being: It majorly deals with the emotions that are event-driven (physical), temporary, conscious experience, and often unleashed by an external or internal stimulus.

Inpatient Department (IPD): A ward in the hospital where a patient diagnosed for a serious condition, such as terminal illness needs to stay there for a short duration or days in order to be treated with state-of-the-art facilities at the hospital for better outcomes.

Outpatient Department (OPD): A ward in the hospital where a patient diagnosed for a serious condition, such as terminal illness comes at a prior taken appointment date and time for the intervention and then after the administration of the appropriate intervention, gets back to their respective homes. Here, the patient need not to stay whole day and night, but have to wait for his turn for intervention to be administered.

Psychological Well Being: These are defined as the learned behaviors predominantly self-perception that deals with the feelings, arising out of sparking and subjective interpretation and expression of emotions that may persist for a long time.

Quality of Life: Quality of life is a general term that integrates patients' physical (movement, physical activities, ability to succeed in work and in family responsibilities), psychological (life satisfaction), social (social activities, being beneficial, body image, anxiety and depression, social support need and role function), economic, spiritual, cognitional &environmental dimensions for wellbeing during the diseased and respective-treatment stage.

Radiotherapy: Radiation therapy uses high doses of radiation to kill cancer cells and shrink tumors. It might cause side-effects such as, skin pigmentation, destruction of salivary glands, severe problems related to eating, swallowing and speech, causing xerostomia, oral infections, dental caries, pain, and discomfort.

Rapport: Ethical Rapport can be defined as a friendly, harmonious relationship characterized by agreement, mutual understanding, or empathy that makes communication possible or easy for collection of data from the potential participants.

Section 6
Post-Disaster Research

Chapter 10
Reflections on Post-Disaster University Research in the Aftermath of Hurricanes Irma and Maria

Kula A. Francis
University of the Virgin Islands, US Virgin Islands

Kenny A. Hendrickson
University of the Virgin Islands, US Virgin Islands

ABSTRACT

This chapter offered a reflection on a research study conducted at the University of the Virgin Islands, a historically black college and university (HBCU), in the aftermath of Hurricanes Irma and Maria. The research study examined student perceptions of authentic university academic care (AUAC) in a post-disaster setting. AUAC can be viewed as genuine humanistic concern and disciplined nurturing in higher education. AUAC is also an important element of examining the quality of academic services. Thus, this chapter shares the experiences of the researchers, their investigative processes, the challenges faced, solutions used, and recommendations.

INTRODUCTION

It is almost impossible to precisely determine the nature, location, time, degree of damage and disruption a hurricane can cause. Unfortunately, the number, intensity and destructiveness of seasonal hurricanes and tropical storms are rapidly increasing

DOI: 10.4018/978-1-7998-5826-3.ch010

in danger and strength (Patel, Min and Lim, 2016). A hurricane's threat to the Caribbean and mid-Atlantic region is significant. The harms caused by hurricanes ultimately affect all aspects of life for residents in societies impacted by these natural disasters. However, natural disaster research is limited due to the unpredictable nature and challenging reality (Taylor, 2016). Furthermore, most of the existing disaster research focus on the geophysical environment, the mental health and psychological consequences of disaster victims. There is not enough research on the significance of a disaster on institutions of higher education.

This chapter revealed the journey of two researchers who unintentionally completed a research on university care following disasters. Our post-disaster research centered on examining authentic university academic care (AUAC) in the aftermath of two category 5 hurricanes. As a reflection, this chapter starts with a summary of the research and context of the study. We also share the challenges faced, solutions used, and provide suggestions for future research.

RESEARCH SUMMARY AND CONTEXT OF THE STUDY

We originally began our research as an investigation of authentic university academic care (AUAC) at the University of the Virgin Islands (UVI). UVI is the only historically black college and university (HBCU) outside of the mainland United States. An HBCU is a college or university founded to educate students of African American descent. As an HBCU, UVI has historically excelled in providing quality academic services to the Virgin Islands, Caribbean, and United States. In our original research, we sought to examine university caring as a measure of academic delivery. The primary question of this research was: *Is UVI's academic care genuine or authentic?* Thus, authentic university academic care (AUAC) became viewed as genuine and humanistic concerns, caring motives, disciplined nurturing and a provision of educational caring necessaries in university's academic services (Hendrickson & Francis, 2020; Addison, 2012). Regarding quality of service, AUAC can be described as a product of students' institutional relationships and organizational systems (e.g., personnel, learning processes, technology). An inherent part of the HBCU experience centers on the term othermothering. Comparable to AUAC, *Othermothering* is defined as a student-centered relational caring that requires nurturing support, motivation and encouragement within academic settings. Othermothering delineates the provision of ethical caring (e.g. empathy, compassion, respect, and communication), "attentive and emotional responses", as well as an institutional culture of academic concern and guardianship (Hendrickson & Francis, 2020, p. 62). Since UVI is an HBCU, we strongly believed that authenticity of this university's academic caring was primarily the result of the othermothering provided by university personnel (i.e., faculty, administration, and service staff).

Our initial research attempted to recognize the components of academic caring, based on students' perceptions. As learners and customers, students' perceptions are useful in providing valuable information on the quality of academic services. Therefore, student's perceptions can be used to measure AUAC. A student-focused survey instrument was created to assess AUAC. *Caring about Academic Care Inventory* (CAACI), the instrument, was disseminated in spring 2017. The distribution of surveys was placed on pause at the end of the spring semester. We planned to continue distributing the survey in fall 2017, once the semester was on its way. Our plans did not go as scheduled, and the events that followed changed the trajectory of our study.

On September 6, 2017, Hurricane Irma struck the U.S. Virgin Islands (USVI), causing wide-spread damage on the islands of St. Thomas, St. John and Water Island. Hurricane Irma is listed as one of the largest hurricanes to ever make landfall in the Caribbean and Atlantic Coast. Less than two weeks later, on September 17, 2017, Hurricane Maria devastated St. Croix, USVI, and further damaged other islands in the USVI and Puerto Rico. Hurricane Maria has been recognized "as the tenth most intense Atlantic hurricane on record and the most intense tropical storm world-wide", as well as "the worst natural disaster on record in the United States" (Cange & McGaw-Césaire, 2019, p. 18). Both hurricanes have been listed among the costliest natural disasters (Cange & McGaw-Césaire, 2019).

Both hurricanes severely undermined important segments of the USVI territory's societal infrastructure including roads, hospitals, seaports, airports, businesses, housing and schools. Substantial damage also occurred at the University of the Virgin Islands (UVI). In his editorial, *UVI is Rising*, Dr. David Hall (2018), President of UVI, provided a post-disaster assessment of UVI. President Hall accounted that many of the critical buildings to the operations of the university across both its St. Croix and St. Thomas campuses as well as its satellite campus on St. John were compromised and rendered inoperable (2018, para 2). UVI's President equated that the damages to UVI ranged from $60 to $80 million (Hall, 2018, para. 2). The unexpected disasters put a temporary stop to all classes, research projects, and all other university business. The community's immediate safety needs did not allow for focus to be placed on returning to business as usual for the territory, businesses, and schools. The same hardships occurred for the University of the Virgin Islands. The basic needs of many were not being met. One of the researchers was forced to temporarily leave the island with a disabled parent, since the living conditions immediately following the hurricanes were life-threatening. Clearly, the status of the study we started in the spring of 2017 was furthest from our minds.

Amid the adversities, UVI was strongly committed to saving the 2017-2018 academic year and rebuilding the university. UVI President, acknowledged the efforts of the campuses' physical plants, operations staff, access and enrollment services (the Office of the Registrar), administrative leadership of the academic components including the Offices of the Provost, Deans of Schools and Colleges, the faculty, and those who took on informal leadership roles. President Hall (2018) noted, in his post-disaster assessment, that the stark reality was that UVI had to resume under conditions such as curfews as well as large portions of the territory lacking permanent power, communications and internet services, including the university campuses (para 5). Over 2,300 students, 500 staff members at UVI, and their families, were also caught in the reality of recovering from their personal damages, losses and the major life disruptions resulting from the hurricanes.

On Monday, October 9, 2017, approximately one month after the Hurricanes occurred, UVI resumed academic operations under less than normal conditions. Although UVI reopened, continuation of our study was still not clear to us. In October of 2017, we began to find our way back into our world of teaching and research. Honestly, this world had disappeared from our priority lists. We realized that our original study, which begun the semester before (Spring 2017) would yield responses from a different mental place for many of the students, still recovering from what was for some, the worst experiences of their lives. With some apprehension, we decided that it was best to resume the study as soon as it was safe for the participants, and at the least, marginally accessible to them. The authors recognized the value of gaining students' perception of university care in the aftermath of the hurricanes. We also recognized that having a study in progress before the hurricanes offered us the opportunity to start right away in completing the post-disaster focused study, and to utilize an established trust built with university students. We resumed the data collection process with the hopes that the effects of the disastrous hurricanes would not deter students from participating in the study.

SURVEY AND DATA COLLECTION PROCESS

We used a cross-sectional exploratory survey research design. This method was chosen because it allowed us to explore university caring using a sample of the population at a specific point in time. We collected data on student's perceptions of Authentic University Academic Care (AUAC) using our original survey, *Caring about Academic Care Inventory* (CAACI). CAACI was a 49-item questionnaire. The CAACI was sectioned within six subscales. The first subscale consisted of five demographic items (age, gender, primary ethnic identity, student status, and principle field of study or discipline), which provided a student narrative to the study. The authors have found that demographic items provide additional (bonus) approaches for extensions to our previous studies. We have also been able to look to the demographic items to unveil angles that we were not expecting to find significance in. For example, students who were majoring in the Natural Sciences (Chemistry, Biology, and Physics) felt the highest level of caring of all other students.

As seen in Figure 1, the next five CAACI subscales were established based on a university academic caring carescapes framework (Framework). This conceptual framework was designed by us to showcase the human relational networks structured to provide academic caregiving to students based on HBCU othermothering ideologies. Academic caregiving can be defined as the delivery of assistance and support to students/learners within academic services. The framework spotlights not only mutually related structures and activities of academic caregiving, but also non-routine and informal caring within university academic components (McQuoid, Jowsey & Talaulikar, 2017, p. 143). Thus, the framework identified academic caregiving points within a university. Academic caregiving-points are conduits that provide students with the necessary caregiving and caring needed for educational achievements. Figure 1 categorized four basic academic caregiving-points: 1.) Institutional operations (administration / support service) provision of care within academic service delivery; 2.) Academic management (colleges / schools) ensuring the necessary level of care was offered within degree curriculum / curricula; 3.) Instruction (teaching faculty) included primary pedagogical academic care within the course learning environment; and, 4.) Academic advisement (faculty and staff advisors) to make certain that students were adjusted to their academic experience and offered curriculum and career grounded guidance. Thirteen caring about academic caregiving (CAAC) criteria were identified and utilized (as seen in Figure 1) to contextualize each academic caregiving point. Table 1 provided the operational definitions for the CAAC criteria. Therefore, the items for the CAACI subscales were constructed based on the following CAAC criteria. The items were evaluated based on a 7-point Likert scale: 7 = Strongly Agree; 6 = Agree; 5 = Somewhat Agree; 4 = Neutral; 3 = Somewhat Disagree; 2 = Disagree; and, 1=Strongly Disagree.

To ensure the validity of the CAACI, we requested the assistance of a panel of experts comprised of five teaching faculty from various academic disciplines (Psychology, English, Business, Natural Sciences and Mathematics), two academic deans (Liberal Arts and Social Sciences and Education), and a vice-provost of student affairs, to evaluate CAACI. Their scrutiny ensured face and content validity. The individuals were chosen based on their involvement in other research projects and / or their participation in a University Research Writing Group, which the authors are also members. Our intent was to utilize as many academic areas, so that their proficient observations were from varied fields. We requested that the panel of experts review the CAACI for the following conditions: 1). the extent to which each item reflected

Figure 1. UACCF

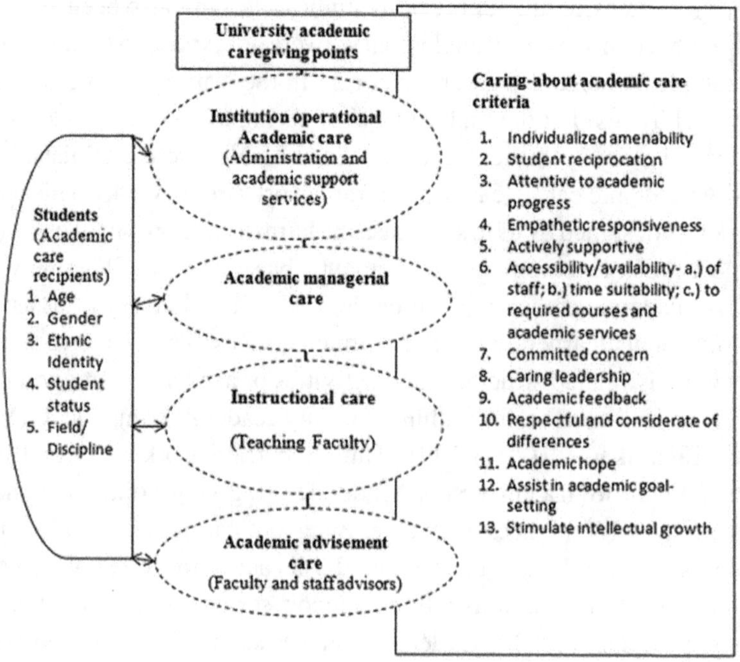

and was consistent with the conceptual framework and the operational definition; and 2). the quality of each item in terms of readability, clarity and simplicity. Based on the panel's evaluation and comments, we made revisions to the CAACI. There were a few items (questions) in the CAACI, where more than one variable was found within the same item. As a clear example, imagine if the item was, 'I like to eat and drink'. In this item, although it discussed the same concept, eating and drinking are separate variables, and should be posed separately. We made the adjustments accordingly and were ready to move on to the next step of the study.

We decided to use a web-based survey, based on our past experiences with it. The CAACI (survey instrument) was converted into Google Forms. Google Forms offered trouble-free accessibility, anonymity and confidentiality. Google Forms also allowed us to streamline the process of CAACI's distribution and data collection. We believed that it may be more beneficial than other options (in person, telephone or mail). We also used this format because of its convenience.

Because of our positions as faculty, we were able to gain access and utilize UVI's student email list serve, which included the official email of all students who were attending UVI. With the entire research plan in place, the authors submitted an application to the university's Institutional Review Board (IRB). After the application was approved, we sent introductory emails to all students' university email addresses.

Figure 2. Operational definitions for a criteria of caring about academic care

1. **Individualized Amenability:** treating learners as individuals and with consideration of their best interests. The offering of institutional connectedness and motivation for students.

2. **Student Reciprocity:** exchanging of institutional and academic support and mutual commitment.

3. **Attentions to academic progress:** establishing personalized relationships with mutual respect and trust along with an open channel communication between academic caregiver and students (Levin, Cox, Cervin & Haberler, 2010 p.34).

4. **Empathetic Responsiveness:** the offering of a consistent, affective and unique professional responses to students' requirements (De Waal, 2008).

5. **Committed Concern:** an unwavering quality of involvement, dedication and obligation to students' academic development (Ujváriné, Zrínyi, Tóth, Zékányné, Szögedi, & József, 2011).

6. **Actively Supportive:** capability of enthusiastically providing necessary resources and helpful encouragement to students.

7. **Accessibility/Availability:** the degree of constancy in attentive availability to students when needed. Reynders, Kerkhof, Molenberghs, and Van Audenhove (2014) spotlights accessibility and availability, as necessary to increase students' willingness to seek assistance.

8. **Needs Amenability:** an accommodative and adaptive nature towards academic services to fulfill students' academic needs, especially those requests that lead to educational accomplishments.

9. **Mindfulness of Differences (Personalization):** awareness of diversity and differentiations of a student population (e.g. ethnicity, culture, age, gender, lifestyle, disabilities or economic status) and its potential bearing on academic development and educational achievement.

10. **Hope:** a belief or assurance of students' abilities to progress through a pathway in achieving their goals of degree completion (Snyder, Shorey, Cheavens Pulvers, Adams and Wiklund, 2002, p. 820).

11. **Academic caring leadership:** authoritative (formal/informal) directing, guiding, promoting and supporting the nurturing of students, inclusive of *active academic caregiving* and an *ethics of care* within a university environment.

12. **Assistance with academic goal setting:** the process of aiding students with academic goal setting and planning, and self-evaluation in their academic advancements (Schunk, 2003, p. 161).

13. **Stimulation of intellectual growth:** ability to stimulate and encourage learning through cognitive development, educational nurturing and disciplinary rigor (Ancess, 2000, p. 595).

Our email included several virtual fieldwork tools, including information about the study, the assurance of complete anonymity and confidentiality, hyperlinks to the survey, clarifying podcast, and a cover letter. We also made it clear that no coercion, or conflicts of interest were involved. The solicitation email was our only interaction with the subjects. The authors realized the importance of this step to the entire study.

Once a student completed and submitted the CAACI survey, the data was captured in an online spreadsheet created in Google Forms. The immediate transfer into a spreadsheet limited data collection errors and allowed for us to quickly analyze that data. Appropriately, the collected data was only accessible via one of the author's university google account. Although our survey did not collect personal or identifiable information, the data was stored on the university's encrypted account to protect for any possible breaches.

After a month, we sent a follow-up email requesting participation for students who had not previously completed the survey. Once the data was collected, we coded and analyzed it using NCSS, which is a statistical software. NCSS is a statistical package, which allows researchers in academic institutions and in businesses the ability to statistically analyze data. NCSS is comparable to both SPSS and SAS. We utilized NCSS only based on the availability of this product on our personal computers.

Although research has previously noted on average, about 10% lower response rate per web-based than other survey modes, the benefits to us seemed to outweigh the harms (Fan & Yan, 2010). Fan and Yan (2010) also noted the advantages included shorter transmission time, lower or no delivery cost, more design options, less data entry time, and little or no need for cleaning or transcribing data. These benefits were especially favorable for us since no funding was available. Another advantage for conducting fieldwork via an online platform was that the participants tended to feel a greater sense of anonymity and confidentiality. This was important for us, since we were conducting research following disasters with students who might be afraid to give their true perception as it related to their professors, support staff, and the university.

CHALLENGES FACED

Conducting this research brought about a number of obstacles with it. Mainly, there were four major challenges: 1. following the hurricanes, the unspecified reopening of the University changed the timing and timeline; 2. the hurricanes caused a high level of friction between the researchers due to differing views on the ethical implications of immediately continuing the study; 3. communications (internet service) and electricity were not widely available or reliable while the islands recovered; and, 4. there were concerns on the survey response rates as it related to the appropriate number of respondents necessary to determine generalizability based on sample size. We realized that these main challenges overlapped, were interconnected, and often led to other smaller snags along the way.

Timing and Changed Timeline

The changes to our timeline were problematic. Nevertheless, these challenges had become inevitable given the disasters that had occurred in September. The hurricanes' timing affected the focus of our study and the data collection process. Not only were there added uncertainty on our expected return to the university, we were also unsure of how the level of participation would affect our outcomes once we resumed the study. Our major objective was to complete the study by May 2018 (the end of the academic year). We had already put in place a rigorous schedule of data analysis and research writing to commence immediately after the end of the data collection process. Because of our timeline, any small change in our timeline could be harmful to our productivity. We also noted that a research timeline "allows researchers to set a higher pace for completing the research study tasks" (Hendrickson & Francis, 2015, p. 233). Thus, it was very important for us to conclude the survey process on time. On time meant that our research study would be concluded at the end of spring 2018. The uncertainty of our schedule was a challenge for us since many households remained without everyday utilities like the internet. We considered the impact of the disaster's timeline on our research project. Xiao, Huang and Wu (2015) identified five phases of natural disaster timeline: (1) Prevention, preparedness, and mitigation; 2. Watches and Warning; (3) impact; (4) response; and, (5) recovery. Phases 1 to 3 had already concluded by the time the hurricanes left our region, so these phases did not affect our study. We were left in the response phase, and was hoping to quickly move to the recovery phase. Government, businesses, and other organizations (Including UVI) rapidly tried to get things back to some level of normalcy. Contingency plans were put in place which aided in the response and recovery process. The immediate response phase occurred in the first four weeks after the impact of the disasters, and the longer-term recovery was to last for years afterwards. Throughout the Virgin Islands, loads of fallen trees, roofs, and debris were removed from the streets and roadways. At the University of the Virgin Islands, assessments were made, and the challenges ahead were discussed with employees. UVI was able to reconvene the fall semester approximately a month after the disasters.

By mid-November, the U.S. Virgin Islands was head long in the process of long-term recovery. During this time, critical societal services, such as electricity, internet, communication services (telephone and cellphone) were steadily resuming into businesses, government offices, and homes in the Virgin Islands. As well, the rebuilding efforts were underway. Since some level of normalcy had been regained, we agreed to start disseminating the emails, which requested participation in our survey. E-notices (via email) were dispersed to the student body on a monthly basis, starting from late November 2017 to early May 2018. We were able to take back some control of our timeline, when we ended the data collection process based on the original schedule of May 2018. By retaining the original deadline, the authors were able to maintain their publications goals. We also realized that there was no guarantee that responses would improve if the study continued on. Although the study lost two crucial months, and responses were not rapidly progressing, we recognized that we had gotten to the data collection finish line.

Differing Views Among Researchers

For some researchers, single-handedly conducting research seems paramount. On the other hand, the authors have found success in our collaborative efforts for more than five years. Although miniscule, individual differences have sometimes been a challenge for us. The proverbial, "there is no I in team" has rarely needed to be mentioned to either member of our partnership. However, with collaborative work, there is always the possibility for differences of opinion and contention. One of the challenges of this research emerged when finding a common ground seemed to be a trying task between the researchers. Both researchers had been knocked down by the events of Hurricanes Irma and Maria. Although we were both hoping to maintain the scheduled completion of the study, we knew that it was going to be a challenge.

As the university reopened, Author Two made it clear that the original schedule for data collection needed to commence as soon as possible. Author One questioned, "How could we ethically carryout a study on students during the recovery of a major, life changing event?" The hurricanes undoubtedly added pressures on everyone. The particulars of living in the aftermath brought about many difficulties. We needed to keep in mind that some of the participants possibly could have recently lost their homes, jobs, belongings and loved ones. The severity was also resounding based on the large number of students that withdrew from classes (approximately 15%).

Additionally, the reality of life in the aftermath of the hurricanes included no running water, no electricity for most of the population. Additionally, there were many homes that were severely damaged. The damages included loss of structure (walls), or more commonly, loss of roofs. In the aftermath of the hurricanes, families could not get in contact with other family members due to the limited or lack of available modes of communication for a timespan that went far beyond a week. A cell phone, even if someone was lucky enough to have one that was charged, hardly had signal to pick up or make a call. Residents were under curfew by dusk until dawn. On top of that, during the day there was major congestion and traffic on roadways. The simple pleasures of refrigeration and lights at night were not available to residents.

The lines into the grocery stores were long. It was common to spend two or more hours in line to purchase a single bag of ice.

Nonetheless, Author One felt that the circumstances facing residents (which included our students) were not optimal for the distribution of surveys. It is important to know that both authors were in agreement that the situation was harsh. Still, Author Two saw a need to immediately resume the study as an opportunity to collect the perceptions of students, while they were consciously concerned about the reopening of school and affected academic school year following the hurricane. Author Two believed the study would find new context in the area of disaster research. Author Two considered the possible assumption of harm, but felt assured that the Institutional Review Board at the University had approved the study, and that all necessary approvals were granted. Particularly, Author Two believed that students who participated in the online survey had to have met some level of safety and security (despite the hurricanes) if they were able to access and complete the survey. Additionally, before beginning the survey, all students would have reviewed the consent e-notice, and had the option to participate only if they chose to.

As a threat to the study, based on the timing and possible ethical concerns, Author One discussed discontinuing the partnership on this research. However, after a long discussion, we quickly realized that we were both entangled in different aspects of this research and needed for both partners to remain on task. For example, only Author Two had access to the Google Forms password and data collected in the spring 2017 semester. On the other hand, Author One had completed a comprehensive literature review on the research. We also reviewed our contingency plan that was prepared at the beginning of the study in early 2017. Unfortunately, it did not consider anything related to natural disasters. Despite the initial differences of opinion, we also came to a compromise that we would wait for an additional month before resuming the surveys.

Communications and Electricity Concerns

With the resumption of the fall 2017 semester in October, it became clear that in just about all aspects of life, residents of the Virgin Islands were doing the best they could with what they had. This was certainly the case with electricity and communications. The situation was not perfect, but citizens were adjusting as much as possible to the inconveniences of daily life. Power and communication outages were constant and expected. The local utilities companies had gotten a large amount of assistance from line crew who flew in from all over the US, especially from New Jersey and New York. They repaired thousands of feet of power and cable lines, which added to the frequent power outages for those who had gotten their power restored. Yet, there remained a large number of residents who had not gotten the services in their home. Could we as researchers expect anyone to complete a survey if the ability to check email, or conduct everyday internet queries required searching for the nearest government or business hotspot?

We knew the power and internet outages might have hampered with our survey returns, but we did not expect what we believe occurred. After the first month of surveys being in the inbox of the students, we started to look at the number of returned surveys. We noted that the responses were low. Given the situation, we were not expecting that the response rate to be extremely high. We also noticed that of the forty-seven surveys attempted in the first month, eleven of them were not fully completed. As researchers, we expected that some students may lose interest and not complete the survey, but we did not expect what we noted. Author Two decided to look more closely at the data to see if there were any patterns with the unfinished responses. Immediately, it was realized that in that first month, six of the eleven incomplete surveys were closed at the exact time, on the same day. This was somewhat of a phenomenon to us.

Although we cannot say with 100% certainty, we believe that there must have been a power or internet outage during the completion of the survey on that specific time and date. With either the power or internet outage, students were unable to complete surveys. This posed to be a challenging concern as we were already anxious about the return rate. All of the eleven invalid surveys in that first month had to be removed. Of the forty-seven surveys in the first month, eleven, or 23% of the surveys were not valid. Accordingly, on that one specific day, in that exact moment, we lost six surveys, or 13% of possible responses collected in November. At the end of the timeframe for the survey (May 2018), we were not able to reach the projected responses that we had hoped for. Of the total 269 attempted surveys, 210 were valid. A total of 59 surveys or 22% were voided for use in the study. We believe the repairing of the infrastructure certainly played a part as one of our major challenges. The connected circumstance of timing did not allow us to sit back and linger. We accepted that an obstacle to disaster research was the unexpected electrical and communication glitches; we moved on.

Concerns on Response Rates

Undoubtedly, the connection between the electricity and communication woes played heavily into the next major challenge – our overall response rates. The enrollment count at the University of the Virgin Islands was 2,300 at the start of the fall 2017 semester. Immediately following the hurricanes, about 350 students (about 15%) withdrew from classes at the university. The withdrawal numbers were unprecedented; however, understanding the effects of the hurricanes made it logical. The enrollment of students following the hurricanes was approximately 1950. Of them, a total of 210 students fully completed the Caring and Academic Caregiving Inventory (CAACI) survey. Of the 269 surveys started, there were 210 completed (valid) surveys. This number represented only 11% of the student body. With a population of 1950, a sample size of at least 322 would be needed in order to obtain a 5% margin of error and 95% confidence level (Islam, 2018). This meant, the findings of our study could not be generalized to the population, which caused some apprehension. We were fearful that the publishing of our findings would suffer from both small study and publication bias. There is generally hesitation that small studies are insignificant, lacking scientific rigor and credibility, unpublishable, or just a waste of time. In fact, small study bias and publication bias has been identified as common threats to small research studies. Small study bias is a belief that the observations and findings of small research studies "are more likely to produce a false positive result than are large research" (Ariyaratnam et al., 2007, p. 736). Likewise, publication bias refers to a situation, when research findings greatly differ from other results that have been published (Stanley, 2008). Stanley (2008) recognized that publication bias and small study bias are used by editors and reviewers of publications to support rejects and exert their preferences of statistical significance. Our major challenge of response rates ultimately was also connected to the issues of publication and small study biases.

The Impact of the Challenges

Not all learning experiences come easy. One of the troublesome situations is the struggle of dealing with challenges. We were faced with several challenges that left us disappointed. Inasmuch as we were in the recovery stage of the hurricane, and understood that this very issue could make or break our research, we tried to remain positive. There is, however, the unwritten feeling of sadness that consumes researchers when challenges become compounded. During the challenges, we both questioned the decisions that had been made. Could one researcher be to blame? What should have been done? Should we have continued the survey a little longer? Should we have forgotten about this study before we even started? Silently, we wondered, but never, at that time, did we ask these questions to each other. It is not until the study had concluded that the authors started to bring their thoughts of despair to the surface. Despite the challenges, it was the solutions that saved us from the experience of complete disaster.

THE SOLUTIONS UTILIZED

We acknowledged the downsides and challenges to this post-disaster research. The difficulties included an uncontrolled changed timeframe, differences in approach among research partners, electricity and communication concerns, and lower than expected response rates. To tackle the challenges of this research, we decided it was best to focus on three important concepts: 1. contextualize the study based on our current situation; 2. plan and create contingencies as we needed them, and, 3. accept the response rates, and gain ground on the completion timeline.

Adding Disaster Context to Research

Within the past decade, earthquakes, tornadoes, floods, tsunamis, wildfires, volcanic eruptions, hurricanes, and other biohazards (such as Covid-19), have been on the rise. In each of these natural (or man-made) disasters, there is greater opportunity for significant research to be conducted. We both realized that there could be great utility if we looked at our research from the perspective of disasters. In our case, hurricanes were not on our research agenda or related to our research inquiry. Nonetheless, this study became a fusion of university caring and natural disasters. Nilsen and Bernhardsson (2019) considered research context as a means of creating variations in studies (p. 190). Adding a new layer of context to our study provided prospects of finding new ideas through modifying the research intent and purpose. As a result, our work evolved! We were able to add the contextual layer of natural disasters while examining the level of care at the University. We were able to use the disaster as an opportunity to complete the research from a new perspective.

There were no thoughts of disaster research or anything related to hurricanes when we started the original study. Our only intention was to conclude the collection process of our ongoing research. The hurricanes led us to rethink our research approach. As cliché as it may sound, we took the lemons that we were given, and made lemonade. The solution to our immediate inability to complete the study turned into a new outlook on caring in the face of a disaster. We needed to change the approach from student's perception of university caring to student's perception of caring following a disaster. We realized that the disaster possibly impacted how students viewed university care. Thus, it was necessary to separate what had been collected prior to the hurricanes with what we collected after.

Creating a Plan and Contingency

We recognized the importance of planning for the successful completion of our research. According to Buswell, Webb, Mitchell, and Leder Mackley (2017), there are five phases to successful research: planning, recruitment, data gathering, data analysis and publications (p. 544). While the recruiting, data collection, data analysis and concluding publications are often the emphasis of research, we realized that planning is often overlooked and perhaps just as important as the other phases. Without a well-conceived plan, the remaining steps will falter. Although we were able to fulfill a worthwhile planning phase for our original research study, it became apparent that we needed additional planning contingencies for the new post-disaster context. While many researchers, especially novice, hope and plan for the best without expecting and planning for the worse, a contingency plan must be constructed. There is a need to ensure that a study will achieve its goals, in the face of unknown conditions and uncertain outcomes. Through our research, it became apparent that it is impossible to plan for all eventualities. After all, considering the effects of a hurricane, snowstorm, or earthquake prior to any research may seem pointless – unfortunately, we never know. In an attempt to circumvent failures, contingency plans should consider that failure is possible. In our research project, we found the need to plan for a likelihood that the distribution of our survey would not be successful due to conditions in the aftermath of the hurricanes. We also considered the possibility that we would not obtain a significant amount of responses to achieve our overall goal of publishing the AUAC data.

Thus, we decided to closely monitor the disaster recovery, along with the university's progress during recovery, in order to modify the project's survey distribution timetable. Following our contingency plan, we also agreed to utilize the initial data collected in the spring 2017, to complete a projected publication on university academic care.

Accepting the Response Rates

A major concern that we contended with was the decision to use a less than projected target sample size. We opted to accept the response rate as is, rather than prolonging the study to gain a greater response rate. Although we were dissatisfied with the response rates, we were determined to ensure statistical significance over the achievement of a larger sample size. We realized that the findings could not be used as a representation of the entire student body. However, internal validity was achieved through our careful data collection and appropriate statistical analysis. Evidence of university caring was apparent based on the 210 student responses. Although the study was not generalizable, there were other factors that worked in our favor. Pearson and Mundform (2010) acknowledged that a sample size of 100 is adequate to attain factor solutions for a preliminary analysis. deWinter, Dodou and Wieringa (2009) accepted that exploratory factor analysis can yield good quality and reliable results even with sample sizes below 50. Regarding correlations,

VanVoorhis and Morgan (2007) offered the general rule of thumb of no less than 50 participants, which would increase given the number independent variables. Based on the recommendations and guidelines, our study hit the target necessary for completion of the study.

Despite our apprehension due to publication and small study biases, we have found previous success in publishing small research studies in the area of higher education (Hendrickson & Francis, 2020; Hendrickson & Francis, 2018). When properly conducted, there are many benefits to completing a small study that outweigh the limitations or biases, especially for researchers early in their career. Small studies can also provide more control, and ease in tracking during the investigation process. Small studies are the best way to explore new hypotheses, as well as pilot research protocols and methods used in analysis. Smaller studies also provide chances to create scientific markers or points for further and future scientific explorations.

A Summary of the Solutions

Although the challenges we faced were demanding, problematic and trying, we were able to modify the research by adding an additional layer to the study that had already been in progress. The emergence of disaster study was a new genre for the researchers. We were able to apply flexibility and uncover a deficiency in disaster research and higher education. A level of confidence allowed us to accept the responses at our intended end date. Rather than stalling for results that very well may not have happened, we moved ahead to maintain our goal. All in all, the solutions preserved and enhanced the research, which provided and opportunity for us to offer recommendations.

RECOMMENDATIONS

At the conclusion of our research study, we reflected on what helped us along the way, and what we would have done differently. Based on our experiences, we offer the following recommendations that may be helpful to researchers conducting similar studies: 1. choose research partners carefully; 2. if possible, include some form of incentive; 3. add a qualitative component; and, 4. include an extended version of the study (such as faculty perspective). The first recommendation emphasizes on the researchers' connection and relationship, while the remaining three suggestions focuses on the design of our research.

Choose Research Partners Carefully

This collaborative research required contributions from both authors. We do not recommend choosing research partners whose sole interest is to increase their publications without putting in the work. Rather, it is important that researchers share a high level of enthusiasm to get through the obstacles of finishing a research project. We also recommend that researchers share a common interest in the research topic.

Researchers should also be willingness to compromise. As we discussed, we encountered research differences that required us to make concessions. Had either author not compromised, the completion of this project may not have concluded. It is rather easy for differences to disturb the strength of the collaboration. In these cases, we recommend a having a plan, which distributes tasks and obligations, which will ensure that the necessary work is completed. The authors both had assigned tasks and due dates. It is also important to realize that each individual will bring their own strengths. Utilizing each member's abilities can help guarantee optimal contribution from all participants.

Incentives for Increased Response Rates

The authors believe that providing an incentive may have positively improved their response rates. Studies have noted that incentives positively increase the number of respondents in web-based survey (Dykema, et al, 2013; Fan & Yan, 2010; van Gelder, Bretveld, & Roelevend, 2010). Monetary incentives yield better responses, even compared to e-gift cards, or electronic codes such as Amazon e-card. It is also noted that, smaller monetary incentives generally had equal returns to larger monetary incentives (Dykema, et al 2013; Fan & Yan, 2010). As it pertains to the size of the incentive, a five-dollar bill would be equally as effective as twenty-dollar bill.

Another option that the researchers would suggest considering is the use of lottery. The use of a prize (albeit a laptop or iPad) might encourage additional survey completions. The lottery system would not be as costly as providing direct incentives to each individual participant. In virtual fieldwork, the bottom line generally is based on the responses of participants. The need to ensure an adequate number of respondents can make a difference in achieving statistical significance and generalizability of the population.

Qualitative Component

We realize that a qualitative component would have allowed us to get the true voice of the respondents. It also would have provided a more in depth and detailed insight into UVI student's opinion of caring within a naturally occurring setting.

By offering an open-ended question, students could have shared their feelings and attitudes towards their experiences with academic caring. Our research could have also benefited from student's unfiltered responses that were written in their own voice. Qualitative research may have allowed us to discover and generate variables. It would have also allowed for us to identify overlooked relationships.

Extend Research to Faculty Perception

Although students can draw conclusions about the quality and efforts of care received, it only offers one side of university caring. We believe the perception of the faculty, as frontline academic caregivers, could have provided vital and objective observations on the actual academic caring. Faculty-student relationships are integral as a key foundation of academic caring. In support, Labrague, McEnroe-Petitte, Papathanasiou, Edet, and Arulappan (2015) noted that faculty interactions with students in learning positively result in academic caring. Thus, faculty perspective would have enhanced the study, and allowed us to draw additional conclusions. Particularly, it would have been interesting to uncover the faculty's opinions of care within academic services offered following disasters.

CONCLUSION

Completing a research project can be an arduous and challenging undertaking, especially for novice and beginner researchers. A research study can become even more difficult to complete, when faced with unforeseen major hazards, crises and disasters (i.e. natural, accidental, man-made disasters, or terrorist incidents). Nonetheless, conducting research in the aftermath of a disaster offered a unique contextual layer to our existing research. We were able to merge the distinct setting of the response and recovery of a natural disaster with the involvement of caring in relations to the resilience of a university.

As in any worthwhile endeavor, the current research required time, commitment and enthusiasm. It also required a willingness to modify our research study to contend with the challenges due to the very nature of a disaster. In the aftermath of Hurricanes Irma and Maria, we had many personal and collective obstacles, including finding the ultimate balance between work, family, students, research, and other life distractions. We also had to address differences within our research partnership. The reality that both disaster research and research partnerships are complicated and filled with contentions. Despite the longstanding research relationship between the authors, we were forced to tackle the pitfalls. We found methods for dealing with the differences of opinions, uncontrollable delays, the absence or inadequate public utilities, and the low responses. It is our hope that readers understand that overcoming all challenges and contentions within our research partnerships added to the richness and productivity of our study. We contend that the success of our research rested in our established relationship with our participants, along with our willingness to incorporate disaster context in our research project plan and timeline. Our research study was a product of collaborative compromises, while maintaining the ethical integrity. We also believe that our research benefited from our virtual environment, and decision to find statistical significance within our timeline, rather than attaining a larger sample size. Lastly, the success of our project was a result of the involvement and concerns provided by each research partner.

REFERENCES

Addison, B. V. (2012). Academic care, classroom pedagogy and the house group teacher: 'making hope practical' in uncertain times. *Pastoral Care in Education*, *30*(4), 303–315. doi:10.1080/02643944.2012.688064

Ancess, J. (2000). The reciprocal influence of teacher learning, teaching practice, school restructuring, and student learning outcomes. *Teachers College Record*, *102*(3), 590–619. doi:10.1111/0161-4681.00069

Ariyaratnam, R., Casas, J. P., Whittaker, J., Smeeth, L., Hingorani, A. D., & Sharma, P. (2007). Genetics of ischaemic stroke among persons of non-European descent: A meta-analysis of eight genes involving~ 32,500 individuals. *PLoS Medicine*, *4*(4), 728–736. doi:10.1371/journal.pmed.0040131 PMID:17455988

Buswell, R., Webb, L., Mitchell, V., & Leder Mackley, K. (2017). Multidisciplinary research: Should effort be the measure of success? *Building Research and Information*, *45*(5), 539–555. doi:10.1080/09613218.2016.1194601

Cange, C. W., & McGaw-Césaire, J. (2019). Long-term public health responses in high-impact weather events: Hurricane Maria and Puerto Rico as a case study. *Disaster Medicine and Public Health Preparedness*, *4*(1), 18–22. doi:10.1017/dmp.2019.82 PMID:31679557

Davison, C. B., Brandt, N. C., & Hua, D. M. (2015). Predictive modeling of organizational resilience utilizing leadership frame orientation. *Research in Higher Education*, *29*, 1–13.

De Waal, F. B. (2008). Putting the altruism back into altruism: The evolution of empathy. *Annual Review of Psychology*, *59*(1), 279–300. doi:10.1146/annurev.psych.59.103006.093625 PMID:17550343

deWinter, J. C. F., Dodou, D., & Wieringa, P. A. (2009). Exploring factor analysis with small sample sizes. *Multivariate Behavioral Research*, *44*(2), 147–181. doi:10.1080/00273170902794206 PMID:26754265

Dykema, J., Jones, N. R., Piche, T., & Stevenson, J. (2013). Surveying Clinicians by Web: Current Issues in Design and Administration. *Evaluation & the Health Professions*, *36*(3), 352–381. doi:10.1177/0163278713496630 PMID:23975760

Fan, W., & Yang, Z. (2010). Factors affecting response rates of the web survey: A systematic review. *Computers in Human Behavior*, *26*(2), 132–139. doi:10.1016/j.chb.2009.10.015

Flowers, A. M. III, Scott, J. A., Riley, J. R., & Palmer, R. T. (2015). Beyond the call of duty: Building on othermothering for improving outcomes at historically black colleges and universities. *Journal of African American Males in Education*, *6*(1), 59–73. Retrieved May 12, 2020, from https://www.semanticscholar.org/paper/Beyond-the-Call-of-Duty%3A-Building-on-Othermothering-Flowers-Scott/4e821a419092edcd7b6aaa180430dd4572a39f0b

Hall, D. (2018). *UVI is Rising*. Retrieved May 12, 2020, from https://www.uvi.edu/news/articles/2018/18_015_uvi_is_rising.aspx

Hendrickson, K. A., & Francis, K. A. (2015). Surviving Senior Seminar Research Helpful Strategies for Improving Overall Project Performance. In Infusing Undergraduate Research into Historically Black Colleges and Universities Curricula (pp. 219-240). Emerald Group Publishing Limited. doi:10.1108/S1479-364420150000017011

Hendrickson, K. A., & Francis, K. A. (2018). Finding professional true to self-authenticity in public university faculty experience. *Studies in Higher Education*, *43*(12), 2398–2411. doi:10.1080/03075079.2017.1333492

Hendrickson, K. A., & Francis, K. A. (2020). Investigating Authentic University Academic Care at a US Non-Mainland HBCU. In P. Blessinger, E. Sengupta, & M. Makhanya (Eds.), *Improving Classroom Engagement and International Development Programs: International Perspectives on Humanizing Higher Education Innovations in Higher Education Teaching and Learning* (pp. 59–74). Emerald Publishing Limited. doi:10.1108/S2055-364120200000027006

Islam, M. R. (2018). Sample size and its role in Central Limit Theorem (CLT). *International Journal of Physics & Mathematics*, *1*(1), 37–47. doi:10.31295/ijpm.v1n1.42

Labrague, L. J., McEnroe-Petitte, D. M., Papathanasiou, I. V., Edet, O. B., & Arulappan, J. (2015). Impact of instructors' caring on students' perceptions of their own caring behaviors. *Journal of Nursing Scholarship*, *47*(4), 338–346. doi:10.1111/jnu.12139 PMID:25959595

Levin, J., Cox, E., Cervin, C., & Haberler, Z. (2010). The recipe for promisig practices in community colleges. *Community College Review*, *38*(1), 31–58. doi:10.1177/0091552110374505

McQuoid, J., Jowsey, T., & Talaulikar, G. (2017). Contextualizing renal patient routines: Everyday space-time contexts, health service access, and wellbeing. *Social Science & Medicine*, *183*, 142–150. doi:10.1016/j.socscimed.2017.04.043 PMID:28482275

Nilsen, P., & Bernhardsson, S. (2019). Context matters in implementation science: A scoping review of determinant frameworks that describe contextual determinants for implementation outcomes. *BMC Health Services Research*, *19*(1), 189–209. doi:10.118612913-019-4015

Patel, N., Min, M., & Lim, S. (2016, April). Accurate evacuation route planning using forward-backward shortest paths. In *2016 Annual IEEE Systems Conference (SysCon)* (pp. 1-6). IEEE.

Pearson, R. H., & Mundform, D. J. (2010). Recommended sample size for conducting exploratory factor analysis on dichotomous data. *Journal of Modern Applied Statistical Methods; JMASM*, *9*(2), 5, 359–368. doi:10.22237/jmasm/1288584240

Reynders, A., Kerkhof, A. J. F. M., Molenberghs, G., & Van Audenhove, C. (2014). Attitudes and stigma in relation to help-seeking intentions for psychological problems in low and high suicide rate regions. *Social Psychiatry and Psychiatric Epidemiology*, *49*(2), 231–239. doi:10.100700127-013-0745-4 PMID:23896893

Schunk, D. H. (2003). Self-efficacy for reading and writing: Influence of modeling, goal setting, and self-evaluation. *Reading & Writing Quarterly*, *19*(2), 159–172. doi:10.1080/10573560308219

Snyder, C. R., Shorey, H. S., Cheavens, J., Pulvers, K. M., Adams, V. H. III, & Wiklund, C. (2002). Hope and academic success in college. *Journal of Educational Psychology*, *94*(4), 820–826. doi:10.1037/0022-0663.94.4.820

Stanley, T. D. (2008). Meta-regression methods for detecting and estimating empirical effects in the presence of publication selection. *Oxford Bulletin of Economics and Statistics*, *70*(1), 103–127. doi:10.1111/j.1468-0084.2007.00487.x

Taylor, H. A. (2016). Review and conduct of human subjects research after a natural or man–made disaster: Findings from a pilot study. *Narrative Inquiry in Bioethics*, *6*(3), 211–222. doi:10.1353/nib.2016.0061

Ujváriné, A. S., Zrínyi, M., Tóth, H., Zékányné, I. R., Szögedi, I., & József, B. (2011). The role of faculty and clinical practice in predicting why nurses graduate in Hungary. *Nurse Education Today*, *31*(1), 94–101. doi:10.1016/j.nedt.2010.04.003 PMID:21094566

van Gelder, M. M. H. J., Bretveld, R. W., & Roeleveld, N. (2010). Web-based Questionnaires: The Future in Epidemiology? *American Journal of Epidemiology*, *172*(11), 1292–1298. doi:10.1093/aje/kwq291 PMID:20880962

VanVoorhis, C. W., & Morgan, B. L. (2007). Understanding power and rules of thumb for determining sample sizes. *Tutorials in Quantitative Methods for Psychology*, *3*(2), 43–50. doi:10.20982/tqmp.03.2.p043

Xiao, Y., Huang, Q., & Wu, K. (2015). Understanding social media data for disaster management. *Natural Hazards*, *79*(3), 1663–1679. doi:10.100711069-015-1918-0

KEY TERMS AND DEFINITIONS

Academic Caregiving: The act of aiding and supporting students/learners in an academic environment.

Authentic University Academic Care (AUAC): Genuine and humanistic concerns, caring motives, disciplined nurturing, and a provision of educational caring necessaries in university's academic services.

Historically Black College and University (HBCU): A college or university founded to educate students of African American descent.

Hurricane: A natural disaster that forms when a tropical storm reaches or exceeds wind speeds of 74 mph.

Natural Disaster: Major destructive event resulting from the Earth's natural processes.

Othermothering: A student-centered relational caring that entails nurturing support, motivation, and encouragement within academic settings, usually associated with HBCUs.

Post-Disaster Research: Research, investigative inquiry and examination conducted by practitioners in the aftermath of a disaster.

University Academic Caring Carescapes Framework: A conceptual framework that showcases the human relational networks structured to provide academic caregiving to students based on HBCU othermothering ideologies.

Section 7
Online Gaming Research

Chapter 11
Multiplayer Online Game Research:
Navigating an Unreal-yet-Real Terrain

Seyed Abdollah Shahrokni
https://orcid.org/0000-0001-9099-8019
Western Oregon University, USA

ABSTRACT

This chapter explains the process of collecting data for an ethnographic case study in a Massively Multiplayer Online Role-playing Game (MMOG) called Stronghold Kingdoms, describing the methodological aspects of the study such as the research site, participants, data collection sources, and analysis. Further, the chapter explains the challenges encountered while conducting the study, including entering and becoming a member of the game community, data collection and analysis in an ever-changing world, being a participant-observer and the risk of going native, and participant attrition. Finally, the chapter provides some solutions and recommendations for researchers interested in conducting studies in MMOG settings.

INTRODUCTION

Logging in to the game one Fall morning, I found that all the villages around me were marked with red swords (see Figure 1). Then, a message from a friend sent a chill down my spine,

Hi Vahid. Hope you good so far. I was wondering why you have swords now – that means enemy.... Can you do something against that????

DOI: 10.4018/978-1-7998-5826-3.ch011

I panicked. I had turned into an enemy overnight! All I knew was that I accepted a faction invite and I was now an enemy to many whom I had befriended since I joined the game. What a world!

Figure 1. Sword-marked villages indicating belligerent terms between me and other players

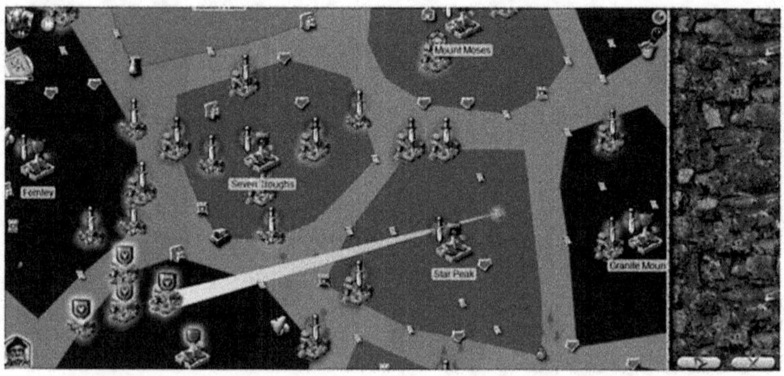

This chapter describes the fieldwork involved in conducting a doctoral study in an online game setting.

My dissertation research, which was defended in September 2018, explored the process of second language socialization (SLS) in a massively multiplayer online game (MMOG) called *Stronghold Kingdoms* (SK; see http://strongholdkingdoms.com), where the situation described above unfolded. SLS is a second language acquisition (SLA) theory which explores how a second language learner gains membership, legitimacy, and language skills through participating in the target language community (Duff & Talmy, 2011). I chose to study SLS in an MMOG setting because I was (and still am) interested in playing MMOGs, and I had personal and anecdotal evidence on how non-native English speaking players (like me) could develop language skills through participating in a multiplayer gaming community. I was particularly interested in learning about 1) what the sociocultural affordances of the community, especially factions (see below), were for SLS, and 2) how the community shaped and supported the non-native English speaker's development of linguistic skills. With these objectives, I used ethnography as an appropriate qualitative research methodology to collect detailed, emic, and holistic data on cultural behaviors (Dörnyei, 2007; Glesne, 2015). Besides, a review of literature showed that SLS had traditionally been studied through ethnographic work for similar reasons. In line with the methodology, then, I used three data sources: 1) observation, 2) artefact analysis, and 3) interview to develop a detailed, emic, and holistic picture of SLS in SK. I started the data collection process by observing the field and looking into chat/forum messages exchanged between and among the participants, a faction of 8 native English speaking (from the US and England; 6 males and 1 female) and 2 non-native English speaking (from Sweden and Egypt; 1 male and 1 female) players. This phase led to generation of codes, categories, and themes through two levels of open and axial coding (Saldaña, 2015), which were then used to discover the sociocultural affordances of the faction community. Then, I conducted an interview with one of the non-native English speakers, Middo. I had originally set out to interview the other non-native English speaker, Tikeno, as well, but she was not responsive to my messages at the time. The results showed that the sociocultural affordances of SK provided a supportive environment for SLS. Further, the results of the interview with Middo suggested that MMOGs in general and SK in particular supported language development (see Shahrokni (in press) for my article based on this study).

BACKGROUND

SK, the game I chose as my field, is a medieval-themed strategy game where players from around the world can join different servers/worlds using a PC, Mac, or portable android/iOS device. In August 2017, when I was still engaged in my research, the game hosted 77 worlds in 9 languages (English, German, Russian, French, Polish, Italian, Turkish, Portoguese, and Chinese). Based on the estimates at the time, SK hosted 5 million active players (Kingdoms Mobile Gameplay | Firefly Studios, 2016), which could be compared to that of a real-world country like Norway at 5,258,317 (Population - SSB, 2017). I played a UK-mapped world for about 5 years before it ended and my fellow players and I had to immigrate to a US-mapped world. Overall, my gameplay for this study spanned 3 years, exploring 4 distinct stages: 1) war against the world, 2) post-war peace and life in exile, 3) end of the world, and 4) immigration to a new world.

To play the game, one needs to create an account, choose a server/world, and join. Entering the game, the player is spawned as a low-rank peasant (Rank 1: Village Idiot) in charge of an empty village. There are no buildings, farms, industries, and/or people in the village and the player is tasked to change this situation. As the player works to develop the village (see Figure 2), they receive honor points, reach milestones, attain achievements, go through ranks, and meet and befriend other players in the proces. This persistent effort to rise, gives players more power, as they can have more villages, armies, gold, vassals, and most importantly, can join factions (at Rank 7: Bondsman), which are groups of 40 players, and play socially. The game's social life can even get more complex with a faction joining a House,

Figure 2. A typical village

a group of 12 factions, and engage in multilateral relations and alliances aimed at ruling the world. This cause, however, requires a lot of communication, coordination, and negotiations for action.

FIELDWORK IN MMOGs

MMOGs are unique places to gather data, as they are both "virtual" and, at the same time, very "real." In other words, they are computer-simulated two- or three-dimensional game worlds inhabited by real people and alive with movements, conversations, negotiations, actions, and emotions. These worlds are "legitimate sites of culture" (Boellstorff, 2008, p. 61), as they depict a dynamic cultural field which could be as real as real life (Kolo & Baur, 2004). Therefore, methodological considerations which are usually observed in physical/real worlds tend to be valid in MMOGs as well (Boellstorff, 2008; Boellstorff, Nardi, Pearce, & Taylor, 2012; Guimarães, 2005; Hine, 2008; Kolo & Baur, 2004). For example, methodologically, an interview protocol designed to interview a participant in either an MMOG or real world may include several categories of items (e.g., structural, experiential, comparison, hypothetical, behavioral, and phenomenological) to comprehensively explore the construct/phenomena under investigation (e.g., Dörnyei, 2007; Glesne, 2015; Saldaña, 2013). However, the virtuality of the medium can also bring about advantages or challenges. For instance, I did not have to physically travel, as real-world ethnographers do, to collect the data about the *Other* (Hine, 2000); I used my login credentials to access the research site, the game world. Therefore, in the remainder of this chapter, I try to describe the data collection process in my dissertation research, recounting the types of challenges I experienced and handled.

Observation

As an ethnographer, I needed to be an observant of the sociocultural phenomena and/or the behavior around me, which required being immersed in the field. As I had two years of gameplay before I actively started collecting data, I was a participant-observer in the study, which provided me with an opportunity to be well-included in the community and in friendly terms with my participants and many others. Likewise, I knew the setting very well. For example, throughout my research, I was on multiple chat threads and Discord (Version 10.0; Discord Inc., 2018), which is an audio chat client gamers use to communicate, allowing me to observe the politics of the game more closely. In other words, I could be both a player and a "fly on the wall" (Howe & Eisenhart, 1990). However, with this full-time residency (Schensul, Schensul, & LeCompte, 1999) came a natural caveat. I had access to too much information that I could not easily process. To cope, I "selectively" observed (Angrosino & DePerez, 2000) the activities in the community (p. 677), focusing on the tenets of the SLS framework. While getting lost in the data is not unique to MMOGs, I have found the multimodal medium of an MMOG so rich in electronic content that could potentially puzzle any researcher; therefore, having a planned observation scheme developed based on the theoretical framework could help guide the process. Another issue the observation scheme helped with was reducing my bais, as I was too immersed in the field and ran the risk of "going native" (Gold, 1958, p. 221; Takyi, 2015), that is, being a resident of the site in a way that I lost my perspective, with my feelings obliterating my ability to observe objectively (Schwartz & Schwartz, 1955; Takyi, 2015). To deal with the issue of bias further, as Bernard (1994) suggests, I removed myself daily from the field at times to remain objective and reflect. Finally, through data triangulation, I was able to diversify the knowledge base of the study as well.

Artefact Analysis

As noted earlier, MMOGs are rich in content; there are so many artefacts which attract one's attention and are generated and added to momentarily. For instance, in SK, some of the artefacts include the avatars with their elegant looks and gears, the various moving and/or static objects on the map (e.g., carts, scouts, armies, enemy siege camps, castles, bandits, stashes, resources, food, wolf lairs, villages, parishes, capitals), the leadership board listings, communication channels (in-game chat and forum and external servers such as Discord and TeamSpeak (Version 3.5; TeamSpeak Systems, Inc., 2018), game reports, quests, ranks, and research. A lot of information could be gleaned from observing these artefacts; however, one could not be certain that accurate data were collected if a careful artefact analysis would not follow. For instance, looking at a player's avatar, one could develop an idea about a player's gender, achievements, completed quests, number of villages, faction, house, and rank (see Figure 3). Although almost all of this information was accurate, I could find cases where this information was misleading. For instance, through my analysis, I learned that some of the male players I had played with chose a female avatar to increase the chances of receiving support from their fellow peers. I had personally fallen for that trick many times. Another surprising finding I gathered was that many players had alternative accounts (ALTs), which rendered the information on their avatar profiles inaccurate, as what had been achieved was through cheating.

Figure 3. My game profile

As part of this analysis, I focused on chat/forum messages as a way to explore the sociocultural norms of the faction community through conversations between and among the members. This phase, coupled with observational data, would not only serve as a formal survey of the field but also inform the design of the interview protocol. There was, however, one problem, the world was ending (it ended in August 2017), and I did not have much time to collect all the data (which were added to momentarily) I needed. In this race against time, I decided to strategically use three free technological tools to record the data. First, I used the screen capturing tool Ezvid (Version 1.0; Ezvid Inc., 2018) to create (103.13 minutes worth of) videos of in-text messages as I loaded them in my Inbox. Then, I used the Free Video to JPG Converter (Version 5.0; DVDVideoSoft, 2018) software to convert videos to images (216,576 images; see Figure 4 for an example). Finally, I used the Free Online OCR (NewOCR.com, 2018) to extract the text from the images. Through this process, I created Word Documents of chat messages, which I could then feed into the statistical package, Atlas.ti (Version 8.2; Scientific Software Development GmbH, 2018), for analysis. Of course, one can analyze the transcripts without a statistical package, but, with a large data set, it would be optimal to use one.

Figure 4. An extracted image of chat exchanges (player names and other identifiers have been blurred to observe participant anonymity)

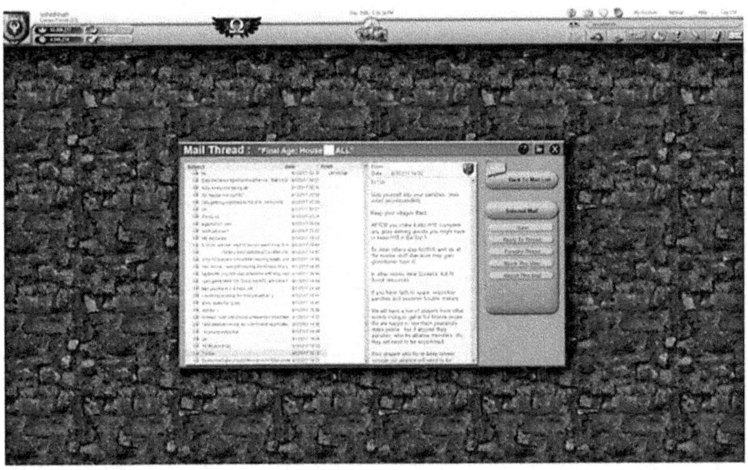

Conversation Analysis

To make sense of the faction's sociocultural and linguistic norms, I needed to analyze the members' conversations. As was mentioned in the previous section, the written discourse needed to be extracted (for which I used the screencast tool Ezvid, the OCR tool Free Online OCR, and the media conversion tool Free Video to JPG Converter) and analyzed (for which I used *Atlas.ti*). In the following excerpt, I present the results of a conversation analysis on the exchanges preceding the war between our faction, Brown Eagle, and House 16. The purpose of this analysis is showing the type of discourse a researcher may encounter in an MMOG as well as the analytic lenses one may use to make sense of the data. For instance, in line with my research questions, in this sample, which relates to the beginning of an all-out war, I was looking for different sociocultural and linguistic norms of the community (see the Appendix for the complete transcript; grammatical and/or spelling mistakes are preserved intentionally).

Debate

After one of our members, Treld, captured a village from an inactive member of House 16, which was the dominant house in the game, Black Dog, the Governor of North West England and a player from House 16, wrote us through the SK's in-game messaging service. In the message, he advanced several language functions: threatening, assuming, encouraging, offering, praising, persuading, assessing, establishing identity, hypothesizing, and evaluating (just to mention a few). The message which had been posted to all of us triggered a long conversation in which everyone was witnessing the development of the topic, deciding for ourselves which side to back.

Black Dog started the debate by addressing the faction and then proclaimed that we had attacked House 16 with no reason. Afterwards, he excluded the members from the address, pointing at our leader, and praised us for our neutrality (as we were not affiliated to any house) and appreciation of peace. Then, he offered us the opportunity to leave our faction, be neutral, and/or join his. Next, he implicitly threatened us, stating that we could have our assets if we left Brown Eagle. Subsequently, he explicitly threatened us that if we did not follow his command, "we would never be given peace," as personal swords would be placed on us (at that point only faction swords had been placed on us). After this argument, he expressed sorrow over occurrences of this incident, stating that he could not tolerate non-aligned players' attacking others. Finally, he threw the ball in our court and, excluding us again from "other players in our faction," and offered us the opportunity to avoid that "pointless conflict":

Black Dog: *Brown Eagle, due to a number of unprovoked attacks on our members . Im sure there are a number of pleyers in the Brown Eagle faction that do not want to be attacked and they is why you are neutral in the first place. We would therefore like to offer you the oppertunity to drop faction and be neutral or even join our alliance. Anyone that drops faction will be left alone and will be able to keep all their land. Those that want to fight stay but know that anyone that sends an attack against us will have personal swords placed on them and will never be given peace. Im sony it has come to this but i cannot have neutral players attacking us without doing anything about it. You all have your chance now to get out this pointless conflict started by some players in your faction. (Regards The Black Dog, General of the League of Swords H16 Governor of North West England. (Records, 2015))*

This message, along with its power-laden signature, was new to us as newbies in the game. We had never seen such a threatening message before, with a governor who was probably required by law to defend the county against our "unprovoked attacks." "Had we really violated the laws of the world?" we thought. "But Treld was only going to capture Chino's village after he went inactive." "Perhaps all the inactive villages in the county belonged to the governor!" We had so many questions to which we needed answers. At this point, Treld, our faction member, wrote back and told our own version of the story, highlighting the fact that he tried to capture a village from another inactive faction member and that he had been attacked first by House 16:

Trled: *I'll summarize the events of today for you all. I left the Brown Eagle faction yesterday to capture from a faction member. While I was sleeping, IsaacU decided that I was neutral. Which according to your definition, I still am whether in a faction or not. Whether they knew I was in Brown Eagle or not I suppose is subjective, though easy enough to prove - and the monking has not stopped. Anyhow, they started to monk me out claiming that I have no rights to the Parish because I'm neutral. P-Jazz presumes I have joined the faction to hide. Which from my perspective is a poorly crafted deception to propagate an agenda. He scouted me, therefore I scouted him. He sends back a msg stating that my actions are of war, when I am the defender, not the aggressor and frankly, kiss my a*s. I know a lot of history. Perhaps P-Jazz should research Hitler, Stalin, Mao, Pot, etc. You know what you're doing. (Records, 2015)*

Weather P-Jazz researched Hitler or not, at this point, Morning Destroyer, our faction leader, replied. We were so happy that he wrote, as we felt completely defenseless. In the message, he questioned the claim, stating that it was unfounded, and that we had responded to House 16's attacks with counterattacks. He also undermined their claims about attacking us and confirmed Treld's story, reminding Black Dog of the rules of engagement (dropping factions to capture an ally's village). Furthermore, he lashed out at Black Dog for claiming that Treld did not belong to our faction, calling it a "blatant lie." Finally, referencing history, he blamed three players (IsaacU, P-Jazz, and Devil) who had sneakingly derailed their collaborations in the past, trying to advance their own "personal agenda" again. Then, he lashed out at the leaders of House 16 for not realizing their agendas and being deceived. Finally, he urged us not to reply to the thread, as he believed that it was a trick to find out who was active in faction:

Morning Destroyer: *Unprovoked attacks? You know that's not true, just as members of the faction know its not true. Scouts and attacks on members of this faction and its parishes by your members and members of House 16 were responded with in kind, thats all. Trying to claim your attacks were "justified" as the Karmageddon members in question had dropped faction to trade flags or villages with other members and were therefore neutral" at the time is a joke. You know full well that* *players have to drop faction to attack their own and you know full well how you would have reacted if we (or anyone) had done the same to any alliance" member that had dropped faction to do same. Trying to also claim members of this faction that have been members for months and held parishes in these lands before H16 even landed in England, were not members of the faction before is just a blatant lie. Your members (IsaacU, P-Jazz, and Devil) started this, just as they caused us to leave that alliance" by betraying H16 and H15 a leaderships word months ago.. Its been them all along with their own little personal agenda they had when they were H10 and are now hiding behind H16 colours after surrendering and flipping to that House without so much as one iota of resistance, trash talking cowards that have suckered you into their whims. Nobody else reply to this, its just a ploy to stir up shit and see who is active and when. The mistakes behind being made here is they think we are fools and woefully underestimate what 3 years experience teaches someone. (Records, 2015)*

After this counterargument, Black Dog responded. In the message, he addressed us, not Morning Destroyer. He said that he had asked Morning Destroyer "a number of times" to provide him with evidence (faction messages and reports) for the claim that Treld was indeed our faction member and he had dropped faction to do a friendly capture, which, accordingly, Morning Destroyer had not complied. Furthermore, he said that Morning Destroyer had tried to show the incident in a different light, pretending that we had been attacked first while, in fact, we were the ones who started it. Finally, excluding us from "a few of [our] members," he restated his claim that we were the ones who declared war:

Black Dog: i have asked morning-destroyer a number of time to provide proof of this. he has not done so. he then tried to say they we were attacking and scouting a member called Cor_rect. he forward me the reports saying that we were in the wrong however i have a report showing that Cor_rect attacked IsaacU first. Morning had been trying to make it seem like my players are being aggressive towards you yet we are only reacting to a few of your members. if you want to be a part of a faction that wants to fight us then thats your will. as for the parish you should have sent the proof before my members were attacked. you guys therefore declared war on us not us on you. (Records, 2015)

Following this message, Black Dog composed another message, but this time addressed to Morning Destroyer. In the message, he ridiculed Morning Destroyer's claims and urged him to forward the reports to him, as his lack of compliance after being asked "3 times" had led to the escalation of this conflict. Furthermore, he did not rule out the possibility of the presence of a rogue player in our faction, which, of course, was not true. Finally, he implicitly threatened that nonobservance of this order would exacerbate the situation:

Black Dog*: Morning are you kidding me?? Trying to claim your attacks were justified as the Brown Eagle members in question had dropped faction to trade flags or villages with other members and were therefore neutral at the time is a joke. "You know full well that players have to drop faction to attack their own and you know full well how you would have reacted if we (or anyone) had done the same to any alliance" member that had dropped faction to do same. Forward me the reports!!!! like ive asked you to do 3 times now. you have caused the problems by not doing as requested. i said 2 days ago any more attacks on our members will be classed as an act of war on against us. you either chose to deliberatly ignore that or you have rouge members. its one of the other. the more you throw your mouth around the worse this is going to get. (Records, 2015)*

After this, Morning Destroyer wrote back, lashing out at Black Dog's propaganda. Drawing on historical happenings, he tried to persuade Black Dog, who was also the king of the world, that House 16 was a not worth defending, as they were "multi-accounting house jumping farmers" that could betray others and "flip House" when threatened. In his devaluing House 16, he also mentioned the historical reason why he, as the leader of a House 15 faction, had to leave the alliance. To further back his claims, he referenced Priest Marco, a veteran player, stating that he had also been forced to abandon the world "disgusted at how H16 was betraying key H15." Finally, he insulted the House 16 players as "trash talking whining bitches" who only cared about their own benefits:

Morning Destroyer*: I forwarded all reports, stop your bullshit propaganda Black dog, you are defending multi-accounting house jumping farmers that done nothing for that Alliance other than flip House the moment they felt threatened. You defend and believe the very people that caused the faction that whilst in H15 had cleared Hertfordshire, Oxfordshire of enemies and pushed into Buckinghamshire, to leave, i.e. us.. Shame Priest Marco isn't still around to clarify historical facts and the real situation in this area... but that's right, he left game, disgusted at how H15 was betraying key H16 factions in favour of trash talking whining bitches that had their own little agenda and had done nothing for that faction or that house. (Records, 2015)*

Right after this message, he also composed another message in response to Black Dog, referencing other historical events backing his claim. He likened Black Dog to a "corrupt politician" who had his agenda and mind set, not willing to look into the evidence provided:

Morning Destroyer: *No I am not kidding you... that's the facts... Your mind's made up though and facts won't change it... you have your agenda and propaganda to stick to, just like a corrupt politician. Honesty and Glory left H16 about the same time as Leo left HM position and H16 started absorbing the multi-accounting, cheating egomaniacs from H12/H1 and H9 in England at the time, letting them dictate a new course for H16 in England... Uncomfortable truths... but truth nonetheless. (Records, 2015)*

As this conversation was unfolding in front of us, we were all amazed at how deep this conflict was and how little we knew of the history of this game. We were all newbies in some way, and I had just joined Brown Eagle. Although we were weighing the evidence provided by both sides, we trusted Morning Destroyer. His skills, knowledge of the game history, language, high rank, and support made us trust him. Besides, we had been part of the operation which led to this conflict; therefore, we had seen for ourselves how deceitful the Black Dog was. When ultimately, after a long debate (see Appendix A), the prospect of peace diminished, and Morning Destroyer offered more evidence on House 16's hypocrisy, the Black Dog declared war:

Black Dog: *I offered you peace multiple times and all you do is attack me. You really want to commit suicide then you'll get your WAR. Having seen the state of you castles in the faction, obviously those two or three players who started this WAR may be ready but the rest of you will die. Any of you not wanting in on this WAR, leave Brown Eagle now or face the consequences. You may have razed one of my castles but I raze ten for your decision to attack me. (Records, 2015)*

We did not say anything on this thread because Morning Destroyer had told us not to, but we shared our support for him in the faction communication thread. He was our leader and we trusted him:

Justin: *I'm all for war. I can also help with monking/excomm. Would be nice to get Morning Destroyer into buckingham. How many votes does Black Dog have?*

Middo: *I'm in.*

Jennifer: *It sounds like a total war. I'm just thinking of heavy defense:-??*

Treld: *Very much defense (probably I overmonk my villages:-D)*

Tikeno: *I am staying in this faction no one are going to force me into something I decide for myself. Even if I will be razed I rather die with proud than leaving when its begin to be unpleasant.*

Golden: *Well, I had invite earlier from H6 knights hospitaller, and now Lord Hellblaze from H12 asks how the faction is, if we are satisfied etc etc. I am with Tikeno on this. I will go down with the ship if I get to choose.*

Treld: *I feel the same way. I'll be speaking with undocumented "allies" on behalf of Morning Destroyer today if I can reach them. (Records, 2015)*

From a linguistic standpoint, in this snippet, we can see examples of language use that are opinion evaluation (e.g., "from my perspective is a poorly crafted deception to propagate an agenda," or "is a blatant lie"), causative (e.g., "he scouted me, therefore I scouted him."), defensive-passive aggressive (e.g., "when I am the defender, not the aggressor and frankly, kiss my a*s."), establishing identity (e.g., "I know a lot of history. Perhaps P-Jazz should research Hitler, Stalin, Mao, Pot, etc."), justification (e.g, "scouts and attacks on members of this faction and its parishes by your members and members of House 16 were responded with in kind"), formulaic (e.g., "you know full well," "is a joke," or "the more you throw your mouth around the worse this is going to get"), hypothesis-conditional (e.g., "how you would have reacted if we (or anyone) had done the same to any alliance"), contradict-contrast (e.g., "however I have a report showing that Cor_rect attached IsaacU first), setting the record straight (e.g., "you guys therefore declared war on us not us on you"), aggressiveness (e.g., "forward me to reports!!!," "you either chose to deliberatly ignore that or you have rouge members. its one of the other."), and metaphoric (e.g., "disgusted at how H15 was betraying key H16 factions in favour of trash talking whining bitches that had their own little agenda and had done nothing for that faction or that house). Of course, this was a small excerpt of this debate (see the Appendix section for the complete transcript), the rest shows a variety of other linguistic moves and speech acts in place.

Interview

After I identified the sociocultural and linguistic norms and patterns through observations and analysis of chat exchanges, I created a protocol to interview my non-native English speaking participants, aiming to find out how the community shaped and supported language development. At this stage, there were two main problems: 1) my Swedish participant, Tikeno, was unresponsive, as she had coincidentally activated the vocation mode (VM), a game feature which allows participants to take a break (of up to 30 days) from the game without losing assets, and 2) my Egyptian participant, Middo, disappeared for almost two months (from July 19, 2018 to September 1, 2018). Of course, attrition is a concern in all research studies working with participants, but, compared to MMOGs, it may be easier to track participants in real-life settings. If the players in MMOGs do not exchange real-life contact information, which is usually the case, the game avatars and communication channels do not provide any real-life connections to players' lives which researchers can use to locate them. As part of safety measures, the game developers advise players to "exercise caution when deciding to disclose personal information in such chat rooms, in-game mail and forums," as all the data exchanged will be considered public data and can be shared with third-party agencies (Privacy Policy - Firefly Studios, 2018). Therefore, players may reasonably abstain from sharing their personal contacts with researchers, which will make it difficult to track them if they go inactive in the middle of the study.

SOLUTIONS AND RECOMMENDATIONS

Provided that you have secured your Institutional Review Board (IRB), as there is no difference between virtual and physical worlds, and you have observed the required methodological considerations, the following will support you in navigating this new terrain.

1. **Explore your field.** Spend time familiarizing yourself with the game environment and rules. Games usually provide tutorials for newcomers; make sure to complete them. These guides will set you on your way to knowing about and using your resources to succeed in both gameplay and research. Additionally, when part of a faction/ clan/ server/ guild, etc., you will be exposed to other learning that may also affect your gameplay and, consequently, research. Using a notebook could help.
2. **Be part of the community**. Socialize, get to know people, help others, collaborate on joined activities, and be a team player. Being an active member in the community will help immensely, as it can give you access to closed circles and resources which are usually hidden from solo players. Besides, the community is there for you in times of need. Although seeing the sword-marked villages was scarry for me at the beginning, there were tens of hands reaching out to me, calming me down and letting me know I was not alone.
3. **Use data logging tools**. Learn what tools you can use to effectuate your data collection. For example, Windows Game Bar (use Windows key+G when running a game) on a PC can record your gameplay. It would be a good idea

to record all your gaming sessions so that you can return to them during your analysis. Of course, securing participant consents and safeguarding your data is vital regardless of field.

4. **Establish a direct communication channel.** Players are advised not to share personal contact information in their gameplay. While not everyone exercises this caution, it would serve you well if you shared an email address (perhaps one specifically designed for the study) with your participants and ask them to email you directly so that you have another means to track them if they go inactive. I also suggest that you design a website for the purpose of your study and share it with your participants so that they learn more about you and the project. This can help secure their trust and encourage them to participate in your study as well as be in touch with you through another non-game communication channel.
5. **Use a statistical package.** It is easy to get lost in the labyrinth of data you can collect. It would be ideal to use a statistical package, such as *Atlas.ti* for qualitative work, to help you with data organization, analysis, and interpretation.
6. **Enjoy.** Researching MMOGs is very engaging and enjoyable. I always refer to my Ph.D. research as a "dream project," which included three years of game play, virtual friendships, friendly conversations, pleasant challenges, collaborations, and writing a dissertation.

CONCLUSION

This chapter explored the fieldwork involved in conducting research in massively multiplayer online games (MMOGs). These games, which are sometimes referred to as "the medium of the twenty-first century" (Pearce, Boellstorff, & Nardi, 2011, p. 66) and "the future of learning" (Shaffer, Squire, Halverson, & Gee, 2005, p. 10), are unique fields to explore. I tried to share a glimpse of these exciting fields through my experiences conducting research in one, *Stronghold Kingdoms*. There is a lot out there you should explore!

REFERENCES

Angrosino, M. V., Mays, D., & Kimberly, A. (2000). Rethinking observation: From method to context. In N. K. Denzin & Y. S. Lincoln (Eds.), *Handbook of qualitative research* (2nd ed., pp. 673–702). Sage.

Atlas.ti. (2018). [Computer software]. Retrieved from https://atlasti.com

Bernard, H. R. (1994). *Research methods in anthropology: Qualitative and quantitative approaches* (2nd ed.). AltaMira Press.

Boellstorff, T. (2008). *Coming of age in Second Life: An anthropologist explores the virtually human*. Princeton University Press.

Boellstorff, T., Nardi, B., Pearce, C., & Taylor, T. L. (2012). *Ethnography and virtual worlds: A handbook of method*. Princeton University Press. doi:10.2307/j.cttq9s20

Discord. (2018). [Computer software]. Retrieved from https://discord.com

Dörnyei, Z. (2007). *Research methods in applied linguistics: Quantitative, qualitative, and mixed methodologies*. Oxford University Press.

Duff, P. A. (2007). Second language socialization as sociocultural theory: Insights and issues. *Language Teaching*, *40*(4), 309–319. doi:10.1017/S0261444807004508

Duff, P. A., & Talmy, S. (2011). Language socialization approaches to second language acquisition. In D. Atkinson (Ed.), Alternative approaches to second language acquisition (pp. 95-116). Routledge.

Ezvid. (2018). [Computer software]. Retrieved from https://www.ezvid.com

Free Online, O. C. R. (2018). [Computer application]. Retrieved from https://www.newocr.com

Free Video to JPG Converter. (2018). [Computer application]. Retrieved from https://www.dvdvideosoft.com/products/dvd/Free-Video-to-JPG-Converter.htm

Glesne, C. (2015). *Becoming qualitative researchers: An introduction*. Pearson.

Gold, R. L. (1958). Roles in sociological field observations. *Social Forces*, *36*(3), 217–223. doi:10.2307/2573808

Guimaraes, M. J. L. J. (2005). Doing anthropology in cyberspace: Fieldwork boundaries and social environments. In C. Hine (Ed.), *Virtual methods: Issues in social research on the Internet* (pp. 141–156). Berg.

Hine, C. (2000). *Virtual ethnography*. Sage Publications. doi:10.4135/9780857020277

Hine, C. (2008). Virtual ethnography: Modes, varieties, affordances. In N. G. Fielding, R. M. Lee, & G. Blank (Eds.), *The Sage handbook of online research methods* (pp. 257–270). Sage Publications. doi:10.4135/9780857020055.n14

Howe, K., & Eisenhart, M. (1990). Standards for qualitative (and quantitative) research: A prolegomenon. *Educational Researcher*, *19*(4), 2–9. doi:10.3102/0013189X019004002

Kingdoms Mobile Gameplay | Firefly Studios. (2016). Retrieved from http://fireflyworlds.com/2016/08/17/kingdoms-mobile-gameplay

Kolo, C., & Baur, T. (2004). Living a virtual life: Social dynamics of online gaming. *Game Studies*, *4*(1), 1–31.

Pearce, C., Boellstorff, T., & Nardi, B. A. (2011). *Communities of play: Emergent cultures in multiplayer games and virtual worlds*. MIT Press.

Population – SSB. (2017). Retrieved from https://www.ssb.no/en/befolkning/nokkeltall/population

Privacy Policy - Firefly Studios. (2018). Retrieved from https://fireflyworlds.com/privacy-policy

Saldaña, J. (2013). *The coding manual for qualitative researchers* (2nd ed.). Sage.

Schensul, S. L., Schensul, J. J., & LeCompte, M. D. (1999). *Essential ethnographic methods: Observations, interviews, and questionnaires*. AltaMira Press.

Schieffelin, B. B., & Ochs, E. (1986). The microgenesis of competence: Methodology in language socialization. In D. Slobin, J. Gerhardt, A. Kyratzis, & J. Guo (Eds.), *Social interaction, social context, and language* (pp. 251–263). Erlbaum.

Schwartz, M. S., & Schwartz, C. G. (1955). Problems in participant observation. *American Journal of Sociology*, *60*(4), 343–353. http://www.jstor.org/stable/2772027. doi:10.1086/221566

Schwartz, M. S., & Schwartz, C. G. (1955). Problems in participant observation. *American Journal of Sociology*, *60*(4), 343–353. doi:10.1086/221566

Shaffer, D. W., Squire, K. R., Halverson, R., & Gee, J. P. (2005). Video games and the future of learning. *Phi Delta Kappan*, *87*(2), 105–111. doi:10.1177/003172170508700205

Shahrokni, S. A. (in press). Evidence for second language socialization in multiplayer online game play. *TESL-EJ, 25*(1).

Takyi, E. (2015). The challenge of involvement and detachment in participant observation. *Qualitative Report*, *20*(6), 864–872.

TeamSpeak. (2018). [Computer software]. Retrieved from https://www.teamspeak.com

KEY TERMS AND DEFINITIONS

Ethnography: A research methodology which provides detailed, emic, and holistic descriptions of cultural norms and behaviors (Dörnyei, 2007).

Massively Multiplayer Online Game (MMOG): Networked online games which can be played on different platforms (e.g., computers and cell phones) by players from around the world.

Second Language Socialization: A process by which second language learners gain membership, legitimacy, and communicative competence in the target community's language and culture (Duff, 2007; Schieffelin & Ochs, 1986).

APPENDIX: DEBATE

Black Dog: *Brown Eagle, due to a number of unprovoked attacks on our members . Im sure there are a number of pleyers in the Brown Eagle faction that do not want to be attacked and they is why you are neutral in the first place. We would therefore like to offer you the oppertunity to drop faction and be neutral or even join our alliance. Anyone that drops faction will be left alone and will be able to keep all their land. Those that want to fight stay but know that anyone that sends an attack against us will have personal swords placed on them and will never be given peace. Im sony it has come to this but i cannot have neutral players attacking us without doing anything about it. You all have your chance now to get out this pointless conflict started by some players in your faction. (Regards The Black Dog, General of the League of Swords H16 Governor of North West England.)*

Trled: *I'II summarize the events of today for you all. I left the Brown Eagle faction yesterday to capture from a faction member. While I was sleeping, IsaacU decided that I was neutral. Which according to your definition, I still am whether in a faction or not. Whether they knew I was in Brown Eagle or not I suppose is subjective, though easy enough to prove - and the monking has not stopped. Anyhow, they started to monk me out claiming that I have no rights to the Parish because I'm neutral. P-Jazz presumes I have joined the faction to hide. Which from my perspective is a poorly crafted deception to propagate an agenda. He scouted me, therefore I scouted him. He sends back a msg stating that my actions are of war, when I am the defender, not the aggressor and frankly, kiss my a*s. I know a lot of history. Perhaps P-Jazz should research Hitler, Stalin, Mao, Pot, etc. You know what you're doing.*

Morning Destroyer: *Unprovoked attacks? You know thats not true, just as members of the faction know its not true. Scouts and attacks on members of this faction and its parishes by your members and members of House 16 were responded with in kind, thats all. Trying to claim your attacks were "justified" as the Karmageddon members in question had dropped faction to trade flags or villages with other members and were therefore neutral" at the time is a joke. You know full well that players have to drop faction to attack their own and you know full well how you would have reacted if we (or anyone) had done the same to any alliance" member that had dropped faction to do same. Trying to also claim members of this faction that have been members for months and held parishes in these lands before H16 even landed in England, were not members of the faction before is just a blatant lie. Your members (IsaacU, P-Jazz, and Devil) started this, just as they caused us to leave that alliance" by betraying H16 and H15 a leaderships word months ago.. Its* *been them all along with their own little personal agenda they had when they were H10 and are now hiding behind H16 colours after surrendering and flipping to that House without so much as one iota of resistance, trash talking cowards that have suckered you into their whims. Nobody else reply to this, its just a ploy to stir up shit and see who is active and when. The mistakes behind being made here is they think we are fools and woefully underestimate what 3 years experience teaches someone.*

Black Dog*: i have asked morning-destroyer a number of time to provide proof of this. he has not done so. he then tried to say they we were attacking and scouting a member called Cor_rect. he forward me the reports saying that we were in the wrong however i have a report showing that Cor_rect attacked IsaacU first. Morning had been trying to make it seem like my players are being aggressive towards you yet we are only reacting to a few of your members. if you want to be a part of a faction that wants to fight us then thats your will. as for the parish you should have sent the proof before my members were attacked. you guys therefore declared war on us not us on you.*

Black Dog*: Morning are you kidding me?? Trying to claim your attacks were justified as the Brown Eagle members in question had dropped faction to trade flags or villages with other members and were therefore neutral at the time is a joke. "You know full well that players have to drop faction to attack their own and you know full well how you would have reacted if we (or anyone) had done the same to any alliance" member that had dropped faction to do same. Forward me the reports!!!! like ive asked you to do 3 times now. you have caused the problems by not doing as requested. i said 2 days ago any more attacks on our members will be classed as an act of war on against us. you either chose to deliberatly ignore that or you have rouge members. its one of the other. the more you throw your mouth around the worse this is going to get.*

Morning Destroyer*: I forwarded all reports, stop your bullshit propaganda Black dog, you are defending multi-accounting house jumping farmers that done nothing for that Alliance other than flip House the moment they felt threatened. You defend and believe the very people that caused the faction that whilst in H15 had cleared Hertfordshire, Oxfordshire of enemies and pushed into Buckinghamshire, to leave, i.e. us.. Shame Priest Marco isn't still around to clarify historical facts and the real situation in this area... but that's right, he left game, disgusted at how H15 was betraying key H16 factions in favour of trash talking whining bitches that had their own little agenda and had done nothing for that faction or that house.*

Morning Destroyer: *No I am not kidding you... that's the facts... Your mind's made up though and facts won't change it... you have your agenda and propaganda to stick to, just like a corrupt politician. Honesty and Glory left H16 about the same time as Leo left HM position and H16 started absorbing the multi-accounting, cheating egomaniacs from H12/H1 and H9 in England at the time, letting them dictate a new course for H16 in England... Uncomfortable truths... but truth nonetheless.*

P-Jazz: *Your member Cor_rect attacked the County Sherrif. apart from being a very stupid thing to do that alone is enough to have WAR declared It's very simple, you have three choices.*

1. Stop interfering with the Alliance taking neutral and enemy parishes and stop trying to undermine those that are in Alliance hands. Basically stop playing your silly little games and you can go back to being peaceful neutral. If anyone wants to drop the faction for any reason and not be seen as neutral then you should inform the County Sherrif. before-hand
2. Rejoin the Alliance and you will be welcomed as an Ally and we can start making the County safer for everyone. If DW wants to take his political ambitions to another County then that is not of my business and I have no objection to him doing so. Just stop the petty games and respect the Alliance decisions. That does not mean that anyone will lose their parishes or villages, just accept the Alliance decision as to who is Sherrif. You can keep all the parishes you currently own, I have no interest in them apart from I want to be in Alliance hands.
3. WAR. As stated by the NW governor, anyone wanting out of this war then just drop Brown Eagle now and you will be spared. Anyone making any form of attack on Alliance positions will be hunted down for-ever.

Black Dog: *LOL i cannot believe im hearing that. H15 factions!!!!!!!!!!! you guys joined after we took england and left because you didnt get your own way. if you were such a key faction then why are you not in our position now?'? I know full well what happened in H18.Capt Earthquake and Bulldog were the HM's of H15 and they were my sister factions generals. All the reports you have sent me prove nothing only that we attacked your player after he attacked us. LOOK AT THE TIME THE ATTACK ON JACOB HIT THEN LOOK AT THE TIMES WE SCOUTED AND A'I'I'ACKED YOUR GUY are you blind orjust ingoring the fact that your players braught this on.*

Morning Destroyer: *Left because we were fighting for something... As Captain Earthquake can confrim... he moved here to ensure what was agreed would be done... then IsaacU-P-Jazz started whining and decided to take the county, which baffled many in H16 who we were working alongside and could not understand why such a dumb move was allowed. I got mails from many in H16 at the time asking why we left and when it had been pointed out what happened supported us in our move and hoped H16 would correct the injustice... it never did... Besides you were away busy or otherwise afk at the time all that happened so you really know nothing iof what happened. by the time you returned you just had 2 member in your faction that nobody had heard anything from bitching endlessly in your faction threads (I saw them) about Buckinghamshire and so you picked sides regardless of prior agreements, alienating not only some in H16 but even some in your own faction!*

Black Dog: *and yet here we are yet again. you have ignored the main bulk of my mail and lumped to the past. I do not live in the past and i dont really care about what 1appened. you have still not forward the eports i have asked for and im not going to ask again. you have not proved aything and are only digging yourself a bigger hole. dont argue anythin unless you have the evidence to back it up. im sure you know that. you seem to be an nteligent person. i see no point in wasting my time talking to you anymore so i still offerthe same peace to all karma members. if you dont want to be attacked then drop.*

P-Jazz: *LOL I have only one account and I don't even think it is possible to cheat at this game, joined H16 when I felt H12 was not working anymore and was the first person in the region to do so. In the N of Bucks there was only two of us and we fought, cajolled and capped the County for H16. You were so busy building your own little fiefdom that you even managed to leave three enemy parishs in the North. Exactly where you claim that you were doing all your work. Still to this day, you allow and even encourage the enemy into your parishs. I have been your neighbour since the beginning and yet I saw none of this fighting you claim to have been involved with. I have openly offered you peace several times and all I get is some childish rants. Get over it. You lost the election for sherrif, it's about time you put aside your own feelings and act like a proper leader and do what is best for your faction. Rejoin the Alliance and we can be allies. I don't ask that you like me, just stop undermining the Alliance. As for my personal agenda, l have little or no influence on Alliance decisions outside of the County of Bucks. I have no way of changing any of the founding principles of the Alliance and just want to see the County of Bucks, totally in Alliance hands (or theirAllies.) You claim to be a man 41, start acting like it then, the decision to rejoin the Alliance is a no-brainer and war is just suicide. Anyone who knows Black Dog will just laugh at your accusations against him.*

Morning Destroyer*: You dont get to choose what is past and when nor what is relevant. It is the past that led to the present and so its very relevant to the current situation. A situation stirred up by the very same people that stirred it up in the past, when we were part of that alliance, said it many times when we were in hat "alliance", and not just in regards to Buckinghamshire, our biggest political and territorial opposition was always some members in H16, far more so than any enemies of the alliance we dealt with at the time. H15 was seen as a pawn to H16... whislt telling people in those "allied" Houses that they would get to be in the glory house, H16, once they proved themselves, instead H16 has busily recruiting and letting in factions from enemies that surrendered. Players that had done nothing whatsoever for the alliance, they just appened to have ill-gotten titles and parishes, a quick way to flip colours. H16 started using its "allies", getting them to clear land and secure it with the promisse of that land becoming theirs. oefore moving H16 players in and claiming it forthemselves instead. H16 has done this to several factions in its "support Houses" and indeed some in its own house and caused them to leave, then calling them traitors for doing so, with no mention of the fact that they had left because they had been lied to, used and betrayed. Thats the facts. again... Many in H16 and that alliance that were in it from when it was solely in Ireland knew how things were meant to work and saw how things increasingly were not working that way... I had mails from them, mails of support and offers of forming breakaway factions there is a discontent within H16 it has sought to suppress and stifle many times. It is a discontent from knowing that the House is no longer what it was, not the house they joined, it is some bastardised form of the houses it abosorbed in England and the egomaniacs in it will eventually turn on each other... or flee to the next new world. to try fk that world up too..*

P-Jazz*: My god, you rants become harderto follow all the time. So now your problem is with the treatment of MS by MS, They are both part of the Alliance. If you just want to be in the House that wins the Glory rounds then join H16, surely a place can be found for you, just to solve this. Forthe record I've only ever been in two factions, h10 for about four weeks from the begginning and h16 since then. I only joined h16 by chance, it was the first Alliance faction to send me an invite. Just stop the tantrums and do the sensible thing, rejoin the Alliance. If you're unhappy with things then try to change it from the inside. Iwouldn't let my personal feelings about you cloud my judgement of what the right thing to do is. As long as you are willing to help make Bucks County completely Alliance territory (or at least not hinderthe process) then I have no problem with you. I don't want to have a any power over you or any of your faction. As long as you abide by Alliance rules, decisions etc. then there will be no problems.*

Black Dog: *morning i do believe i offerend you the chance to join my faction a number of times when you were join gthe alliance. you turned the offer down. The strongest factions get to join and stay in H16. you left before your faction could be recognised. thats your fault. Anyhow. you still have not provided any jusitfication as to why my members were attacked. you have used the attacks from our allies to strike P-Jazz and IsaacU because of your grudges against them that is the reason this has all kicked off so stop blming us forthis skirmish. YOU started it and after several warnings you continued to attack. again your fault. The alliance doesn't have room for players that dont respect the rules set before them. you wanted a fight.*

Black Dog: *I offered you peace multiple times and all you do is attack me. You really want to commit suicide then you'll get your WAR. Having seen the state of you castles in the faction, obviously those two or three players who started this WAR may be ready but the rest of you will die. Any of you not wanting in on this WAR, leave Brown Eagle now or face the consequences. You may have razed one of my castles but I raze ten for your decision to attack me.*

Compilation of References

Acar, F. (1993). Women and university education in Turkey. *Higher Education in Europe*, *18*(4), 65–77. doi:10.1080/0379772930180406

Addison, B. V. (2012). Academic care, classroom pedagogy and the house group teacher: 'making hope practical' in uncertain times. *Pastoral Care in Education*, *30*(4), 303–315. doi:10.1080/02643944.2012.688064

Agee, J. (2009). Developing qualitative research questions: A reflective process. *International Journal of Qualitative Studies in Education: QSE*, *22*(4), 431–447. doi:10.1080/09518390902736512

Akyol, B., & Tanrısevdi, F. (2018). Academicians in Turkey: An Evaluation of Current Status of Academic Staff in Higher Education. *Higher Education Studies*, *8*(4), 129–138. doi:10.5539/hes.v8n4p129

Al Ghamdi, A., Samarji, A., & Watt, A. (2016). Essential considerations in distance education in KSA: Teacher immediacy in a virtual teaching and learning environment. *International Journal of Information and Education Technology, 6*(1), 17–22. doi:10.7763/ijiet.2016.v6.651

Allen, I. E., & Seaman, J. (2013). Changing course: Ten years of tracking online education in the United States. Newburyport, MA: Babson Survey Research Group and Quahog Research Group, LLC: Pearson.

All-free-download.com. (2020). *China political map.* Retrieved May 1, 2020 from https://all-free-download.com/free-vector/download/china-political-map-vector_587825.html

Alshenqeeti, H. (2014). Interviewing as a data collection method: A critical review. *English Linguistics Research*, *3*(1), 39–45. doi:10.5430/elr.v3n1p39

American Council on the Teaching of Foreign Languages. (2012). *The ACTFL performance descriptors for language learners 2012*. Retrieved June 4, 2018 from https://www.actfl.org/sites/default/files/pdfs/PerformanceDescriptorsLanguageLearners.pdf

American University. (2020, May 24). *Human subjects research IRB*. https://www.american.edu/irb/

Anand, P., Chung, S., & Wagers, M. (2010). Widening the net: Challenges for gathering linguistic data in the digital age. *NSF SBE 2020: Future Research in the Social, Behavioral & Economic Sciences,* 1-5.

Ancess, J. (2000). The reciprocal influence of teacher learning, teaching practice, school restructuring, and student learning outcomes. *Teachers College Record, 102*(3), 590–619. doi:10.1111/0161-4681.00069

Andrews, D., Nonnecke, B., & Preece, J. (2003). Electronic survey methodology: A case study in reaching hard-to-involve internet users. *International Journal of Human-Computer Interaction, 16*(2), 185–210. doi:10.1207/S15327590IJHC1602_04

Angrosino, M. V., Mays, D., & Kimberly, A. (2000). Rethinking observation: From method to context. In N. K. Denzin & Y. S. Lincoln (Eds.), *Handbook of qualitative research* (2nd ed., pp. 673–702). Sage.

APA. (2013). Posttraumatic Stress Disorder. In *Diagnostic and statistical manual of mental disorders: DSM-5*. American Psychiatric Association.

Aparicio, E. M., Birmingham, A., Rodrigues, E. N., & Houser, C. (2018). Dual experiences of teenage parenting and homelessness among Native Hawaiian youth: A critical interpretative phenomenological analysis. *Child & Family Social Work*. Advance online publication. doi:10.1111/cfs.12618

Appadurai, A. (1996). *Modernity at large: Cultural dimensions of globalization*. University of Minnesota Press.

Ariyaratnam, R., Casas, J. P., Whittaker, J., Smeeth, L., Hingorani, A. D., & Sharma, P. (2007). Genetics of ischaemic stroke among persons of non-European descent: A meta-analysis of eight genes involving~ 32,500 individuals. *PLoS Medicine, 4*(4), 728–736. doi:10.1371/journal.pmed.0040131 PMID:17455988

Arnold, N., Ducate, L., & Lomicka, L. (2007). Virtual communities of practice in teacher education. In M. A. Kassen, R. Z. Lavine, K. Murphy-Judy, & M. Peters (Eds.), *Preparing and developing technology-proficient L2 teachers* (pp. 103–132). CALICO.

Atlas.ti. (2018). [Computer software]. Retrieved from https://atlasti.com

Aufderheide, P. (2016, August 17). Does this have to go through the IRB? *The Chronicle of Higher Education.* https://www.chronicle.com/article/Does-This-Have-to-Go/237476

Aycan, Z., Kanungo, R. N., Mendonca, M., Yu, K., Deller, J., Stahl, G., & Khursid, A. (2000). Impact of Culture on Human Resource Management Practices: A Ten Country Comparison. *Applied Psychology, 49*(1), 192–220. doi:10.1111/1464-0597.00010

Ayyıldız Ünnü, N. A., & Şentürk, B. (2020). All-in-One Academics: Mental Workload in Turkish Academic Employment. In Evaluating Mental Workload for Improved Workplace Performance (pp. 69-87). IGI Global.

Baer, R. A. (2003). Mindfulness training as a clinical intervention: A conceptual and empirical review. *Clinical Psychology: Science and Practice, 10*(2), 125–143. doi:10.1093/clipsy.bpg015

Baez, B. (2002). Confidentiality in qualitative research: Reflections on secrets, power and agency. *Qualitative Research*, *2*(1), 35–58. doi:10.1177/1468794102002001638

Baker, A., & Lee, J. (2011). Mind the gap: Unexpected pitfalls in doing classroom research. *Qualitative Report*, *16*(5), 1435–1447.

Ballamingie, P. C., & Johnson, S. (2011). The vulnerable researcher: Some unanticipated challenges of doctoral fieldwork. *Qualitative Report*, *16*(3), 711–729.

Bar-On, D. (1996). Ethical Issues in Biographical Interviews and Analysis. In *Ethics and Process in the Narrative Study of Lives* (Vol. 4). Sage. doi:10.4135/9781483345451.n2

Bar-On, D. (Ed.). (2000). *Bridging the gap. Storytelling as a way to work through political and collective hostilities*. Körber-Stiftung.

Bate, S. P. (1997). Whatever happened to organizational anthropology? A review of the field of organizational ethnography and anthropological studies. *Human Relations*, *50*(9), 1147–1175. doi:10.1177/001872679705000905

Bedford, O., & Hwang, K. K. (2003). Guilt and shame in Chinese culture: A cross-cultural framework from the perspective of morality and identity. *Journal for the Theory of Social Behaviour*, *33*(2), 127–144. doi:10.1111/1468-5914.00210

Bell, C. (2020). The Tourism Researcher: Ethical Dilemmas During Fieldwork in Africa, Bali and Myanmar. *Critical Arts*, 1–6. doi:10.1080/02560046.2020.1712445

Bell, H. (2003). Strengths and Secondary Trauma in Family Violence Work. *Social Work*, *48*(4), 513–522. doi:10.1093w/48.4.513 PMID:14620108

Bengry-Howell, A., & Griffin, C. (2012). Negotiating access in ethnographic research with 'hard to reach'young people: Establishing common ground or a process of methodological grooming? *International Journal of Social Research Methodology*, *15*(5), 403–416. doi:10.1080/13645579.2011.600115

Benson, P., Chik, A., Gao, X., Huang, J., & Wang, W. (2009). Qualitative research in language teaching and learning journals, 1997–2006. *Modern Language Journal*, *93*(1), 79–90. doi:10.1111/j.1540-4781.2009.00829.x

Berg, B. L. (2001). *Qualitative research methods for social sciences*. Allyn & Bacon.

Berger, R., & Weiss, T. (2009). The Posttraumatic Growth Model: An Expansion to the Family System. *Traumatology*, *15*(1), 63–74. doi:10.1177/1534765608323499

Bernard, H. R. (1994). *Research methods in anthropology: Qualitative and quantitative approaches* (2nd ed.). AltaMira Press.

Bernhardt, E., & Hammadou, J. (1987). A decade of research in foreign language teacher education. *Modern Language Journal*, *71*(3), 289–299. doi:10.1111/j.1540-4781.1987.tb00369.x

Berno, T., & Ward, C. (2005). Innocence abroad: A pocket guide to psychological research on tourism. *The American Psychologist, 60*(6), 593–600. doi:10.1037/0003-066X.60.6.593 PMID:16173892

Bildirici, M., Anafarta, N., & Aykaç, E. (2003). Türkiye üniversitelerinde kadın emek gücü. *İktisat, İşletme ve Finans, 18*(206), 92-108.

Birnbach, M. (1961). Neo-Freudian Social Philosophy. Stanford University Press.

Bittner, E. (1965). The concept of organization. *Social Research*, 239–255.

Bloomquist, K. R., Wood, L., Friedmeyer-Trainor, K., & Kim, H. W. (2015). Self-care and Professional Quality of Life: Predictive Factors among MSW Practitioners. *Advances in Social Work, 16*(2), 292–311. doi:10.18060/18760

Boden, D. (1994). *The business of talk: Organizations in action.* Polity Press.

Bodur, M., & Kabasakal, H. (2002). Türkiye-Arap Kümesinde Kurumsal Kültür. *Journal of Management Research, 2*(1), 2–22.

Boellstorff, T. (2008). *Coming of age in Second Life: An anthropologist explores the virtually human.* Princeton University Press.

Boellstorff, T., Nardi, B., Pearce, C., & Taylor, T. L. (2012). *Ethnography and virtual worlds: A handbook of method.* Princeton University Press. doi:10.2307/j.cttq9s20

Borasky, D., Cooper, J., & FitzGerald, K. (2019). Paying subjects to take part in research: A new perspective on coercion and undue influence. *Clinical Research, 33*(3). https://acrpnet.org/2019/03/12/paying-subjects-to-take-part-in-research-a-new-perspective-on-coercion-and-undue-influence/

Bouchard, M. A., Normandin, L., & E-Hélène Séguin, M. (1995). Countertransference as instrument and obstacle: A comprehensive and descriptive framework. *The Psychoanalytic Quarterly, 64*(4), 717–745. doi:10.1080/21674086.1995.11927473 PMID:8584597

Bradbury-Jones, C. (2007). Enhancing rigor in qualitative health research: Exploring subjectivity through Peshkin's I's. *Journal of Advanced Nursing, 59*(3), 290–298. doi:10.1111/j.1365-2648.2007.04306.x PMID:17590210

Brady, J. (2009). From research idea to workplace project: Gaining access to the corporate world. In W. C. V. Bhatia, B. Du-Babcock, & J. Lung (Eds.), *Language for professional communication: Research, practice and training* (pp. 161–174). The Hong Kong Polytechnic University.

Brahier, B. R. (2006). *Examining a model of teachers' technology adoption decision making: An application of diffusion of innovations theory.* University of Minnesota.

Breaking News English. (2020). Retrieved from https://breakingnewsenglish.com/

Brend, D., Krane, J., & Saunders, S. (2019). *Exposure to trauma in intimate partner violence human service work: A scoping review. Traumatology.* doi:10.1037/trm0000199

Brewin, C., Andrews, B., & Valentine, J. (2000). Meta-analysis of risk factors for posttraumatic stress disorder in trauma-exposed adults. *Journal of Consulting and Clinical Psychology*, *68*(5), 748–766. doi:10.1037/0022-006X.68.5.748 PMID:11068961

Bride, B. (2007). Prevalence of Secondary Traumatic Stress among Social Workers. *Social Work*, *52*(1), 63–70. doi:10.1093w/52.1.63 PMID:17388084

Bride, B. E., Jones, J. L., & MacMaster, S. A. (2007). Correlates of secondary traumatic stress in child protective services workers. *Journal of Evidence-Based Social Work*, *4*(3-4), 69–80. doi:10.1300/J394v04n03_05

Broadhead, R. S., & Rist, R. C. (1976). Gatekeepers and the social control of social research. *Social Problems*, *23*(3), 325–336. doi:10.2307/799778

Brockington, D., & Sullivan, S. (2003). Qualitative research. In R. Scheyvens & D. Storey (Eds.), *Development Fieldwork: A Practical Guide* (pp. 57–74). Sage. doi:10.4135/9781849208864.n4

Brown, H. (2020). *Researchers Awarded Grant to Study Secondary Traumatic Stress in Health Workers*. Retrieved August 5, 2020, from http://uknow.uky.edu/research/researchers-awarded-grant-study-secondary-traumatic-stress-health-workers

Brown, J. S., Collins, A., & Duguid, P. (1989). Situated cognition and the culture of learning. *Educational Researcher*, *18*(1), 32–42. doi:10.3102/0013189X018001032

Bryman, A. (2016). *Social research methods* (5th ed.). Oxford University Press.

Buchanan, D., Boddy, D., & McCalman, J. (1986). *Getting in, getting on, getting out, getting back*. University of Glasgow Department of Management Studies.

Buchmann, M. (1984). The use of research knowledge in teacher education and teaching. *American Journal of Education*, *92*(4), 421–439. doi:10.1086/443768

Burns, P. M., & Novelli, M. (2006). Tourism and social identities: Introduction. In P. M. Burns & M. Novelli (Eds.), *Tourism and social identities: Global frameworks and local realities* (pp. 1–12). Elsevier.

Buswell, R., Webb, L., Mitchell, V., & Leder Mackley, K. (2017). Multidisciplinary research: Should effort be the measure of success? *Building Research and Information*, *45*(5), 539–555. doi:10.1080/09613218.2016.1194601

Butler, J. (1997). *The psychic life of power: theories in subjection*. Stanford University Press.

Calvi, M. V. (2001). El léxico del turismo. In Á. M. Felices Lago (Ed.), *ELE para contextos profesionales*. Cultura e intercultura en la enseñanza del español a extranjeros. Retrieved from http://www.ub.es/filhis/culturele/cont_pro.html

Campbell, S., Greenwood, M., Prior, S., Shearer, T., Walkem, K., Young, S., ... Walker, K. (2020). Purposive sampling: Complex or simple? Research case examples. *Journal of Research in Nursing*, *174498712092720*. Advance online publication. doi:10.1177/1744987120927206

Canadian Institutes of Health Research, N. S. a. E. R. C. o. C., & and Social Sciences and Humanities Research Council of Canada. (2014, December). Tri-Council Policy Statement: Ethical Conduct for Research Involving Humans. Author.

Cange, C. W., & McGaw-Césaire, J. (2019). Long-term public health responses in high-impact weather events: Hurricane Maria and Puerto Rico as a case study. *Disaster Medicine and Public Health Preparedness*, *4*(1), 18–22. doi:10.1017/dmp.2019.82 PMID:31679557

Caretta, M. A., & Jokinen, J. C. (2017). Conflating privilege and vulnerability: A reflexive analysis of emotions and positionality in postgraduate fieldwork. *The Professional Geographer*, *69*(2), 275–283. doi:10.1080/00330124.2016.1252268

Cerwonka, A., & Malkki, L. H. (2008). *Improvising theory: Process and temporality in ethnographic fieldwork*. University of Chicago Press.

Chege, N. (2019). 'If you give me time I can love you': A Pregnant Researcher among Male Beach Workers on Kenya's Liminal South Coast Beaches. *Anthropology Matters*, *19*(1), 1–34. doi:10.22582/am.v19i1.494

Cheng, C. C. (1998, May). *Extra-linguistic data for understanding dialect mutual intelligibility*. Paper presented at the meeting of Annual Conference of the Pacific Neighborhood Consortium (PNC). Taipei, Taiwan.

Chen, Y., & Feng, S. (2013). Access to public schools and the education of migrant children in China. *China Economic Review*, *26*, 75–88. doi:10.1016/j.chieco.2013.04.007

Chism, R., & Graff, C. (2020). The impact of online lower-level courses on world language learners' self-perceptions, mindset and willingness to communicate. In M. Freiermuth & N. Zarrinabadi (Eds.), *Technology and the Psychology of Second Language Learners and Users: New Language Learning and Teaching Environments* (pp. 543–569). Palgrave Macmillan. doi:10.1007/978-3-030-34212-8_21

Choi, G. (2011a). Organizational Impacts on the Secondary Traumatic Stress of Social Workers Assisting Family Violence or Sexual Assault Survivors. *Administration in Social Work*, *35*(3), 225–242. doi:10.1080/03643107.2011.575333

Choi, G. (2011b). Secondary Traumatic Stress of Service Providers who Practice with Survivors of Family or Sexual Violence: A National Survey of Social Workers. *Smith College Studies in Social Work*, *81*(1), 101–119. doi:10.1080/00377317.2011.543044

Christopher, J. C., & Maris, J. A. (2010). Integrating mindfulness as self-care into counseling and psychotherapy training. *Counselling & Psychotherapy Research*, *10*(2), 114–125. doi:10.1080/14733141003750285

Chronicle Forums. (2011, September 2). *The Chronicle of Higher Education*. Retrieved May 20, 2020 from https://www.chronicle.com/forums/index.php?topic=81783.0

Clarke, M. (1975). Survival in the field: Implications of personal experience in fieldwork. *Theory and Society*, *2*(1), 95–123. doi:10.1007/BF00212729

Clark, T. (2011). Gaining and maintaining access: Exploring the mechanisms that support and challenge the relationship between gatekeepers and researchers. *Qualitative Social Work: Research and Practice*, *10*(4), 485–502. doi:10.1177/1473325009358228

CoHE. (2018, June). *Mission Differentiation and Specialization Aimed at Regional Development Program.* Council of Higher Education. https://bolgeselkalkinma.yok.gov.tr/hakkinda/proje-hakkinda

Cole, A. (2000). Academic freedom and the publish or perish paradox in schools of education. *Teacher Education Quarterly*, *27*(2), 33–48.

Collis, J., & Hussey, R. (2003). *Business research: A practical guide for undergraduate and postgraduate students* (2nd ed.). Palgrave Macmillan.

Cordingley, P., & Needham, K. (2010). Chapter 15: School leaders using inquiry and research. In Campbell & S. Groundwater-Smith (Ed.), Connecting Inquiry and Professional Learning in Education: International perspectives and Practical solutions (pp. 192-199). Routledge.

Corr, C. A., Corr, D. M., & Doka, K. J. (2019). *Death & dying, life & living* (8th ed.). Cengage.

Coughlin, S., Smith, S., & Fernández, M. (2017). *Handbook of community-based participatory research.* doi:10.1093/acprof:oso/9780190652234.001.0001

Cox, K., & Steiner, S. (2013). Preserving commitment to social work service through the prevention of vicarious trauma. *Journal of Social Work Values and Ethics*, *10*(1), 52–60.

Creese, A., Baraç, T., Bhatt, A., Blackledge, A., Hamid, S., Wei, L., Yağcıoğlu-Ali, D. (2008). *Investigating multilingualism in complementary schools in four communities.* Final Report for the ESCR for Project No: RES-000-23, 1180.

Creese, A., & Blackledge, A. (2012). Voice and meaning-making in team ethnography. *Anthropology & Education Quarterly*, *43*(3), 306–324. doi:10.1111/j.1548-1492.2012.01182.x

Creswell, J. W. (2007). *Qualitative inquiry & research design: choosing among five approaches.* Sage.

Creswell, J. W. (2007). *Qualitative inquiry and research design: Choosing among five approaches* (2nd ed.). SAGE.

Creswell, J. W. (2013). *Research design: Qualitative, quantitative, and mixed methods approaches.* Sage publications.

Cross-Disciplinary Learning. (2012). In N. M. Seel (Ed.), *Encyclopedia of the Sciences of Learning* (pp. 858–858). Springer US.

Crossley, É. (2019). Deep reflexivity in tourism research. *Tourism Geographies*, •••, 1–22. doi:10.1080/14616688.2019.1571098

Crowhurst, I., & Kennedy-Macfoy, M. (2013). Troubling gatekeepers: Methodological considerations for social research. *International Journal of Social Research Methodology, 16*(6), 457–462. doi:10.1080/13645579.2013.823281

Cunliffe, A. L., & Alcadipani, R. (2016). The politics of access in fieldwork: Immersion, backstage dramas, and deception. *Organizational Research Methods, 19*(4), 535–561. doi:10.1177/1094428116639134

D'Cruz, H., Gillingham, P., & Melendez, S. (2007). Reflexivity, its meaning and relevance for social work: A critical review of the literature. *British Journal of Social Work, 37*(1), 73–90. doi:10.1093/bjsw/bcl001

Dalenberg, C. (2000). *Countertransference and the treatment of trauma*. American Psychological Association. doi:10.1037/10380-000

Dalenberg, C. J. (2004). Maintaining the safe and effective therapeutic relationship in the context of distrust and anger: Countertransference and complex trauma. *Psychotherapy (Chicago, Ill.), 41*(4), 438–447. doi:10.1037/0033-3204.41.4.438

Davidson, L. (2003). *Livening outside mental illness: Qualitative studies of recovery in schizophrenia*. New York University Press.

Davison, C. B., Brandt, N. C., & Hua, D. M. (2015). Predictive modeling of organizational resilience utilizing leadership frame orientation. *Research in Higher Education, 29*, 1–13.

De Neve, G., & Unnithan-Kumar, M. (2016). *Critical journeys: The making of anthropologists*. Routledge. doi:10.4324/9781315575032

De Waal, F. B. (2008). Putting the altruism back into altruism: The evolution of empathy. *Annual Review of Psychology, 59*(1), 279–300. doi:10.1146/annurev.psych.59.103006.093625 PMID:17550343

Deckert, S. K., & Vickers, C. H. (2011). *An introduction to Sociolinguistics*. Continuum International Publishing Group.

Denzin, N. K., & Lincoln, Y. S. (2005) Methods of collecting and analyzing empirical materials. In The Sage Qualitative Research, (3rd edition (pp.641-642). Thousand Oaks, CA: Sage.

Devlet Personel Daire Başkanlığı. (2015). *Üniversitelere ilişkin sayısal veriler*. Retrieved May, 5, 2017, from http://www.dpb.gov.tr/trtr/istatistik/universitelere-ilskin-sayisal-veriler

deWinter, J. C. F., Dodou, D., & Wieringa, P. A. (2009). Exploring factor analysis with small sample sizes. *Multivariate Behavioral Research, 44*(2), 147–181. doi:10.1080/00273170902794206 PMID:26754265

Dikmen N., & Maden, D. (2012). Kadın akademisyenlerin görünmeyen emeği üzerine bir araştırma: Ordu Üniversitesi örneği. *İstanbul Ticaret Üniversitesi Sosyal Bilimler Dergisi, 11*(21), 257-288.

Discord. (2018). [Computer software]. Retrieved from https://discord.com

Dörnyei, Z. (2007). *Research methods in applied linguistics: Quantitative, qualitative, and mixed methodologies*. Oxford University Press.

Douglas, J. D. (1976). *Investigative social research: Individual and team field research*. Sage.

Drame, E. R., & Irby, D. J. (2016). *Black participatory research: power, identity, and the struggle for justice in education*. Academic Press.

Drexler, W. (2010). The networked student model for construction of personal learning. *Language Learning & Technology*, *26*(3), 369–385.

Driscoll, L., Parkes, K., Tilley-Lubbs, G., Brill, J., & Bannister, V. (2009). Navigating the lonely sea: Peer mentoring and collaboration among aspiring women scholars. *Mentoring & Tutoring*, *17*(1), 5–21. doi:10.1080/13611260802699532

Duarte, J., & Pinto-Gouveia, J. (2016). Effectiveness of a mindfulness-based intervention on oncology nurses' Burnout and compassion fatigue symptoms: A non-randomized study. *International Journal of Nursing Studies*, *64*(1), 98–107. doi:10.1016/j.ijnurstu.2016.10.002 PMID:27744228

Dubetsky, A. (1976). Kinship, primordial ties, and factory organization in Turkey: An Anthroplogical View. *International Journal of Middle East Studies*, *7*(3), 433–451. doi:10.1017/S0020743800025745

Dudley, J. R. (2005). *Research methods for social work: becoming consumers and producers of research*. Pearson.

Dudovskiy, J. (2019). *Snowball sampling*. Retrieved May 1, 2020 from https://research-methodology.net/sampling-in-primary-data-collection/snowball-sampling/

Duff, P. A., & Talmy, S. (2011). Language socialization approaches to second language acquisition. In D. Atkinson (Ed.), Alternative approaches to second language acquisition (pp. 95-116). Routledge.

Duff, P. A. (2007). Second language socialization as sociocultural theory: Insights and issues. *Language Teaching*, *40*(4), 309–319. doi:10.1017/S0261444807004508

Duncombe, J., & Jessop, J. (2012). Doing rapport and the ethics of faking friendship. In Ethics in qualitative research (pp. 108–121). London: Sage. doi:10.4135/9781473913912.n8

Dunya, A., Lewando, H. G., & Blackburn, C. (2011). Issues of gender, reflexivity and positionality in the field of disability: Researching visual impairment in an Arab society. *Qualitative Social Work: Research and Practice*, *10*(4), 467–484. doi:10.1177/1473325010370188

Dwyer, S. C., & Buckle, J. L. (2009). The space between: On being an insider-outsider in qualitative research. *International Journal of Qualitative Methods*, *8*(1), 54–63. doi:10.1177/160940690900800105

Dykema, J., Jones, N. R., Piche, T., & Stevenson, J. (2013). Surveying Clinicians by Web: Current Issues in Design and Administration. *Evaluation & the Health Professions*, *36*(3), 352–381. doi:10.1177/0163278713496630 PMID:23975760

Eastwood, C. D., & Ecklund, K. (2008). Compassion fatigue risk and self-care practices among residential treatment center childcare workers. *Residential Treatment for Children & Youth*, *25*(2), 103–122. doi:10.1080/08865710802309972

Edmondson, A. C., & McManus, S. E. (2007). Methodological fit in management field research. *Academy of Management Review*, *32*(4), 1246–1264. doi:10.5465/amr.2007.26586086

Edwards, R., & Holland, J. (2013). *What is qualitative interviewing?* Bloomsbury.

Egbert, J. (2006). Learning in context: Situating language teacher learning in CALL. In P. Hubbard & M. Levy (Eds.), *Teacher Education in CALL* (pp. 167–181). John Benjamins. doi:10.1075/lllt.14.15egb

Egbert, J., & Salsbury, T. (2009). Out of complacency and into action: An exploration of professional development experiences in school/home literacy engagement. *Teaching Education*, *20*(4), 375–393. doi:10.1080/10476210902998474

El Shaban, A. (2017). I*n-Service ELL Teachers and Technology Adoption: Exploring Diffusion of Innovations in Language Education* [Doctoral dissertation, Washington State University]. Retrieved from https://research.libraries.wsu.edu/xmlui/bitstream/handle/2376/13005/ElShaban_wsu_0251E_12122.pdf?sequence=1&isAllowed=y

El Shaban, A., & Egbert, J. (2018). Diffusing education technology: A model for language teacher professional development in CALL. *System*, *78*, 234–244. doi:10.1016/j.system.2018.09.002

Elo, S., Kääriäinen, M., Kanste, O., Pölkki, T., Utriainen, K., & Kyngäs, H. (2014). Qualitative Content Analysis: A Focus on Trustworthiness. *SAGE Open*, *4*(1). doi:10.1177/2158244014522633

EnglishCentral. (2020). Retrieved from https://www.englishcentral.com/videos

Ergöl, Ş., Koç, G., Eroğlu, K., & Taşkın, L. (2012). Türkiye'de kadın araştırma görevlilerinin ev ve iş yaşamlarında karşılaştıkları güçlükler. *Yükseköğretim Dergisi*, *2*(1), 43–49.

Eriksson, P., & Kovalainen, A. (2008). Ethnographic research. In P. Eriksson & A. Kovalainen (Eds.), *Introducing qualitative methods: Qualitative methods in business research* (pp. 138–153). Sage. doi:10.4135/9780857028044.d77

Ernø-Kjølhede, E., Husted, K., Mønsted, M., & Wenneberg, S. B. (2001). Managing university research in the triple helix. *Science & Public Policy*, *28*(1), 49–55. doi:10.3152/147154301781781679

Etchegoyen, R. H. (2018). *The fundamentals of psychoanalytic technique*. Routledge. doi:10.4324/9780429481864

European Commission. (2016). *She Figures 2015 Report.* Retrieved October, 5, 2014, from https://ec.europa.eu/research/swafs/pdf/pub_gender_equality/she_figures_2015-final.pdf

Ezvid. (2018). [Computer software]. Retrieved from https://www.ezvid.com

Facebook. (2020). Retrieved from https://m.facebook.com/?refsrc=http%3A%2F%2Fwww.google.com%2Fsearch&_rdr

Factile. (n.d.). Retrieved from https://www.playfactile.com/

Falzon, M.-A. (2009). Introduction: Multi-sited ethnography: Theory, praxis and locality in contemporary research. In M.-A. Falzon (Ed.), *Multi-sited ethnography: Theory, praxis and locality in contemporary research*. Ashgate Publishing, Ltd.

Fan, W., & Yang, Z. (2010). Factors affecting response rates of the web survey: A systematic review. *Computers in Human Behavior*, *26*(2), 132–139. doi:10.1016/j.chb.2009.10.015

Far, P. K. (2018). Challenges of recruitment and retention of university students as research participants: Lessons learned from a pilot study. *Journal of the Australian Library and Information Association*, *67*(3), 278–292. doi:10.1080/24750158.2018.1500436

Feiman-Nemser, S., & Remillard, J. (1996). Perspectives on learning to teach. In F. Murray (Ed.), *The teacher educator's handbook: Building a knowledge base for the preparation of teachers* (pp. 63–91). Jossey-Bass.John.

Feldman, M. S., Bell, J., & Berger, M. T. (2003). *Gaining access: A practical and theoretical guide for qualitative researchers*. AltaMira.

Félix-Brasdefer, J. C. (2015). *The language of service encounters: A pragmatic-discursive approach*. Cambridge University Press. doi:10.1017/CBO9781139565431

Ferro, A. (2006). *Psychoanalysis as therapy and storytelling*. Routledge.

Fetterman, D. M. (2010). *Ethnography: Step-by-step*. Sage.

Figley, C. (1995). *Compassion fatigue: Coping with secondary traumatic stress disorder in those who treat the traumatized*. Bruner/ Mazel.

Fikret Pasa, S., Kabasakal, H., & Bodur, M. (2001). Society, organisations, and leadership in Turkey. *Applied Psychology*, *50*(4), 559–589. doi:10.1111/1464-0597.00073

Fine, G. A. (1993). Ten lies of ethnography: Moral dilemmas of field research. *Journal of Contemporary Ethnography*, *22*(3), 267–293. doi:10.1177/089124193022003001

Fineman, S., Gabriel, Y., & Sims, D. (2010). *Organizing and organizations* (4th ed.). Sage.

Finn, M., Elliott-White, M., & Walton, M. (2000). *Tourism and leisure research methods: Data collection, analysis and interpretation*. Longman.

Flaherty, C. (2014, April 9). So much to do, so little time. *Inside Higher Ed*. Retrieved June 26, 2020, from https://www.insidehighered.com/news/2014/04/09/research-shows-professors-work-long-hours-and-spend-much-day-meetings

Flaherty, C. (2017). Should grad students publish? *Inside Higher Ed.* Retrieved June 26, 2020, from https://www.insidehighered.com/news/2017/08/23/renewed-debate-over-whether-graduate-students-should-publish

Flores, G. M. (2016). Discovering a hidden privilege: Ethnography in multiracial organizations as an outsider within. *Ethnography*, *17*(2), 190–212. doi:10.1177/1466138115575660

Flowers, A. M. III, Scott, J. A., Riley, J. R., & Palmer, R. T. (2015). Beyond the call of duty: Building on othermothering for improving outcomes at historically black colleges and universities. *Journal of African American Males in Education*, *6*(1), 59–73. Retrieved May 12, 2020, from https://www.semanticscholar.org/paper/Beyond-the-Call-of-Duty%3A-Building-on-Othermothering-Flowers-Scott/4e821a419092edcd7b6aaa180430dd4572a39f0b

Fojtik, R. (2017). The Ozobot and education of programming. *New Trends and Issues Proceedings on Humanities and Social Sciences*, *5*, 8–16.

Free Online, O. C. R. (2018). [Computer application]. Retrieved from https://www.newocr.com

Free Video to JPG Converter. (2018). [Computer application]. Retrieved from https://www.dvdvideosoft.com/products/dvd/Free-Video-to-JPG-Converter.htm

Freud, S. (1957). The future prospects of psycho-analytic therapy. In J. Strachey (Ed. & Trans.), The standard edition of the complete psychological works of Sigmund Freud, (Vol. 11, pp. 141-151). London: Hogarth Press. (original work was published in 2010)

Freud, S., & Bonaparte, P. M. (1954). *The origins of psychoanalysis* (Vol. 216). Imago.

Fromm, E. (2013). *The revision of psychoanalysis*. Open Road Media.

Fullan, M. (2007). *The new meaning of educational change*. Routledge.

Gal, S. (2019) *Risk and protective factors for the mental health consequences of childhood political trauma (Argentina 1976-1983) among adult Jewish Argentinian immigrants to Israel* (Unpublished doctoral dissertation). McGill University, Montreal, Canada.

Gal, S. (2020). "Israel is not Switzerland": The impact of the political violence in Israel on historically traumatized Argentinian immigrants. *Journal of International Migration and Integration.*

Gal, S. (2020a). Childhood experiences of Anti-Semitism in Argentina (1976-1983): Stories of trauma, resilience and long term outcome. In H. S. Moffic, J. Peteet, A. Hankir, & M. V. Seeman (Eds.), *Psychiatry and Anti-Semitism - Recognition, Prevention, and Interventions.* Springer.

Gal, S., & Hanley, J. (2019). The long-term impact of political activism on mental wellbeing in the context of adult Jewish immigrants who experienced childhood political trauma. *Journal of Immigrant and Minority Health.*

Ganassin, S., & Holmes, P. (2013). Multilingual research practices in community research: The case of migrant/refugee women in North East England. *International Journal of Applied Linguistics*, *23*(3), 342–356. doi:10.1111/ijal.12043

Gangopadhyay, J. (2019). The ethical challenges of field research: A personal narrative. *Indian Journal of Medical Ethics*, *5*(1), 34–38. doi:10.20529/IJME.2019.081 PMID:32103818

Gerstl-Pepin, C., & Patrizion, K. (2009). Learning from Dumbledore Pensive: Metaphor as an aid in teaching reflexivity in qualitative research. *Qualitative Research*, *9*(3), 299–308. doi:10.1177/1468794109105029

Glesne, C. (2015). *Becoming qualitative researchers: An introduction.* Pearson.

Glogster. (2020). Retrieved from http://edu.glogster.com/?ref=com

Gobrin, A. (2020) *The cultural sofa: Psychoanalysis in the Spaces of Literature, Philosophy and Society*. Narkisus Publication.

Gold, R. L. (1958). Roles in sociological field observations. *Social Forces*, *36*(3), 217–223. doi:10.2307/2573808

Goldstein, H. (1994). Ethnography. Critical inquiry and social work practice. In E. Sherman & W. Ried (Eds.), *Qualitative research in social work* (pp. 42–51). Columbia University Press.

Gole, N. (2004). Visible women: Actress in the public sphere. *New Perspectives Quarterly*, *21*(2), 12–13. doi:10.1111/j.1540-5842.2004.00653.x

Google Docs. (2020). Retrieved from https://www.google.com/docs/about/

Google. (2020). Retrieved from https://www.google.com/

Gray, D. E. (2014). *Doing research in the real world* (3rd ed.). Sage.

Grönroos, C. (2001). The perceived service quality concept – a mistake. *Managing Service Quality*, *11*(3), 150–152. doi:10.1108/09604520110393386

Grove, J. (2013). *Global Gender Index, 2013.* Retrieved May, 10, 2012, from https://www.timeshighereducation.com/cn/features/global-gender-index-2013/2003517.article

Guimaraes, M. J. L. J. (2005). Doing anthropology in cyberspace: Fieldwork boundaries and social environments. In C. Hine (Ed.), *Virtual methods: Issues in social research on the Internet* (pp. 141–156). Berg.

Guiyang Education Bureau. (2010). *Guiyang Education Bureau on further strengthening the private primary and secondary schools (kindergartens) safety prevention work of the relevant provisions*. Retrieved May 1, 2020 from http://www.fsou.com/html/text/lar/172412/17241244.html

Guiyang Education Bureau. (2018). *Measures for the management of private schools in Guiyang city (Trail)*. Retrieved May 1, 2020 from http://www.xifeng.gov.cn/zxfw/bmlqfw/jypx/xxgl/201811/t20181130_1953320.html

Guiyang Education Bureau. (2019). *Table of basic information of education in Guiyang in 2019.* Guiyang Open Government Data Platform. Retrieved May 1, 2020 from https://data.guiyang.gov.cn/city/toListResContent.htm

Guiyang Local Chronicles Compilation Committee. (2018). *Guiyang Yearbook 2018*. Retrieved May 1, 2020 from http://www.guiyang.gov.cn/ztzl/ztzlsjtj/ztzlsjtjgynj/

Gul, R. B., & Ali, P. A. (2010). Clinical trials: The challenge of recruitment and retention of participants. *Journal of Clinical Nursing*, *19*(1–2), 227–233. doi:10.1111/j.1365-2702.2009.03041.x PMID:20500260

Gupte, M. (2015). Does information and communication technology have an important role to play in Indian higher education? *Abhinav International Monthly Refereed Journal of Research in Management & Technology*, *4*(1), 41–46.

Gürüz, K. (2012). Remuneration of Academic Staff in Turkish Universities. In P. G. Altbach, L. Reisberg, M. Yudkevich, G. Androushchak, & I. F. Pacheco (Eds.), *Paying the Professoriate: A Global Comparison of Compensation and Contracts* (pp. 297–307). Routledge.

Hafernik, J. J., Messerschmitt, D. S., & Vandrick, S. (1997). Research news and comment: Collaborative research: Why and how? *Educational Researcher*, *26*(9), 31–35.

Hall, D. (2018). *UVI is Rising*. Retrieved May 12, 2020, from https://www.uvi.edu/news/articles/2018/18_015_uvi_is_rising.aspx

Hall, M. C. (2010). Island destinations: a natural laboratory for tourism: introduction. *Asia Pacific Journal of Tourism Research*, *15*(3), 245–249. doi:10.1080/10941665.2010.503613

Hammersley, M., & Atkinson, P. (2007). *Ethnography: Principles in practice* (3rd ed.). Routledge. doi:10.4324/9780203944769

Hamzeh, M. Z., & Oliver, K. (2010). Gaining research access into the lives of Muslim girls: Researchers negotiating muslimness, modesty, inshallah, and haram. *International Journal of Qualitative Studies in Education: QSE*, *23*(2), 165–180. doi:10.1080/09518390903120369

Hannerz, U. (2003). Being there... and there... and there! Reflections on multi-site ethnography. *Ethnography*, *4*(2), 201–216. doi:10.1177/14661381030042003

Hansson, F., & Mønsted, M. (2006). *Leadership Dilemmas and Dynamics in research projects – entrepreneurial leadership?* MPP Working Paper, No. 6.

Harrison, R. L., & Westwood, M. J. (2009). Preventing vicarious traumatization of mental health therapists: Identifying protective practices. *Psychotherapy (Chicago, Ill.)*, *46*(2), 203–219. doi:10.1037/a0016081 PMID:22122619

Harvey, M. R., Mischler, E. G., Koenen, K., & Harney, P. A. (2000). In the aftermath of sexual abuse: Making and remaking meaning in narratives of trauma and recovery. *Narrative Inquiry*, *10*(2), 291–311. doi:10.1075/ni.10.2.02har

Heidegger, M., Macquarrie, J., & Robinson, E. (1962). *Being and time*. Blackwell.

Hellawell, D. (2006). Inside–out: Analysis of the insider–outsider concept as a heuristic device to develop reflexivity in students doing qualitative research. *Teaching in Higher Education, 11*(4), 483–494. doi:10.1080/13562510600874292

Hellsten, L., Martin, S., McIntyre, L., & Kinzel, A. (2011). Women on the academic tenure track: An autoethnographic inquiry. *International Journal for Cross-Disciplinary Subjects in Education, 2*(1), 271–275. doi:10.20533/ijcdse.2042.6364.2011.0039

Helm, F. (2015). The practices and challenges of telecollaboration in higher education in Europe. *Language Learning & Technology, 19*(2), 197–217. http://llt.msu.edu/issues/june2015/helm.pdf

Hendrickson, K. A., & Francis, K. A. (2015). Surviving Senior Seminar Research Helpful Strategies for Improving Overall Project Performance. In Infusing Undergraduate Research into Historically Black Colleges and Universities Curricula (pp. 219-240). Emerald Group Publishing Limited. doi:10.1108/S1479-364420150000017011

Hendrickson, K. A., & Francis, K. A. (2018). Finding professional true to self-authenticity in public university faculty experience. *Studies in Higher Education, 43*(12), 2398–2411. doi:10.1080/03075079.2017.1333492

Hendrickson, K. A., & Francis, K. A. (2020). Investigating Authentic University Academic Care at a US Non-Mainland HBCU. In P. Blessinger, E. Sengupta, & M. Makhanya (Eds.), *Improving Classroom Engagement and International Development Programs: International Perspectives on Humanizing Higher Education Innovations in Higher Education Teaching and Learning* (pp. 59–74). Emerald Publishing Limited. doi:10.1108/S2055-364120200000027006

Herrington, J., & Oliver, R. (2000). An instructional design framework for authentic learning environments. *Educational Technology Research and Development, 48*(3), 23–48. doi:10.1007/BF02319856

Herskovits, M. J. (1950). The hypothetical situation: A technique of field research. *Southwestern Journal of Anthropology, 6*(1), 32–40. doi:10.1086outjanth.6.1.3628688

Hevezi, J. A. (2016). Evaluation of a meditation intervention to reduce the effects of stressors associated with compassion fatigue among nurses. *Journal of Holistic Nursing, 34*(4), 343–350. doi:10.1177/0898010115615981 PMID:26598000

Hine, C. (2000). *Virtual ethnography*. Sage Publications. doi:10.4135/9780857020277

Hine, C. (2008). Virtual ethnography: Modes, varieties, affordances. In N. G. Fielding, R. M. Lee, & G. Blank (Eds.), *The Sage handbook of online research methods* (pp. 257–270). Sage Publications. doi:10.4135/9780857020055.n14

Hirvi, L., & Snellman, H. (Eds.). (2012). *Where is the field? The experience of migration viewed through the prism of ethnographic fieldwork.* Studia Fennica Ethnologica. doi:10.21435fe.14

Hobbs, V., & Kubanyiova, M. (2008). The challenges of researching language teachers: What research manuals don't tell us. *Language Teaching Research, 12*(4), 495–513. doi:10.1177/1362168808097162

Hofstede, G. (1980). *Culture's Consequences: International Differences in Work-Related Values*. Sage Publications.

Hofstede, G. (2001). *Culture's Consequences: Comparing Values, Behaviors, Institutions, and Organizations Across Nations* (2nd ed.). Sage Publications.

Hofstede, G., & Bond, M. H. (1988). The Confucius Connection: From Cultural Roots to Economic Growth. *Organizational Dynamics*, *16*(4), 4–21. doi:10.1016/0090-2616(88)90009-5

Hogan, K., Hegarty, J., Ward, T., & Dodd, L. (2012). Counsellors' experiences of working with male victims of female-perpetrated domestic abuse. *Counselling & Psychotherapy Research*, *12*(1), 44–52. doi:10.1080/14733145.2011.630479

Holmes, J. (2014). *How do we research workplace talk?* Paper presented at the 17th Warwick International Postgraduate Conference in Applied Linguistics, University of Warwick.

Holmes, J. (2007). Monitoring organisational boundaries: Diverse discourse strategies used in gatekeeping. *Journal of Pragmatics*, *39*(11), 1993–2016. doi:10.1016/j.pragma.2007.07.009

Holmes, J. (2014). Countertransference in qualitative research: A critical appraisal. *Qualitative Research*, *14*(2), 166–183. doi:10.1177/1468794112468473

Homan, R. (2001). The principle of assumed consent: The ethics of gatekeeping. *Journal of Philosophy of Education*, *35*(3), 329–343. doi:10.1111/1467-9752.00230

Hooker, J. (2012). Cultural differences in business communication. In C. B. Paulston, S. F. Kiesling, & E. S. Rangel (Eds.), *The handbook of intercultural discourse and communication* (pp. 389–407). John Wiley & Sons Ltd. doi:10.1002/9781118247273.ch19

Horwood, J., & Moon, G. (2003). Accessing the research setting: The politics of research and the limits to enquiry. *Area*, *35*(1), 106–109. doi:10.1111/1475-4762.00116

House, R. J., Hanges, P. J., Ruiz-Quintanilla, S. A., Dorfman, P. W., Javidan, M., Dickson, M. W., Gupta, V., & … . (1999). Cultural Influences on Leadership and Organizations: Project GLOBE. In W. H. Mobley, M. J. Gessner, & V. Arnold (Eds.), *Advances in Global Leadership* (pp. 171–233). JAI Press.

Howe, K., & Eisenhart, M. (1990). Standards for qualitative (and quantitative) research: A prolegomenon. *Educational Researcher*, *19*(4), 2–9. doi:10.3102/0013189X019004002

Høyland, S., Hollund, J. G., & Olsen, O. E. (2015). Gaining access to a research site and participants in medical and nursing research: A synthesis of accounts. *Medical Education*, *49*(2), 224–232. doi:10.1111/medu.12622 PMID:25626753

Hsieh, H.-F., & Shannon, S. E. (2005). Three Approaches to Qualitative Content Analysis. *Qualitative Health Research*, *15*(9), 1277–1288. doi:10.1177/1049732305276687 PMID:16204405

Hu, H., Lu, S., & Huang, C. C. (2014). The psychological and behavioral outcomes of migrant and left-behind children in China. *Children and Youth Services Review*, *46*, 1–10. doi:10.1016/j.childyouth.2014.07.021

Hyndman, B. (2018, August 13). Ten reasons teachers can struggle to use technology in the classroom. *The Conversation*. Retrieved from http://theconversation.com/ten-reasons-teachers-can-struggle-to-use-technology-in-the-classroom-101114/

Iliffe, G., & Steed, L. (2000). Exploring the Counselor's Experience of Working With Perpetrators and Survivors of Domestic Violence. *Journal of Interpersonal Violence*, *15*(4), 393–412. doi:10.1177/088626000015004004

International Religious Freedom Report. (2018). *Executive Summary Turkey*. Retrieved 03, November, 2019, from https://www.state.gov/wp-content/uploads/2019/05/TURKEY-2018-INTERNATIONAL-RELIGIOUS-FREEDOM-REPORT.pdf

Islam, M. R. (2018). Sample size and its role in Central Limit Theorem (CLT). *International Journal of Physics & Mathematics*, *1*(1), 37–47. doi:10.31295/ijpm.v1n1.42

Ivanova, M., Buda, D.-M., & Burrai, E. (2020). Creative and disruptive methodologies in tourism studies. *Tourism Geographies*, 1–10. doi:10.1080/14616688.2020.1784992

Jackson, J. M. (1977). The organization and its communication problems. In L. W. Porter & K. H. Roberts (Eds.), *Communication in organizations* (pp. 84–134). Penguin Books.

Januszewski, A., & Molenda, M. (Eds.). (2013). *Educational technology: A definition with commentary*. Routledge. doi:10.4324/9780203054000

Johl, S. K., & Renganathan, S. (2010). Strategies for gaining access in doing fieldwork: Reflection of two researchers. *Electronic Journal of Business Research Methods*, *8*(1), 42–50.

Johnson, R. B., Onwuegbuzie, A. J., & Turner, L. A. (2007). Toward a definition of mixed methods research. *Journal of Mixed Methods Research*, *1*(2), 112–133. doi:10.1177/1558689806298224

Jones, A. (2014). *TEDxESL*. Retrieved from https://tedxesl.com/

Josselson, R. (1996). On writing other people's lives: Self-analytic reflections of a narrative researcher. In R. Josselson (Ed.), The narrative study of lives, Vol. 4. Ethics and process in the narrative study of lives (p. 60–71). Sage Publications, Inc.

Jung, C. G. (1969). *The psychology of the transference* (Vol. 20). Bollingen Foundation.

Kabasakal, H., & Bodur, M. (2002). Arabic cluster: A bridge between East and West. *Journal of World Business*, *37*(1), 40–54. doi:10.1016/S1090-9516(01)00073-6

Kabasakal, H., & Bodur, M. (2007). Leadership and Culture in Turkey: A Multi-Faceted Phenomenon. In J. Chhokar, F. Brodbeck, & R. House (Eds.), *Managerial Cultures of the World: A GLOBE Re- port of In-depth Studies of the Cultures of 25 Countries* (pp. 835–874). Lawrence Erlbaum Associates.

Kabat-Zinn, J. (1990). *Full Catastrophe Living: Using the Wisdom of Your Body and Mind to Face Stress, Pain, and Illness*. Bantam Doubleday Dell.

Kadın Adayları Destekleme Derneği. (2015). *Kadın İstatistikleri*. Retrieved January, 10, 2014 from http://www.ka-der.org.tr/tr- TR/Page/Show/400/kader-istatistikleri.html

Kahoot. (2020). Retrieved from https://getkahoot.com/

Kahraman, L., Kahraman, A. B., Ozansoy, N., Akıllı, H., Kekillioğlu, A., & Özcan, A. (2014). Nevşehir Hacı Bektaş Veli Üniversitesi toplumsal cinsiyet algısı araştırması. *Turkish Studies*, *9*(2), 811–831. doi:10.7827/TurkishStudies.6244

Kaler, A., & Beres, M. (2016). *Essentials of field relationships*. Routledge. doi:10.4324/9781315429335

Kalir, B. (2006). The field of work and the work of the field: Conceptualising an anthropological research engagement. *Social Anthropology*, *14*(2), 235–246.

Kapelovitz, L. H. (1976). *To love and to work: A demonstration and discussion of psychotherapy*. Grune & Stratton.

Kerbrat-Orecchioni, C. (2006). Politeness in small shops in France. *Journal of Politeness Research*, *2*(1), 79–103. doi:10.1515/PR.2006.005

Kessler, G. (2006). Assessing CALL teacher training: What are we doing and what could we do better. *Teacher Education in CALL*, 23-42.

Kessler, R. C., Chiu, W., Demler, O., & Walters, E. E. (2005). Prevalence, severity, and comorbidity of 12-month dsm-iv disorders in the national comorbidity survey replication. *Archives of General Psychiatry*, *62*(6), 617–627. doi:10.1001/archpsyc.62.6.617 PMID:15939839

Kilb, A., & Herzig, K. (2016). Increasing the efficiency of data collection with a research participation night. *Teaching of Psychology*, *43*(1), 53–58. doi:10.1177/0098628315620881

Killian, K. D. (2008). Helping till it hurts? A multimethod study of compassion fatigue, burnout, and self-care in clinicians working with trauma survivors. *Traumatology*, *14*(2), 32–44. doi:10.1177/1534765608319083

Kingdoms Mobile Gameplay | Firefly Studios. (2016). Retrieved from http://fireflyworlds.com/2016/08/17/kingdoms-mobile-gameplay

Kirkpatrick, H. (2008). A narrative framework for understanding experiences of people with severe mental illnesses. *Archives of Psychiatric Nursing*, *22*(2), 61–68. doi:10.1016/j.apnu.2007.12.002 PMID:18346562

Knaifel, E., & Mirsky, J. (2015). Interplay of identities: A narrative study of self-perceptions among immigrants with severe mental illness from the former Soviet Union. *Transcultural Psychiatry*, *52*(1), 74–95. doi:10.1177/1363461514552585 PMID:25315487

Kolo, C., & Baur, T. (2004). Living a virtual life: Social dynamics of online gaming. *Game Studies*, *4*(1), 1–31.

Konokman, G. Y., & Yelken, T. Y. (2016). The development of the resistance scale towards technology supported instruction: Exploratory and confirmatory factor analysis. *International Online Journal of Educational Sciences*, *8*(4), 199–214. doi:10.15345/iojes.2016.04.016

Korthagen, F. (2010). Situated learning theory and the pedagogy of teacher education: Towards an integrative view of teacher behavior and teacher learning. *Teaching and Teacher Education*, *26*(1), 98–106. doi:10.1016/j.tate.2009.05.001

Kosygina, L. V. (2005). Doing gender in research: Reflection on experience in field. *Qualitative Report*, *10*(1), 87–95.

Kubanyiova, M. (2008). Rethinking research ethics in contemporary applied linguistics: The tension between macroethical and microethical perspectives in situated research. *Modern Language Journal*, *92*(4), 503–518. doi:10.1111/j.1540-4781.2008.00784.x

Kubler-Ross, E. (1969). *On death and dying*. Macmillan Publishing.

Labrague, L. J., McEnroe-Petitte, D. M., Papathanasiou, I. V., Edet, O. B., & Arulappan, J. (2015). Impact of instructors' caring on students' perceptions of their own caring behaviors. *Journal of Nursing Scholarship*, *47*(4), 338–346. doi:10.1111/jnu.12139 PMID:25959595

Largent, E., Grady, C., Miller, F. G., & Wertheimer, A. (2013). Misconceptions about coercion and undue influence: Reflections on the views of IRB members. *Bioethics*, *27*(9), 500–507. http://healthpolicy.ucla.edu/chis/tac2015/Documents/SDSM%20TAC/Langert_etal_2013.pdf. doi:10.1111/j.1467-8519.2012.01972.x PMID:22493972

Lave, J., & Wenger, E. (1991). *Situated learning: Legitimate peripheral participation*. Cambridge University Press. doi:10.1017/CBO9780511815355

Lavrakas, P. (Ed.). (2008). *Encyclopedia of survey research methods*. Sage Publications. doi:10.4135/9781412963947

Laws, E., Prideaux, B., & Moscardo, G. (2006). Quality and service management perspectives. In B. Prideaux, G. Moscardo, & E. Laws (Eds.), *Managing tourism and hospitality services: Theory and international applications* (pp. 1–14). CABI. doi:10.1079/9781845930127.0001

Letherby, G. (2003). *Feminist Research in Theory and Practice*. Open University Press.

Levin, J., Cox, E., Cervin, C., & Haberler, Z. (2010). The recipe for promisig practices in community colleges. *Community College Review*, *38*(1), 31–58. doi:10.1177/0091552110374505

Lev-Wiesel, R., & Amir, M. (2006). Growing out of ashes: Posttraumatic growth among Holocaust child survivors—is it possible. In L. G. Calhoun & R. G. Tedeschi (Eds.), *Handbook of posttraumatic growth: Research and practice* (pp. 248–263). Psychology press.

Lichterman, P. (2017). Interpretive reflexivity in ethnography. *Ethnography*, *18*(1), 35–45. doi:10.1177/1466138115592418

Lin, Q., Ren, M., & Yang, M. (2019). Identity crisis among rural-to-urban migrant children in China: A proposal for school and government interventions. *Journal of Asian Public Policy*, *12*(2), 144–159. doi:10.1080/17516234.2019.1588841

Lofland, J., Snow, D. A., Anderson, L., & Lofland, L. H. (1971). *Analyzing social settings* (4th ed.). Thomson Wadsworth Pub.

Mack, N., Woodson, C., MacQueen, K. M., Guest, G., & Namey, E. (2005). *Qualitative research methods: A data collector's field guide*. Family Health International.

Madden, R. (2017). *Being ethnographic: A guide to the theory and practice of ethnography*. Sage. doi:10.4135/9781529716689

Maner, J. K. (2016). Into the wild: Field research can increase both replicability and real-world impact. *Journal of Experimental Social Psychology*, *66*, 100–106. doi:10.1016/j.jesp.2015.09.018

Marcus, G. E. (1995). Ethnography in/of the world system: The emergence of multi-sited ethnography. *Annual Review of Anthropology*, *24*(1), 95–117. doi:10.1146/annurev.an.24.100195.000523

Mason, J. (2002). *Qualitative researching* (2nd ed.). Sage.

McCabe, S., & Marson, D. (2006). Tourist constructions and consumptions of space: Place, modernity and meaning. In P. M. Burns & M. Novelli (Eds.), *Tourism and social identities: Global frameworks and local realities* (pp. 91–108). Elsevier. doi:10.1016/B978-0-08-045074-2.50011-4

McCann, I., & Pearlman, L. (1990). Vicarious Traumatization: A framework for understanding the psychological effects of working with victims. *Journal of Traumatic Stress*, *3*(1), 131–149. doi:10.1007/BF00975140

McFarlane-Morris, S. (2020). "Home sweet home?" Struggles of intracultural "betweenness" of doctoral fieldwork in my home country of Jamaica. *Area*, *52*(2), 394–400. doi:10.1111/area.12580

McGrail, E. (2005). Teachers, technology and change: English teachers' perspectives. *Journal of Technology and Teacher Education*, *13*(1), 5–25.

McKinley, J. (2019). Evolving the TESOL teaching-research nexus. *TESOL Quarterly*, *53*(3), 875–884. doi:10.1002/tesq.509

McNeil, L. (2018). Situated learning in CALL teacher education. In J. Liontas (Ed.), *The TESOL encyclopedia of English language teaching*. John Wiley & Sons. doi:10.1002/9781118784235.eelt0438

McQuoid, J., Jowsey, T., & Talaulikar, G. (2017). Contextualizing renal patient routines: Everyday space-time contexts, health service access, and wellbeing. *Social Science & Medicine*, *183*, 142–150. doi:10.1016/j.socscimed.2017.04.043 PMID:28482275

Mehan, H. (1991). The school's work of sorting students. In D. Boden & D. H. Zimmerman (Eds.), *Talk and social structure: Studies in ethnomethodology and conversation analysis* (pp. 71–90). Polity Press.

Mehra, B. (2002). Bias in qualitative research: Voices from an online classroom. *Qualitative Report, 7*(1), 1–19. https://nsuworks.nova.edu/tqr/vol7/iss1/2

Miller, J. (2015). Seven Self-Care Strategies. *Reflections: Narratives of Professional Helping, 21*(1), 52–58.

Milton, J., Polmear, C., & Fabricius, J. (2011). A short introduction to psychoanalysis. *Sage (Atlanta, Ga.).*

Mligo, E. S. (2013). *Doing effective fieldwork: A textbook for students of qualitative field research in higher learning institutions*. Resource Publications.

MOE. (2017). *Education in China*. Retrieved May 1, 2020 from http://www.moe.gov.cn/jyb_sjzl/s5990/201810/t20181018_352057.html

Monahan, T., & Fisher, J. A. (2010). Benefits of 'observer effects': Lessons from the field. *Qualitative Research, 10*(3), 357–376. doi:10.1177/1468794110362874 PMID:21297880

Moncrief, W. C., & Marshall, G. W. (2005). The evolution of the seven steps of selling. *Industrial Marketing Management, 34*(1), 13–22. doi:10.1016/j.indmarman.2004.06.001

Morran, D. (2008). Firing up and burning out: The personal and professional impact of working in domestic violence offender programmes. *Probation Journal, 55*(2), 139–152. doi:10.1177/0264550508090272

Morrow, S. L., & Smith, M. L. (2000). Qualitative research for counseling psychology. Handbook of Counseling Psychology, 3, 199-230.

Murchison, J. (2010). *Ethnography essentials: Designing, conducting, and presenting your research* (Vol. 25). John Wiley & Sons.

Nagington, M. (2016). Judith Butler's theories: Reflections for nursing research and practice. *Nursing Philosophy, 17*(4), 307–316. doi:10.1111/nup.12134 PMID:27321751

National People's Congress. (2000). *Law of the People's Republic of China on the standard spoken and written Chinese language* (final rule Jan. 1, Asian Legal Information Institute, Trans). Retrieved May 1, 2020 from http://www.asianlii.org/cn/legis/cen/laws/ssawcll414/

NBS, UNFPA, & UNICEF. (2019). *Domestic migrants in China (2018): Trends, challenges and recommendations*. Retrieved May 1, 2020 from https://china.unfpa.org/zh-Hans/node/42245

Newell, J., & MacNeil, G. (2010). Professional burnout, vicarious trauma, secondary traumatic stress, and compassion fatigue: A review of theoretical terms, risk factor, and preventative methods for clinicians and researchers. *Best Practices in Mental Health, 6*(2), 57–68.

NGOCN. (2018). *Beijing's largest Privately-run migrant school may be demolished after its 20th anniversary*. Retrieved May 1, 2020 from https://chinadigitaltimes.net/chinese/2018/08/ngocn-%E4%BA%8C%E5%8D%81%E5%91%A8%E5%B9%B4%E6%A0%A1%E5%BA%86%E5%88%9A%E8%BF%87%EF%BC%8C%E5%8C%97%E4%BA%AC%E6%9C%80%E5%A4%A7%E6%89%93%E5%B7%A5%E5%AD%90%E5%BC%9F%E5%AD%A6%E6%A0%A1%E6%88%96%E5%B0%86/

Nielsen, M. F. (2009). Interpretative management in business meetings: Understanding managers' interactional strategies through conversation analysis. *Journal of Business Communication*, *46*(1), 23–56. doi:10.1177/0021943608325752

Nielsen, M. F., Nielsen, S. B., Gravengaard, G., & Due, B. (2012). Interactional functions of invoking procedure in institutional settings. *Journal of Pragmatics*, *44*(11), 1457–1473. doi:10.1016/j.pragma.2012.06.007

Nilsen, P., & Bernhardsson, S. (2019). Context matters in implementation science: A scoping review of determinant frameworks that describe contextual determinants for implementation outcomes. *BMC Health Services Research*, *19*(1), 189–209. doi:10.118612913-019-4015

Nobahar, N., Nobahar, M., & Hamidi, M. (2014). Investigation of Research Barriers and Challenges in University from the Perspective of Faculty Members of Kermanshah City. *IAU International Journal of Social Sciences*, *4*(4), 15–21.

Norris, N. (1997). Error, bias and validity in qualitative research. *Educational Action Research*, *5*(1), 172–176. doi:10.1080/09650799700200020

O'Connor, K., Greene, H. C., Good, A., & Zhang, G. (n.d.). Finding balance: A challenge for untenured faculty. *International Education Studies*, *4*(4), 3–12.

OECD. (2019). *OECD Labour Force Statistics 2019*. Advance online publication. doi:10.1787/g2g9fb3e-en

Ogle, A. (2009). Making sense of the hotel guestroom. *Journal of Retail & Leisure Property*, *8*(3), 159–172. doi:10.1057/rlp.2009.7

Oğuz, H. Ş. (2012). Alanda Bir Kadın. *Fe Dergi*, *4*(1), 64–78. doi:10.1501/Fe0001_0000000064

Okumus, F., Altinay, L., & Roper, A. (2007). Gaining access for research: Reflections from experience. *Annals of Tourism Research*, *34*(1), 7–26. doi:10.1016/j.annals.2006.07.006

Opdenakker, R. (2006). Advantages and disadvantages of four interview techniques in qualitative research. *Forum Qualitative Social Research*, *7*(4), 1–13.

ÖSYM. (2013a). *Öğretim elemanları özet tablosu*. Retrieved December, 3, 2012, from http://www.osym.gov.tr/Eklenti/5190,2ogretimelemanlarisayozettablosupdf.pdf?0

ÖSYM. (2013b). *Yüksek Öğretim istatistikleri, 2012-2013 öğretim yılı lisans eğitimi veren yükseköğretim programlarında görevli öğretim elemanlarının öğretim alanlarına göre sayıları.* Retrieved December, 3, 2012, from http://www.osym.gov.tr/dosya/1-69402/h/13ogretimalanlisansogrencisay.pdf

Özbilgin, M. F., & Healy, G. (2004). The Gendered Nature of Career Development of University Professors: The Case of Turkey. *Journal of Vocational Behavior*, *64*(2), 358–371. doi:10.1016/j.jvb.2002.09.001

Özçetin, H. (2009). 'Breaking the silence': The religious Muslim women's movement in Turkey. *Journal of International Women's Studies*, *11*(1), 106–119.

Ozer, E., Best, S., Lipsey, T., & Weiss, D. (2003). Predictors of posttraumatic stress disorder and symptoms in adults: A meta-analysis. *Psychological Bulletin*, *129*(1), 52–73. doi:10.1037/0033-2909.129.1.52 PMID:12555794

Özgüç, N. (1998). Türkiye Üniversitelerinde Kadın Coğrafyacılar. In N. Arat (Ed.), *Aydınlanmanın Kadınları* (pp. 177–201). Cumhuriyet Kitap Kulübü.

Özkanlı, Ö., & White, K. (2008). Leadership and strategic choices: Female professors in Australia and Turkey. *Journal of Higher Education Policy and Management*, *30*(1), 53–63. doi:10.1080/13600800701745051

Ozobot & Evollve, Inc. (2020). *Ozoblockly.* Ozobot. https://ozobot.com/create/ozoblockly

Ozobot & Evollve, Inc. (2020). *Ozobot: Robots to code and create with.* Ozobot. https://ozobot.com/

Padgett, D. K. (2008). *Qualitative Methods in Social Work Research.* Sage.

Padilla-Walker, L. M., Thompson, R. A., Zamboanga, B. L., & Schmersal, L. A. (2005). Extra credit as incentive for voluntary research participation. *Sage Journals Teaching of Philosophy*, *32*(3), 150–153. doi:10.1207/ s15328023top3203_2

Pagura, J., Stein, M. B., Bolton, J. M., Cox, B. J., Grant, B., & Sareen, J. (2010). Comorbidity of borderline personality disorder and posttraumatic stress disorder in the U.S. population. *Journal of Psychiatric Research*, 44. PMID:20537660

Paluck, E. L., & Cialdini, R. B. (2014). Field research methods. Handbook of research methods in social and personality psychology, 81-97.

Partington, G. (2001). Qualitative research interviews: Identifying problems in technique. *Issues in Educational Research*, *11*(2), 32–44.

Patel, N., Min, M., & Lim, S. (2016, April). Accurate evacuation route planning using forward-backward shortest paths. In *2016 Annual IEEE Systems Conference (SysCon)* (pp. 1-6). IEEE.

Patton, M. Q. (2002). *Qualitative research & evaluation methods* (3rd ed.). Sage.

Pearce, C., Boellstorff, T., & Nardi, B. A. (2011). *Communities of play: Emergent cultures in multiplayer games and virtual worlds.* MIT Press.

Pearlman, L., & MacIan, P. (1995). Vicarious Traumatization: An empirical study on the effects of trauma work on trauma therapists. *Professional Psychology, Research and Practice, 26*(6), 558–565. doi:10.1037/0735-7028.26.6.558

Pearlman, L., & Saakvitne, K. (1995). *Trauma and the therapist: countertransference and vicarious traumatization in psychotherapy with incest survivors*. W.W. Norton.

Pearson, R. H., & Mundform, D. J. (2010). Recommended sample size for conducting exploratory factor analysis on dichotomous data. *Journal of Modern Applied Statistical Methods; JMASM, 9*(2), 5, 359–368. doi:10.22237/jmasm/1288584240

Pegg, J., Adams, A., Risser, H., Bottoms, S., Kern, A., & Wu, K. (2014). Finding Friends: Creating a community of support for early career academics. *Brock Education, 24*(1), 47–54.

Pellegrini, E., & Scandura, T. (2006). Leader-Member Exchange, Paternalism and Delegation in the Turkish Business Culture: An Empirical Investigation. *Journal of International Business Studies, 37*(2), 264–279. doi:10.1057/palgrave.jibs.8400185

People's Daily Online. (2014). *Beijing proposes to control the size of the population to and strengthen the residing management.* Retrieved May 1, 2020 from http://news.eastday.com/eastday/13news/auto/news/china/u7ai865159_K4.html

Peshkin, A. (1988). In search of subjectivity—one's own. *Educational Researcher, 17*(7), 17–21.

Peticca-Harris, A., deGama, N., & Elias, S. R. (2016). A dynamic process model for finding informants and gaining access in qualitative research. *Organizational Research Methods, 19*(3), 376–401. doi:10.1177/1094428116629218

Pew Research Center. (n.d.). *Pew Research Center Methods. Questionnaire Design.* Pew Research Center. https://www.pewresearch.org/methods/u-s-survey-research/questionnaire-design/

Piller, I. (2012). Intercultural communication: An overview. In C. B. Paulston, S. F. Kiesling, & E. S. Rangel (Eds.), *The handbook of intercultural discourse and communication* (pp. 3–18). John Wiley & Sons Ltd. doi:10.1002/9781118247273.ch1

Pillow, W. S. (2003). Confession, catharsis, or cure? Rethinking the uses of reflexivity as methodological power in qualitative research. *International Journal of Qualitative Studies in Education: QSE, 16*(2), 175–196. doi:10.1080/0951839032000060635

Pini, B. (2002). Focus Groups, Feminist Research and farm women: Opportunities for empowerment in rural social research. *Journal of Rural Studies, 18*(3), 339–351. doi:10.1016/S0743-0167(02)00007-4

Pole, C., & Hillyard, S. (2016). *Doing Fieldwork*. Sage. doi:10.4135/9781473966383

Polkinghorne, D. E. (1991). Narrative and self-concept. *Journal of Narrative and Life History, 1*(2-3), 135–153. doi:10.1075/jnlh.1.2-3.04nar

Popplet. (2018). Retrieved from http://popplet.com/

Population – SSB. (2017). Retrieved from https://www.ssb.no/en/befolkning/nokkeltall/population

Porter, W. W., & Graham, C. R. (2016). Institutional drivers and barriers to faculty adoption of blended learning in higher education. *British Journal of Educational Technology*, *47*(4), 748–762. doi:10.1111/bjet.12269

Poyraz, B. (2013). Akademi kadınların cenneti mi? Ankara Üniversitesi örneği. *Ankara Üniversitesi Sosyal Bilimler Enstitüsü Dergisi*, *4*(2), 1–18. doi:10.1501beder_0000000057

Privacy Policy - Firefly Studios. (2018). Retrieved from https://fireflyworlds.com/privacy-policy

Pultz, S. (2018). *Flexibility in research design: How unexpected events can improve learning and research.* SAGE Research Methods Cases., doi:10.4135/9781526435217

Putnam, R., & Borko, H. (2000). What do new views of knowledge and thinking have to say about research on teacher learning? *Educational Researcher*, *29*(1), 4–15. doi:10.3102/0013189X029001004

Rahman, M. S. (2017). The advantages and disadvantages of using qualitative and quantitative approaches and methods in language "testing and assessment" research: A literature review. *Journal of Education and Learning*, *6*(1), 102–112. doi:10.5539/jel.v6n1p102

Rebora, A. (2016). Teachers still struggling to use tech to transform instruction, survey finds. *Education Week*, *35*(35), 4–5. https://www.edweek.org/ew/articles/2016/06/09/teachers-still-struggling-to-use-tech-to.html/

Reeves, C. L. (2010). A difficult negotiation: Fieldwork relations with gatekeepers. *Qualitative Research*, *10*(3), 315–331. doi:10.1177/1468794109360150

Regehr, C., Hemsworth, D., Leslie, B., Howe, P., & Chau, S. (2004). Predictors of post-traumatic distress in child welfare workers: A linear structural equation model. *Children and Youth Services Review*, *26*(4), 331–346. doi:10.1016/j.childyouth.2004.02.003

Reyes, V. (2018). Ethnographic toolkit: Strategic positionality and researchers' visible and invisible tools in field research. *Ethnography*. Advance online publication. doi:10.1177/1466138118805121

Reynders, A., Kerkhof, A. J. F. M., Molenberghs, G., & Van Audenhove, C. (2014). Attitudes and stigma in relation to help-seeking intentions for psychological problems in low and high suicide rate regions. *Social Psychiatry and Psychiatric Epidemiology*, *49*(2), 231–239. doi:10.100700127-013-0745-4 PMID:23896893

Richards, K. (2009). Trends in qualitative research in language teaching since 2000. *Language Teaching*, *42*(2), 147–180. doi:10.1017/S0261444808005612

Riese, J. (2019). What is 'access' in the context of qualitative research? *Qualitative Research*, *19*(6), 669–684. doi:10.1177/1468794118787713

Riessman, C. K. (1993). *Narrative analysis*. Sage.

Robson, C. (2002). *Real world research: A resource for social scientists and practitioner-researchers* (Vol. 2). Blackwell Oxford.

Rodham, K., & Gavin, J. (2006). The ethics of using the internet to collect qualitative research data. *Research Ethics Review*, *2*(3), 92–97. doi:10.1177/174701610600200303

Rogers, E. M. (2003). *Diffusion of innovations* (5th ed.). Free Press.

Rosen, M. (1991). Coming to terms with the field: Understanding and doing organizational ethnography. *Journal of Management Studies*, *28*(1), 1–24. doi:10.1111/j.1467-6486.1991.tb00268.x

Rosenthal, G. (1993). Reconstruction of life stories: Principles of selection in generating stories for narrative biographical interviews. *The Narrative Study of Lives*, *1*(1), 59-91.

Rosenthal, G. (2003). The healing effects of storytelling: On the conditions of curative storytelling in the context of research and counseling. *Qualitative Inquiry*, *9*(6), 915–933. doi:10.1177/1077800403254888

Rothenberg, C. E. (2016). *On doing fieldwork in Palestine: Advice, fieldnotes, and other thoughts*. Springer. doi:10.1007/978-3-319-34201-6

Roulston, K., & Shelton, S. (2015). Reconceptualizing bias in teaching qualitative research methods. *Qualitative Inquiry*, *21*(4), 332–342. doi:10.1177/1077800414563803

Rudd, M. D., & Joiner, T. (1997). 'Countertransference and the therapeutic relationship: A cognitive perspective. *Journal of Cognitive Psychotherapy*, *11*(4), 231–250. doi:10.1891/0889-8391.11.4.231

Saba, F. (2000). Research in distance education: A status report. *International Review of Research in Open and Distance Learning*, *1*(1), 1–9. doi:10.19173/irrodl.v1i1.4

Saldana, J. (2009). *The coding manual for qualitative researchers*. Sage Publishing.

Salih, S. (2002). *Judith Butler*. Academic Press.

Sandler, J. J. (1976). Countertransference and role-responsiveness. *The International Review of Psycho-Analysis*, *3*, 43–47.

Sarangi, S., & Roberts, C. (1999). The dynamics of interactional and institutional orders in work-related settings. In S. Sarangi & C. Roberts (Eds.), *Talk, work and institutional order: Discourse in medical, mediation and management settings* (pp. 1–60). Mouton de Gruyter. doi:10.1515/9783110208375.1.1

Sareen, J., Cox, B., Stein, M., Afifi, T., Fleet, C., & Asmundson, G. (2007). Physical and mental comorbidity, disability, and suicidal behavior associated with posttraumatic stress disorder in a large community sample. *Psychosomatic Medicine*, *69*(3), 242–248. doi:10.1097/PSY.0b013e31803146d8 PMID:17401056

Sargut, S. (2001). *Kültürlerarası Farklılaşma ve Yönetim*. Verso Yayıncılık.

Sarsby, J. (1984). The Fieldwork Experience. In R. Ellen (Ed.), *Ethnographic research: A guide to general conduct*. Academic Press.

Sayar, K. (2003). *Kültür ve Ruh Sağlığı*. Metis Yayınevi.

Schensul, S. L., Schensul, J. J., & LeCompte, M. D. (1999). *Essential ethnographic methods: Observations, interviews, and questionnaires*. AltaMira Press.

Schieffelin, B. B., & Ochs, E. (1986). The microgenesis of competence: Methodology in language socialization. In D. Slobin, J. Gerhardt, A. Kyratzis, & J. Guo (Eds.), *Social interaction, social context, and language* (pp. 251–263). Erlbaum.

Schleiermacher, F. (1998). *Hermeneutics and criticism and other writings* (A. Bowie, Trans.). CUP.

Schunk, D. H. (2003). Self-efficacy for reading and writing: Influence of modeling, goal setting, and self-evaluation. *Reading & Writing Quarterly*, *19*(2), 159–172. doi:10.1080/10573560308219

Schwartz, M. S., & Schwartz, C. G. (1955). Problems in participant observation. *American Journal of Sociology*, *60*(4), 343–353. http://www.jstor.org/stable/2772027. doi:10.1086/221566

Şentürk, B. (2016). Çokuz Ama Yokuz: Türkiye'deki Akademisyen Kadınlar Üzerine Bir Analiz. *ViraVerita E-Dergi*, (2), 1-22.

Shaffer, D. W., Squire, K. R., Halverson, R., & Gee, J. P. (2005). Video games and the future of learning. *Phi Delta Kappan*, *87*(2), 105–111. doi:10.1177/003172170508700205

Shahrokni, S. A. (in press). Evidence for second language socialization in multiplayer online game play. *TESL-EJ*, *25*(1).

Sharma, G., & Good, D. (2013). The work of middle managers: Sensemaking and sensegiving for creating positive social change. *The Journal of Applied Behavioral Science*, *49*(1), 95–122. doi:10.1177/0021886312471375

Sheehan, T. (1988). Heideggar and the Nazis. *The New York Review of Books*, *35*(10), 1–12.

Sherman, R. (2005). Producing the superior self: Strategic comparison and symbolic boundaries among luxury hotel workers. *Ethnography*, *6*(2), 131–158. doi:10.1177/1466138105057546

Sherry, E. (2013). The vulnerable researcher: Facing the challenges of sensitive research. *Qualitative Research Journal*, *13*(3), 278–288. doi:10.1108/QRJ-10-2012-0007

Shi, J., & Sercombe, P. G. (2020). Poverty and inequality in rural education: Evidence from China. *Education as Change*, *24*, 1–26. doi:10.25159/1947-9417/4965

Silverman, D. (2013). *A very short, fairly interesting and reasonably cheap book about qualitative research* (2nd ed.). SAGE. doi:10.4135/9781526402264

Sim, J., & Waterfeld, J. (2019). Focus group methodology: Some ethical challenges. *Quality & Quantity*, *53*(6), 3003–3022. doi:10.100711135-019-00914-5

Singer, E., & Bossarte, R. M. (2006). Incentives for survey participation: When are they 'coercive'? *American Journal of Preventive Medicine, 31*(5), 411–418. doi:10.1016/j.amepre.2006.07.013 PMID:17046413

Slattery, S., & Goodman, L. (2009). Secondary traumatic stress among domestic violence advocates: Workplace risk and protective factors. *Violence Against Women, 15*(11), 1358–1379. doi:10.1177/1077801209347469 PMID:19809098

Smith, D. (2012). *Essays on Deleuze*. Academic Press.

Smith, J., Flowers, P., & Larkin, M. (2009). *Interpretative phenomenological analysis: theory, method and research*. SAGE.

Snyder, C. R., Shorey, H. S., Cheavens, J., Pulvers, K. M., Adams, V. H. III, & Wiklund, C. (2002). Hope and academic success in college. *Journal of Educational Psychology, 94*(4), 820–826. doi:10.1037/0022-0663.94.4.820

Spencer, G. (1973). Methodological issues in the study of bureaucratic elites: A case study of West Point. *Social Problems, 21*(1), 90–103. doi:10.2307/799979

Sprang, G., Clark, J. J., & Whitt-Woosley, A. (2007). Compassion fatigue, compassion satisfaction, and burnout: Factors impacting a professional's quality of life. *Journal of Loss and Trauma, 12*(3), 259–280. doi:10.1080/15325020701238093

Stamm, B. (1995). *Secondary Traumatic Stress: Self-care issues for clinicians, researchers and educators*. Sidran Press.

Standing, K. (1998). Writing the voices of the less powerful: research on lone mothers in Feminist dilemmas in qualitative research: Public knowledge and private lives. In J. Ribbens & R. Edwards (Eds.), (pp. 186–202). Sage Publications.

Stanley, G. (2013). *Language learning with technology: Ideas for integrating technology in the classroom*. Cambridge University Press.

Stanley, T. D. (2008). Meta-regression methods for detecting and estimating empirical effects in the presence of publication selection. *Oxford Bulletin of Economics and Statistics, 70*(1), 103–127. doi:10.1111/j.1468-0084.2007.00487.x

Stavretis, L., & Boucher, C. (2001). *Researching the researchers: Comments on the experience of doing research in universities*. RMIT Business, School of Management. Working Paper Series.

Stockwell, A. (2017). *Effectiveness of Self-Care in Reducing Symptoms of Secondary Traumatic Stress*. St. Catherine University. Retrieved from https://sophia.stkate.edu/msw_papers/797

StoryJumper. (2017). Retrieved from https://www.storyjumper.com/

Stronach, I., Garratt, D., Pearce, C., & Piper, H. (2007). Reflexivity, the picturing of selves, the forging of method. *Qualitative Inquiry, 13*(2), 179–203. doi:10.1177/1077800406295476

Sui, H., & Song, Y. (2017). Does the left-behind experience affect the exam behavior, drop-out behavior, and intention to colleges of secondary vocational students? Based on longitudinal data from Zhejiang and Shaanxi Provinces. *Chinese Education & Society: The education of China's rural children: Chinese migrant children and left-behind children, 50*(4), 350-367.

Sullivan, D., Mitchell, M., & Uhl-Bien, M. (2003). The New Conduct of Business: How LMX can Help Capitalize on Cultural Diversity. In G. Graen (Ed.), *Dealing with Diversity* (pp. 183–218). Information Age Publishing.

Sultana, F. (2007). Reflexivity, positionality and participatory ethics: negotiating fieldwork dilemmas in international research. *ACME, 6*(3).

Takyi, E. (2015). The challenge of involvement and detachment in participant observation. *Qualitative Report*, *20*(6), 864–872.

Taylor, H. A. (2016). Review and conduct of human subjects research after a natural or man–made disaster: Findings from a pilot study. *Narrative Inquiry in Bioethics*, *6*(3), 211–222. doi:10.1353/nib.2016.0061

TeamSpeak. (2018). [Computer software]. Retrieved from https://www.teamspeak.com

Tedlock, B. (1991). From participant observation to the observation of participation: The emergence of narrative ethnography. *Journal of Anthropological Research*, *47*(1), 69–94. doi:10.1086/jar.47.1.3630581

Terenzini, P. T. (1993). On the nature of institutional research and the knowledge and skills it requires. *Research in Higher Education*, *34*(1), 1–10. doi:10.1007/BF00991859

Terrell, S. S. (2013). Integrating online tools to motivate young English language learners to practice English outside the classroom. In *Explorations of Language Teaching and Learning with Computational Assistance* (pp. 184–192). IGI Global. doi:10.4018/978-1-4666-1855-8.ch012

Teulier, R., & Rouleau, L. (2013). Middle managers' sensemaking and interorganizational chage initiation: Translation spaces and editing practices. *Journal of Change Management*, *13*(3), 308–337. doi:10.1080/14697017.2013.822674

Thummapol, O., Park, T., Jackson, M., & Barton, S. (2019). Methodological Challenges Faced in Doing Research With Vulnerable Women: Reflections From Fieldwork Experiences. *International Journal of Qualitative Methods*, *18*(1), 1–11.

Tilley, S. (2016). *Doing respectful research: power, privilege and passion*. Fernwood Publishing.

Touchstone, E. E., Koslow, S., Shamdasani, P. N., & D'Alessandro, S. (2017). The linguistic servicescape: Speaking their language may not be enough. *Journal of Business Research*, *72*, 147–157. doi:10.1016/j.jbusres.2016.10.008

Ujváriné, A. S., Zrínyi, M., Tóth, H., Zékányné, I. R., Szögedi, I., & József, B. (2011). The role of faculty and clinical practice in predicting why nurses graduate in Hungary. *Nurse Education Today*, *31*(1), 94–101. doi:10.1016/j.nedt.2010.04.003 PMID:21094566

United Nations Development Programme. (2019). *Human Development Report, Gender Inequality Index.* http://hdr.undp.org/en/composite/GII

Ünnü, N. A. A., & Şentürk, B. (2020). All-in-One Academics: Mental Workload in Turkish Academic Employment. In *Evaluating Mental Workload for Improved Workplace Performance* (pp. 69–87). IGI Global. doi:10.4018/978-1-7998-1052-0.ch004

Urry, J. (2005). The 'consuming' of place. In A. Jaworski & A. Pritchard (Eds.), *Discourse, communication and tourism* (pp. 19–27). Channel View Publications. doi:10.21832/9781845410216-004

Vajda, J. (2007). Two survivor cases: Therapeutic effect as side product of the biographical narrative interview. *Journal of Social Work Practice, 21*(1), 89–102. doi:10.1080/02650530601173664

Van Cott, D. L. (2005). A graduate student's guide to publishing scholarly journal articles. *PS, Political Science & Politics, 38*(4), 741–743. http://www.jstor.com/stable/30044359. doi:10.1017/S1049096505050237

Van der Kolk, B. A. (2005). Developmental trauma disorder: Toward a rational diagnosis for children with complex trauma histories. *Psychiatric Annals, 35*(5), 401–408. doi:10.3928/00485713-20050501-06

van Gelder, M. M. H. J., Bretveld, R. W., & Roeleveld, N. (2010). Web-based Questionnaires: The Future in Epidemiology? *American Journal of Epidemiology, 172*(11), 1292–1298. doi:10.1093/aje/kwq291 PMID:20880962

Van Maanen, J. (1979). The fact of fiction in organizational ethnography. *Administrative Science Quarterly, 24*(4), 539–550. doi:10.2307/2392360

Van Maanen, J. (2011). *Tales of the field: On writing ethnography* (2nd ed.). The University of Chicago Press. doi:10.7208/chicago/9780226849638.001.0001

VanVoorhis, C. W., & Morgan, B. L. (2007). Understanding power and rules of thumb for determining sample sizes. *Tutorials in Quantitative Methods for Psychology, 3*(2), 43–50. doi:10.20982/tqmp.03.2.p043

Vatanartiran, S., & Karadeniz, S. (2015). A needs analysis for technology integration plan: Challenges and needs of teachers. *Contemporary Educational Technology, 6*(3), 206–220. Retrieved from https://files.eric.ed.gov/fulltext/EJ1105647.pdf

Veal, A. J. (1997). Research methods for leisure and tourism: A practical guide (2nd ed.). London: Ilam.

Venkatesh, S. A. (2006). *Off the books: The underground economy of the urban poor*. Harvard University Press.

Wacquant, L. (2004). Following Pierre Bourdieu into the field. *Ethnography, 5*(4), 387–414. doi:10.1177/1466138104052259

Wanat, C. L. (2008). Getting past the gatekeepers: Differences between access and cooperation in public school research. *Field Methods*, *20*(2), 191–208. doi:10.1177/1525822X07313811

Wasti, A. S. (2003). The influence of cultural values on antecedents of organisational commitment: An individual-level analysis. *Applied Psychology*, *52*(4), 533–554. doi:10.1111/1464-0597.00150

WHO Department of Reproductive Health and Research. London School of Hygiene and Tropical Medicine, & South African Medical Research Council. (2013). Global and regional estimates of violence against women: Prevalence and health effects of intimate partner violence and non-partner sexual violence. WHO.

Wilkinson, S., & Kitzinger, C. (2013). Representing our own experience: Issues in "insider" research. *Psychology of Women Quarterly*, *37*(2), 251–255. doi:10.1177/0361684313483111

Williams, P., Barclay, L., & Schmied, V. (2004). Defining Social Support in Context: A Necessary Step in Improving Research, Intervention, and Practice. *Qualitative Health Research*, *14*(7), 942–960. doi:10.1177/1049732304266997

Windchief, S., & San Pedro, T. (2019). *Applying indigenous research methods: storying with peoples and communities*. doi:10.4324/9781315169811

Winnicott, D. (1960). The theory of the parent-child relationship. *The International Journal of Psycho-Analysis*, *41*, 585–595. PMID:13785877

Woodcock, J. (2015). *Coding with Scratch: Workbook*. New York, NY: DK. https://scratch.mit.edu/

World Economic Forum. (2019). *Global Gender Gap Report 2020*. http://www3.weforum.org/docs/WEF_GGGR_2020.pdf

Wright, K. B. (2017). Researching internet-based populations: Advantages and disadvantages of online survey research, online questionnaire authoring software packages, and web survey services. *Journal of Computer-Mediated Communication*, *10*(3), 00. Advance online publication. doi:10.1111/j.1083-6101.2005.tb00259.x

Xiao, Y., Huang, Q., & Wu, K. (2015). Understanding social media data for disaster management. *Natural Hazards*, *79*(3), 1663–1679. doi:10.100711069-015-1918-0

Yağcı, Y. (2010). A different view of the Bologna process: The case of Turkey. *European Journal of Education*, *45*(4), 588–600. doi:10.1111/j.1465-3435.2010.01456.x

Yücel, A. G., & Demir, S. B. (2018). Academic Incentive Allowance: Scientific Productivity, Threats, Expectations. *International Online Journal of Educational Sciences*, *10*(1), 39–55.

Yu, L., & Gatua, M. W. (2014). Methodological considerations for qualitative research with immigrant populations: Lessons from two studies. *Qualitative Report*, *19*(30), 1–16.

Zarrinabadi, N. (2017, June 1). *Call for chapters: Technology and language learner psychology* [Msg 3]. Message posted to https://listserv.dartmouth.edu/scripts/wa.exe?A1=ind1706A&L=LLTI

Zhang, F. (1989). The initials of Chengdu speech as compared with Standard Chinese. *Journal of the International Phonetic Association*, *15*(2), 59–68. doi:10.1017/S002510030000298X

Zimmerman, D. H., & Boden, D. (1991). Structure-in-action: An introduction. In D. Boden & D. H. Zimmerman (Eds.), *Talk and social structure: Studies in ethnomethdology and conversation analysis* (pp. 3–21). Polity Press.

Zuberi, T., & Bonilla-Silva, E. (2008). *White logic, white methods: Racism and methodology.* Rowman & Littlefield Publishers.

About the Contributors

Reima Abobaker is a Professor of ESL at Savannah College of Art and Design. Her research and publications mainly focused on incorporating engaging tasks with technology into ESL and ELL classroom settings.

Nazlı Ayşe Ayyıldız Ünnü has been working as an Associate Professor in Ege University, Faculty of Economics and Administrative Sciences since 2015. She is the Chair of International Management Department and Coordinator of her home university's Human Resources Unit.She has been the Departmental Coordinator for Erasmus Program since 2014. She has been part of many consultancy projects in her home country (reorganization of FOB's, Designing HRM Systems). Her scholarly interests range widely from management & organization generally to HRM, Organizational Behavior and Gender Roles specifically.She has many articles published in national and international refereed journals.She got Young Academician Incentive Award 2015, which is given based on annual academic performance by a Scientific Committee at Ege University. Her research, conducted in collaboration with two colleagues, titled "Gender of Turkish Academia" was awarded by Turkish Researchers' Association in 2016. She was awarded a three-month scholarship, by DAAD to conduct a research in Trier University, entitled "The Dynamics of Academic Employment in The Light of Gender Roles: A Comparative Perspective on German and Turkish Universities".

Geraldine Bengsch is a professor at School of Education, Communication and Society, King's College London.

Denise Brend completed her PhD at McGill University, her postdoctoral fellowship at the Université de Sherbrooke, and is currently an Assistant Professor at Concordia University. She is a researcher and consultant on the implementation and evaluation of an adaptation of the Attachment, Regulation, and Competency (ARC) model for children and youth placed in residential child welfare settings. Dr.

Brend's research focusses on the impacts of complex, interpersonal, and systemic trauma; the implementation of trauma-informed care; and, work-related threats to the well-being of helping professionals, including secondary traumatic stress and moral injury. She is a professional social worker, licenced psychotherapist, and a member of the American Psychoanalytic Association (APsaA) College of Research Fellows and the International Society for the Study of Trauma and Dissociation.

Rebecca Chism is an Associate Professor of Foreign Language Pedagogy and Pedagogy Coordinator in Modern and Classical Language Studies at Kent State University in Kent, Ohio, U.S.A. She teaches undergraduate and graduate courses in second and foreign language teaching methods and approaches. Her research interests include best practices in teaching pre-and in-service teacher preparation, differentiated instruction, and computer mediated communication.

Joy Egbert is Professor of ELL and Education Technology at WSU's Pullman campus. She has published and presented widely on the topics of task engagement, CALL, and teacher education.

Kula Francis is an Associate Professor of Political Science in the College of Liberal Arts and Social Sciences at the University of the Virgin Islands. She holds a Doctorate of Philosophy in Urban Studies and Public Affairs from the University of Akron in Ohio. Her current research interests are social equity in public organizations, evaluation, international relations, history, politics, and public administration.

Sigalit Gal has completed her doctoral studies at McGill University, Montreal, Canada, achieving a cumulative GPA of 3.80. Given her background in counseling with Jewish immigrants in Montreal, Dr. Gal is interested in exploring the intersection between childhood political trauma with immigration stress. After her doctoral study examined the long-term impact of childhood trauma in adult Jewish immigrants to Israel who experienced the military dictatorship in Argentina, her postdoctoral study expands her inquiry by looking at intergenerational trauma transmission, never before studied in this context. Since her graduation Dr. Gal has submitted and published book chapters and articles in peer-reviewed journals, and presented papers in national and international conferences. In light of the COVID-19 pandemic, Dr. Gal was asked to contribute her research expertise and to co-author a special report for the Prime Minister of Israel. This report focused on the specific mental health and welfare issues experienced by women from diverse cultures in Israel. Changes in policy were suggested. In addition, she initiated and conducted a study, focused on the impact of COVID-19 on mental health among Israeli orthodox Jews. Following Dr. Gal's academic performance, the originality and potential contribution of her research to the fields of Trauma and mental health among culturally diverse communities, she has been offered a full-time research appointment as a Visiting Researcher at Harvard University, in Cambridge, Boston.

Carine Graff teaches French and translation at the University of North Texas. Her research interests include translation studies, foreign language acquisition, writing, and pedagogy.

Kavita Gupta is an Assistant Professor in the Department of Psychology, Faculty of Education and Psychology at The Maharaja Sayajirao University of Baroda, Vadodara (Gujarat). She is Pursuing her Ph.D. in Clinical Psychology from Sardar Patel University, Anand (Gujarat). Besides, she has her Doctoral thesis on 'Quality of Life in Cancer Patients undergoing Chemotherapy and Radiotherapy'. She is working with the university students targeting their Personal, Emotional, Psychological problems and enhancement of their mental health and well-being. Her area of interest includes Positive Psychology, Clinical Psychology, Social Psychology, Research Methodology, Educational Psychology. She is dedicating to serving Undergraduate, Postgraduate and PG Diploma students. She further hopes to apply the knowledge and skills garnered from her various disciplines towards improving the welfare and happiness of people of the world. Her research findings have been published through more than 15 research papers in referred/indexed journals and National & International conferences.

Kenny Hendrickson is an Associate Professor of Public Administration and Social Sciences, and the Director of the Master of Public Administration Program in the College of Liberal Arts and Social Sciences at the University of the Virgin Islands. He holds a Doctorate of Philosophy in Urban Studies and Public Affairs from the University of Akron in Ohio. His current research interests are centered on organizational authenticity and care, workplace performativity, discretion and professional decision making.

Peter Sercombe is Senior Lecturer in Applied Linguistics at Newcastle University, UK. His research interests include sociolinguistics in relation to Austronesian languages, particularly effects of language contact.

Seyed Abdollah Shahrokni is an Instructional Designer in the Center for Academic Innovation at Western Oregon University. His research interests include computer assisted language learning (CALL), second language socialization (SLS) in online social spaces, and Task Engagement.

Yao Wang is a doctoral student at Newcastle University. She received a bachelor's degree in Philosophy from Beijing Normal University and a master's degree in Philosophy of Education from the Institute of Education, University College London. Her doctoral research is about how privately-run migrant school operate to cater to the education need of China's internal migrant children and their educational and social circumstances. She is also interested in intercultural communication and education policy.

Index

A

academic care 219-223, 225, 233, 237-238, 240
academic caregiving 223, 231, 240-241
academics 58, 83, 87, 92, 149, 151-157, 159-161, 164, 166-167, 172, 213
access 18, 25-26, 30, 32-33, 36, 39, 48, 51-52, 54-58, 62-63, 65, 67, 69-70, 79, 98, 100-101, 109, 112, 118-123, 125-126, 128-134, 136-137, 140-143, 145-147, 152, 158-159, 162, 166, 188, 196, 199-200, 203-204, 214, 222, 224, 229, 239, 246-247, 256
ACTFL Performance Descriptors 77, 89, 93
actionable 118, 123
Authentic University Academic Care (AUAC) 219-220, 222, 240
authenticity 2, 220

B

barriers 23, 31-32, 34, 36, 40, 42, 46-48, 50-51, 59, 68, 108, 118, 121, 131, 134, 137, 150, 158, 162, 167, 170, 183
burnout 199, 208-210, 212, 214-216
Butler 96, 110-117

C

calibrate 15, 20, 24
call 22-23, 37, 43, 52-53, 56, 62, 73, 75-76, 78, 81, 93, 102, 158, 164, 187, 228, 238
cancer 199-204, 206-214, 216-217
challenges 1-3, 5-9, 11-14, 18, 20-23, 25-29, 31-35, 42, 46-51, 55, 64, 66-68, 71, 73-74, 79, 81-82, 87-91, 93, 97, 107, 110, 118, 120-122, 132-133, 138, 149-151, 154, 158, 163, 165-166, 170, 175, 199-200, 202-204, 207-210, 212-216, 219-220, 226-228, 230-232, 234, 236-237, 243, 246-247, 257
chemotherapy 200-203, 205-206, 209, 213, 216
coding 2-5, 10, 12-13, 16-17, 19, 23-24, 43, 245, 259
cold canvassing 123, 129, 131, 147
collaboration 11, 40, 48, 80, 90, 201, 235
compassion fatigue 98, 115, 199, 207-210, 212-217
Computer-Mediated Communication (CMC) 93
confidentiality 101, 104, 107-108, 112, 114, 160, 165, 177, 199-200, 203, 207, 214, 217, 224, 226
context 2-3, 7, 11, 20-22, 26, 28, 31, 36, 42, 48-49, 52-53, 58-59, 62, 67-68, 86-87, 97, 101, 105, 107-108, 110, 117-118, 128, 130, 132, 136, 140, 145, 150-152, 155, 158-159, 161-162, 164-165, 167, 175-179, 182, 184, 188, 191-192, 194-196, 198-199, 207, 220, 229, 232-233, 237, 239, 257, 259
Council of Higher Education 152, 168, 172
countertransference 116, 175-176, 178-181, 183-184, 187, 189, 191-196, 198
creative solutions 118, 139

D

Debugging 13, 24
disaster research 219-220, 229-230, 232, 234, 236

E

education 1-4, 14, 16, 21-23, 25-29, 31, 33-35, 42-43, 49-54, 56-57, 59-60, 63-64, 69-71, 74, 78, 83, 89-93, 108, 115, 143, 146, 149-155, 158, 161, 164, 167-168, 171-172, 193, 196-197, 203, 219-220, 223, 234, 237-238, 240
education technology 22, 25-29, 34-35, 43
elementary 1-4, 11, 14, 16, 21, 24
elementary school 1-2, 4, 11, 21, 24
Emotional Well Being 210, 217
ethnography 47, 50-51, 69-71, 123, 133, 140, 142-146, 196, 244, 258, 260

F

fieldwork 21, 35, 42, 46-53, 57, 59-62, 66-68, 70-71, 74, 76, 89-90, 92, 118-119, 121-124, 129, 131, 133-134, 139-145, 149-150, 159, 161, 199-203, 205-214, 216, 226, 235, 244, 246, 257-258
fieldwork challenges 42, 46-47, 149, 200, 202-203, 212
flexibility 11, 46-47, 68, 74, 82, 85, 88, 92, 138, 234

G

gatekeeper 25, 32, 62-63, 102
gatekeepers 33, 63, 68, 97-98, 100, 102, 107, 109, 111, 125, 128-130, 137-138, 141-142, 145-146
gatekeeping 129-130, 132, 134, 143, 147
growth mindset 76, 78-79, 93
Guiyang City 49, 54, 58, 70

H

higher education 22-23, 74, 78, 83, 89-90, 108, 146, 149-150, 152-154, 158, 161, 164, 167-168, 171-172, 196, 219-220, 234, 237-238
Historically Black College and University (HBCU) 219-220, 223, 238, 240-241
Hurricane 219-221, 229, 231, 233, 237, 240

I

in-depth interviews 98, 149, 152, 156-158, 161, 163-164
Information-Gap Activity 93
Inpatient Department (IPD) 217
Institutional Review Board 6, 24, 76, 78, 82, 201, 203-204, 224, 229, 256
instructional design 3-4, 22, 24
Integrated Performance Assessment (IPA) 93
Internal Review Board (IRB) 94
Interpretive Phenomenological Analysis (IPA) 96-97
intimate partner violence 96-97, 114, 117

L

language 4, 7, 16-17, 19, 22-29, 31-35, 37-38, 43-44, 46-48, 51, 59-62, 66-68, 71, 73-76, 78-79, 85, 89-93, 105-106, 124, 141-142, 146-147, 244-245, 250, 254-255, 258-260
Learning Management System (LMS) 94

M

Massively Multiplayer Online Game (MMOG) 243, 257
MAXQDA 157, 172
migrant children 46, 49-54, 58, 64, 69-71
Ministry of Development 153, 172
Mission Differentiation and Specialization Aimed at Regional Development 153, 168, 172
mixed methods research 166, 170, 172-173
mixed methods study 149

MMOG 243-244, 246-247, 249, 260

N

narrative research 175-176, 178, 180, 184, 191-192, 194
natural disaster 220-221, 227, 236, 240
New Delhi 200, 203, 213
Norm Cadre 153, 173
novice researchers 25, 29, 32, 35, 42, 67, 75, 118, 123, 127, 150, 158-160, 162, 164, 167, 199-200, 204, 207, 210, 212-213

O

online learning 73, 79
organization 15, 35, 86, 98-99, 102, 104, 106, 118-140, 143, 147, 160, 163, 168, 172, 257
othermothering 220-221, 223, 238, 240-241
Outpatient Department (OPD) 217
Ozoblokly 4, 19, 24
Ozobot 5, 10, 14-15, 22-24

P

positionality 62, 67, 71, 141, 175-176, 180, 184, 189, 191, 193, 195
post-disaster 219-222, 232-233, 241
post-disaster research 220, 232, 241
power 18, 32, 41, 96-97, 107, 109-115, 117, 130, 134-135, 163-164, 166, 189, 197, 205, 209, 222, 229-230, 240, 245, 265
Privately-run migrant school 71
professional development 1-3, 11, 13, 21-22, 24-25, 27-28, 36, 38-39, 43, 74
psychoanalysis 176, 178-179, 194-197
psychological well being 203, 206-207, 209-211, 213-214, 217

Q

qualitative research 48, 71, 73-74, 89-90, 92-93, 96, 100, 108, 114, 121, 127, 144-146, 166, 171, 173, 175-176, 191-192, 194-197, 215-216, 236, 244, 257
Qualitative research methods 93, 96, 144, 194
quality of life 199-202, 204, 206, 208-209, 214, 216-217

R

radiotherapy 200-203, 205-206, 209, 213, 217
rapport 29, 33, 65, 133, 147, 150, 158-159, 199-200, 204-205, 214-215, 217
research 1-4, 6-9, 11-17, 19-23, 25-36, 39-43, 46-59, 61-64, 66-71, 73-76, 78-94, 96-98, 100-105, 107-121, 123-147, 149-151, 153-173, 175-182, 184-185, 187-204, 207-217, 219-224, 226-241, 243-247, 249, 251, 255-258, 260-261
Research in Higher Education 146, 149, 237
research methods 47, 68, 91-93, 96, 104, 117, 123, 129, 141-146, 165, 194-195, 258
research process 26-27, 46-48, 67-68, 82, 88, 140, 149, 160, 164-165, 167, 173, 191-192, 200-202, 210, 213
research proposal 124, 126, 131, 202-203, 213
research site 46-49, 51-52, 54, 57-58, 66-68, 70, 138, 243, 246
risk 32, 80-81, 83, 96-103, 105-106, 108-109, 111, 115, 117, 120, 132, 155, 163, 175, 185, 196, 207, 215-216, 243, 247

S

Sales Pitch 123, 125, 147
Sampling and Recruitment 46, 55, 67
Scratch 3, 9, 19, 23-24
second language socialization 244, 258-260
secondary traumatic stress 98, 115, 117, 199, 207-210, 212-214, 216
site 18, 20, 46-52, 54, 57-58, 66-68, 70, 102, 118-122, 124, 126, 131-139, 147, 199-200, 203-204, 214, 243, 246-247
SLS 244-245, 247
Social Work 98, 101-102, 114-115, 141, 195-198, 214-215
Stratified Sampling Method 166, 173

T

teacher professional development 2-3, 21, 28, 39, 43
teaching 11-14, 20, 22-24, 26-27, 29, 31-35, 37-39, 44, 60, 74-76, 78, 80, 87, 89-93, 151, 153-154, 157, 160-161, 196, 222-223, 237-238, 258
technology 1-3, 6, 13, 18-20, 22-29, 31-37, 39-40, 42-44, 73-76, 90, 93, 104, 220
time 5-9, 11-21, 28-32, 34-39, 41, 50, 52-53, 58, 61, 63-65, 73, 76-78, 80-89, 91, 99, 104-105, 109, 111, 113, 116, 122, 128-129, 131-135, 137-138, 154-155, 157-158, 160-161, 167-168, 180-182, 184-191, 200, 202, 204, 206, 210-213, 217, 219, 222, 226-227, 230-231, 236, 245-246, 248, 251-254, 256, 261-265
trauma 98-99, 114, 116, 175-176, 180-181, 184-185, 187-190, 192, 195-196, 198, 208, 211, 215-216
triangulation 166, 173, 177, 247
Turkey 149-152, 155-156, 162-163, 166-172
Turkish higher education system 150, 152-153, 161
Turkish universities 149-150, 158, 164, 167, 169

U

University Academic Caring Carescapes Framework 223, 241
US Virgin Islands 219

W

web-based survey 219, 224, 235
Willingness to Communication (WTC) 94

www.ingramcontent.com/pod-product-compliance
Lightning Source LLC
LaVergne TN
LVHW061547290926
849162LV00004B/930